ACID RAIN: ARE THE PROBLEMS SOLVED?

MAY 2-3, 2001 WASHINGTON D.C.
CONFERENCE PROCEEDINGS

Sponsored and Organized by the

Center for Environmental Information Inc.
55 St. Paul St.
Rochester, NY 14604
585-262-2870
ceiroch@frontiernet.net
http://www.ceinfo.org

Edited by
James C. White, Ph.D
Professor Emeritus, Cornell University

Production Editor
John Terninko
Center for Environmental Information

These proceedings were funded by grants from the
Ecological Society of America (ESA)
National Acid Precipitation Assessment Program (NAPAP)
and the
National Oceanic and Atmospheric Administration (NOAA)

The Executive Summary was supported by funds from the U.S. Geological Survey and U.S. Fish and Wildlife Service.

American Fisheries Society
Trends in Fisheries Science and Management 2

American Fisheries Society
Bethesda, Maryland
2003

Suggested citation format:

American Fisheries Society. 2003. Acid rain: are the problems solved? American Fisheries Society Trends in Fisheries Science and Management 2, Bethesda, Maryland.

Library of Congress Control Number 2003100678
ISBN 1-888569-47-6

Printed in the United States of America on acid-free paper.

American Fisheries Society
5410 Grosvenor Lane, Suite 110
Bethesda, Maryland 20814-2199
USA

Cover image: Hydrogen ion concentration as pH from measurements made at the Central Analytical Laboratory, 2001. National Atmospheric Deposition Program (NRSP-3)/National Trends Network. (2003). NADP Program Office, Illinois State Water Survey, 2204 Griffith Dr., Champaign, IL 61820.

Table of Contents

Table of Contents

Preface

Acid rain was first recognized as a serious threat to the environment in the 70s, but public concern about the issue decreased with passage of the Clean Air Act Amendments of 1990 and the emerging awareness of global climate change. As the papers in these proceedings attest, however, the problem is still of major importance. This conference was needed to highlight the continued damage to our environment, documented by monitoring and research findings since the 1990 Clean Air Act amendments, and to reveal the changing face of the problem as regulation, science, and technology have or have not been applied.

The conference brought together those whose research, policymaking , enforcement , and implementation actions are affecting acid deposition. New knowledge and appreciation of the significant role of nitrogen, and especially ammonia, largely from food production, is presented from several points of view, along with the story of the unexpectedly cost-effective success of the emissions trading program.

Many speakers emphasized that, based on research and monitoring, the Clean Air Act Amendments of 1990 were inadequate to control the adverse effects of acid deposition and that additional reductions in emissions are a pressing and immediate need if damage is to be slowed and recovery possible. Others emphasized the severe and costly public health effects associated with acid rain precursors.

We were especially fortunate to have several members of the U.S. Senate and House of Representatives on the program, and active in the hearing of the House Science Committee held in conjunction with the conference.

The Steering Committee's excellent work in designing the program and selecting speakers who were authorities in their field is reflected in the outstanding quality of the presentations.

We gratefully acknowledge the support of the 54 co-sponsors of the conference and the proceedings.

Special thanks go to John Terninko who acted as conference and publication coordinator and to Shirley Sherman who was responsible for the registration and support services. The conference and these proceedings would not have happened without the initiative and organizational skills of Liz Thorndike.

James C. White, Ph.D
Editor
July, 2002

About the Center

The **Center for Environmental Information** (CEI) is a private, nonprofit, educational organization, founded in Rochester, New York, in 1974. CEI provides information and communication services, publications, and educational programs in order to:

- Advance public understanding of environmental issues;
- Act as a communication link among scientists, educators, decision makers and the public;
- Advocate informed action based on the free exchange of information and ideas.

CEI and Acid Rain

In 1980, New York Senator Patrick Moynihan sponsored legislation creating the National Acid Precipitation Assessment Program (NAPAP), a cooperative federal research, monitoring and assessment effort. The Program was intended to address the acid rain problem which scientists had been documenting for some years (1979, "Acid Rain", *Scientific American*, Likens, Wright, Galloway, Butler, Vol. 241, No.4, pp.43-51).

In 1982, with New York State's Adirondack Park the bullseye of the acid rain problem, CEI established the Acid Rain Information Clearinghouse (ARIC). For eight years ARIC served as a central information, education and communication link among scientists, decision makers, environmental organizations, educators and the public. With a distinguished international advisory council, ARIC initiated, edited and disseminated the monthly bulletin, *Acid Precipitation Digest*; organized conferences and educational programs in New York and Washington, D.C.; published and disseminated conference proceedings; served as Documentation Center for the New York-Quebec Acid Rain Agreement; helped create a bi-lingual database (ACIDOC) on acid rain research and policy; identified 708 research, monitoring and assessment projects, in addition to the NAPAP program, being carried out by 25 states and 56 private and non-governmental organizations, thus verifying the scope and magnitude of attention being given to this problem; and established a library and information and referral service known internationally. By 1990 the ARIC acid rain collection included over 6000 publications.

In 1984, ARIC's first major conference , "Acid Rain: A New York State Agenda", held in Albany, was followed by passage of the first state Acid Deposition Control Act. Subsequent conferences in Albany and in Washington, D.C., organized until passage of the 1990 Clean Air Act Amendments, were entitled "Acid Rain: Economic Assessment"; "Liming Acidic Waters: Environmental and Policy Concerns"; "Acid Rain: The Relationship Between Sources and Receptors"; "Acid Rain: The View From The States"; "Global Climate Change Linkages: Acid Rain, Air Quality, and Stratospheric Ozone"; "Acid Rain Research in New York State".

Passage of the 1990 Clean Air Act Amendments and a shift of attention to ozone depletion and global climate change, diminished acid deposition as an issue in the public eye. Important new research information from regions around North America has brought the acid rain issue to the forefront again.

For 26 years the Center for Environmental Information has acted as a catalyst to advance the public agenda toward soundly conceived, scientifically-informed environmental policies. The 2001 acid rain conference is another milestone along that path.

Conference Steering Committee

Ed Bennett – New York State Department of Environmental Conservation
Christopher Bernabo – Collaborative Solutions, Inc.
Rona Birnbaum – U.S. EPA
Richard Haeuber – U.S. EPA
Kathleen Fallon Lambert – Hubbard Brook Research Foundation
Leonard Levin – EPRI
Bernard Melewski – Adirondack Council
Mark Nilles – U.S. Geological Survey
Elizabeth Thompson – Environmental Defense
Elizabeth Thorndike – Center for Environmental Information
Michael Uhart – National Acid Precipitation Assessment Program
James C. White – Center for Environmental Information

Conference Co-Sponsors

- * Adirondack Council
- Adirondack Mountain Club
- Air & Waste Management Association
- American Fisheries Society
- American Gas Association
- American Hiking Society
- American Institute of Biological Sciences
- American Lung Association
- American Sportfishing Association
- Appalachian Mountain Club
- Appalachian Voices
- Association for the Protection of the Adirondacks
- Center for Clean Air Policy
- Citizens Campaign for the Environment
- Clean Air Network
- Clean Air Task Force
- Conference of New England Governors and Eastern Canadian Premiers
- * Cornell University Center for the Environment
- * Department of Foreign Affairs and International Trade Canada / Ministère des Affaires Étrangères et du Commerce International du Canada
- Ecological Society of America
- * Edison Electric Institute
- Environment & Energy Study Institute
- * Environment Canada / Environnement Canada
- * Environmental Defense
- * EPRI
- Hubbard Brook Research Foundation
- Izaak Walton League of America
- * Keyspan Foundation
- National Acid Precipitation Assessment Program
- National Audubon Society
- National Council for Science and the Environment
- National Oceanic & Atmospheric Administration, Air Resources Laboratory
- National Trust for Historic Preservation
- * National Wildlife Federation
- Natural Resources Defense Council
- * New York Power Authority
- * New York State Adirondack Park Agency
- * New York State Department of Environmental Conservation, Division of Air Resources
- * New York State Energy Research and Development Authority
- * Niagara Mohawk
- Northeast States for Coordinated Air Use Management
- RAND Corporation
- Residents' Committee to Protect the Adirondacks
- Resources for the Future
- Save Outdoor Sculpture!
- * State University of New York College of Environmental Science & Forestry
- Trout Unlimited
- * U.S. Department of the Interior, Fish and Wildlife Service
- * U.S. Department of the Interior, National Park Service, Air Resources Division
- * U.S. Department of the Interior, National Park Service, PRIMENet
- * U.S. Environmental Protection Agency, Clean Air Markets Division
- * U.S. Geological Survey
- Union of Concerned Scientists
- * World Resources Institute

* Funding co-sponsor

Conference Program

May 2, 2001

Welcome
Elizabeth Thorndike – Center for Environmental Information

Session I. Addressing the Acid Rain Problem – Twenty Years in Retrospect.
Chair: Christopher Bernabo – Collaborative Solutions, Inc.
Brian McLean – U.S. EPA: Federal Legislative and Regulatory Record
Richard Valentinetti – STAPPA: State/Local Legislative and Regulatory Role
John McManus – American Electric Power: Compliance Experience of Regulated Entities

Session II. Research And Analysis: What the Reports Say.
Chair: Bernard Melewski – Adirondack Council
Charles Driscoll – Syracuse University: Acid Rain Revisited: Sources, Effects and Recovery in the Northeastern U.S.
Michael Uhart – NAPAP: National Acid Precipitation Assessment Program Analysis
Rona Birnbaum – Clean Air Markets Division, U.S. EPA: Beyond Title IV: Perspectives on Additional Reductions

Luncheon Speaker
Senator Charles Schumer

Session III. Acid Rain Impacts: State of the Science.
Chair: Mary Barber – Ecological Society of America
Van Bowersox – Illinois State Water Survey: Sources and Receptors: Monitoring the Data
Arthur Bulger – University of Virginia: Freshwater Aquatic Systems
Robert Howarth – Environmental Defense: Coastal Ecosystems
Greg Lawrence – US Geological Survey: Forests and Terrestrial Systems
Morton Lippmann – New York University School of Medicine: Human Health
Mary F. Striegel – National Center for Preservation Training and Technology: Materials and Cultural Resources

Special Program: Acid Rain Information Resources
Fred Stoss – SUNY Buffalo Biological Sciences Librarian: Review significant printed matter, abstracting and indexing services, and internet resources related to acid rain.

Conference Program

May 3, 2001

Session IV. Acid Rain Linkages: Ozone, Hazardous Air Pollutants, Particulate Matter, Eutrophication, and Climate Change.

Chair: Tamara Blett – National Park Service

Dr. Paulette Middleton – RAND Corporation: Overview of the Sources

James Galloway – University of Virginia: Overview of the Impacts

Session V. North American Regional Impacts Panel.

Chair: Ellis Cowling – North Carolina State University

Richard Poirot – Conference of New England Governors and Eastern Canadian Premiers: Northeast

Douglas Knauer – Wisconsin Department of Natural Resources: Mid-West

Don Campbell – U.S. Geological Survey: Intermountain/West

Brent Takemoto – California Air Resources Board: Southwest

Guy Fenech – Environment Canada: Canada

Ellis Cowling – North Carolina State University: Southeast

Luncheon Speaker

Representative John E. Sweeney

Session VI. Economics of Acid Rain

Chair: John D. Kinsman – Edison Electric Institute

Jeremy Platt – EPRI: Compliance Costs and Markets

Dallas Burtraw – Resources for the Future: Economic Benefits of Controls

Guest Speaker

Senator Hillary Rodham Clinton

Session VII. Hearing of the House Science Committee – In conjunction with the conference (held in the Rayburn Building)

Chair: Representative Sherwood L. Boehlert

Charles Driscoll – Syracuse University: The latest science on air pollution that causes acid rain, ozone and particulates

Ellis Cowling – North Carolina Sate University: Impacts of air pollution on forests, fish, and surface and ground water quality in the Southeast and the importance of monitoring networks

Jill Baron – Colorado State University: The impacts of nitrogen deposition in the Colorado Front Range

Jerry Keeler – University of Michigan: Our evolving understanding of the sources of mercury pollution, mercury's fate in the environment, and future research and monitoring needs

The response to acid rain has thus far been a story of the system succeeding—a story of Congress acting on the basis of the best research, even in the face of uncertainty, and then trying to update policy on the basis of continuing research. That's a pattern we need to perpetuate. We need to keep funding the research and heeding its results.

Congressman Sherwood Boehlert (R-NY)
Opening Statement, Acid Rain Hearing
House Science Committee, May 3, 2001

Introductory Message

The 29 papers and presentations in these proceedings include the most comprehensive and most authoritative information on the topic of acid rain and its linkages presented to date in a public forum. Included are the results of new research findings since enactment of the 1990 Clean Air Act Amendments, recent monitoring data, economic analyses, and political perspectives. The magnitude of the health benefits alone, compared to the costs of controls, makes the case for the value of reduced emissions.

The conference was convened in May, 2001 to put the acid rain problem – continental in scope – squarely back on the forefront of the public agenda. Conference participants came from 24 states, the District of Columbia and Canada. Conference co-sponsors included 54 federal and state agencies, organizations, associations, institutions, and companies.

The Executive Summary of the conference was sent to members of Congress and key administration officials in July, 2001. Since then, various legislative bills and initiatives have been put forward by members of Congress and the Administration to address acid rain and linked problems such as mercury accumulation, particulate matter, over-fertilization of ecosystems (nitrogen saturation), ozone formation, and climate change.

Each of these mounting problems has a common source--combustion of fossil fuels used in transportation and for electric generation – joined by emissions of ammonia from food production. The impact of these growing emissions is now regarded by scientists as a cascade of effects on health, visibility, forests, soils, groundwater, streams, rivers, and coastal ecosystems, as well as a cause of increased concentrations of potent greenhouse gases globally. As Dr. James Galloway of the University of Virginia notes, "Linkages among effects require a multi-pollutant strategy for control."

Mobile and point sources, both parts of the problem, must now become part of the solution. Deployment of existing and emerging energy efficiency technologies and strategies, along with alternative energy sources, can lead the inevitable transition to cleaner sustainable energy resources which are critical to our economic vitality, competitive edge, and quality of life.

Elizabeth Thorndike
Center for Environmental Information
July, 2002

SESSION I: Addressing the Acid Rain Problem – Twenty Years in Retrospect

Christopher Bernabo – Collaborative Solutions, Inc.

This lead-off session is entitled a 20-year retrospective. I think it is quite amazing that we are here 20 years later doing this. I would like to particularly thank Dr. Thorndike and the Center for Environmental Information for having the vision to bring us back together and discuss this in a timely fashion. Again, the issue, which has long been off the front pages, is getting some attention.

The reason we are choosing 20 years is fairly obvious. That is when National Acid Precipitation Assessment Program (NAPAP) began. It wasn't the beginning of the acid rain issue. It was the beginning of a really focused national effort.

I have been asked to make a few remarks to put the session in context. I am going to do that by saying some things about NAPAP. There are programs that preceded this, such as the National Acid Deposition Program (NADP), for which there is some literature out front, as it still goes on.

NAPAP was not the beginning of studying or understanding the acid rain program. It was the beginning of a nation-wide consciousness to not just do research, but to try to harness that research in service to society, to try to help us make some decisions. Science, in itself, does not do that, unless you harness it in some manner.

To this day, I would argue that NAPAP is the benchmark of such efforts. While there have been a number of other efforts going on – notably climate change, marine pollution, weather modification and national programs of multi-agency participation – none of them have achieved what NAPAP has achieved, for all of its warts and problems, in terms of actually effectively helping us move on to take action.

There were times when the efforts were going on and people said, my God, it was $500 million over 10 years. That was a lot of money.

It took five years to do the first assessment. It could have happened quicker, but it stood the test of time. The climate effort, a major issue that eclipsed acid rain, actually began two years before acid rain. That is not well known, but with turnover on the Hill, memory leaves.

1978 was the year of the National Climate Program Act. The National Climate Program Act in 1978 required that the climate program do assessments every two years. For those of you familiar with the situation, 22 years later, last year, the first climate assessment was done. So, I don't feel bad about five years.

When people talk about $500 million, we have spent $15 billion on climate research so far, are still spending at a rate of about $1.6 billion a year, and I don't see a framework for decision making. I don't see policy from the program.

NAPAP is still the benchmark to beat. There are many reasons, by the way – I won't get into all of them – why that is the case. Part is the issue, part is the politics. Indeed, they have yet to be bested in terms of attacking these problems.

Secondly, I think sometimes we forget how effective the program was, directly and indirectly, in bringing the science to bear on the issue.

Mostly attention was focused on the assessments, the interim assessment and the final assessment. In point of fact, the program was designed to release results throughout.

NAPAP produced annual reports to the President and Congress, dozens of congressional hearings, newspaper articles, and scientific papers. In fact, throughout the effort, it was affecting the policy debate.

Acid rain emerged in the 1980s as a large concern about agriculture but, through research and time this was supplanted with concern about forestry.

The focus was entirely on sulfur, which later expanded to include nitrogen. There were major changes from the science into legislation.

There has been a misconception, propagated by some articles, when people were debating this at the end of the program, that it was ineffective for the following reason, that Congress didn't wait the full 10 years to take action, which is ludicrous.

That is true, that Congress took action before the final report, but Congress did not say that it was creating a 10-year program and was going to wait to take action. It was the opposite.

They said, we should be able to take action within 10 years. So, we will set the program out longer than it will take and we will listen to the results of the program, wait until the political will is there, and act.

In fact, the whole notion that the program had an end that was supposed to be a policy end is fallacious. The reason for the 10 years, (just for historical reasons the act passed in the Carter administration) was that was something then – for those of you old enough to know it – called sunset laws, that you don't just create programs and let them go on forever. They typically chose a 10-year time interval and said, if the program is still needed after 10 years, renew it. If not, we don't have to kill something with an infinite life span.

In fact, another interesting thing about the program is, it still exists twenty years later. Its director is here. That is an amazing feat of foresight on, largely, Senator Moynihan's part.

The typical thing our country does, and most countries do, is we study something, we get upset about it, we pass legislation. We say it is solved. Very, very rarely do we go back and measure whether we got what we thought the nation wants. But that is the reason we are here today.

Finally, I would like to mention something about the important things we learned about the use of assessment and science as the underpinning for control programs.

As you know, and we are going to hear in a minute, that one of the most important things to result from this is a market-based program to address the issue. What is less widely understood is that a market-based program could not be effective if we did not understand the source receptor relationship. Let me explain that.

If the most cost effective sources to control were all on the east coast and if the most sensitive resources were in the Midwest, a market-based system would result in most of the benefit being out in the Atlantic Ocean. It wouldn't have done much good for the sensitive receptors.

It is a happy coincidence that the place with the most bang for the buck controls was in the same place that had the most bang for the buck in terms of results and receptors.

To show why this has sort of just passed us by and we don't think about it, China has recently been developing an acid rain program, and Brian and myself and others have been involved in that.

The Chinese are very interested. They know they have a problem, they are interested in solving it and right away having a market program, which is interesting for a former Communist country. Anyway, they want to use the market approach to get more bang for the buck.

They didn't want to do an assessment. Yet, they really haven't figured out if, where their sources are, they can control most effectively the environmental benefit.

So, these things go hand in hand. I think we have learned a lot and are going to hear a lot. We have three retrospectives today, one from an official in the government who has been through all this, one from the state perspective and one from a private industry perspective.

Federal Legislative and Regulatory Record

Brian McLean – U.S. EPA

I am going to look back over the last 20 years to help, as part of this session, set the stage for a lot of the information you are going to hear in the presentations in the next two days.

As I look back over the acid rain story over the past 20 years, I really see two intertwined story lines, one about the science and the other about the policy and program. The decade of the 1980s opened with rather polarized views about the issue of acid rain. One view held that acid rain was a serious environment problem. It was killing our aquatic and terrestrial life. It was getting rapidly worse and might be irreversible. It was caused primarily from emissions from power plants, and primarily power plants in the Midwest.

The other view held that it was not a problem at all, that it was actually a political issue. If it were a problem, it was not a serious one. It had existed for decades it wasn't getting any worse. Besides, you couldn't prove that emissions from the Midwest led to acidic deposition in the east, and that deposition was actually responsible for the effects that we were seeing on the lakes and streams in the east.

Therefore, it was premature to do anything about the issue. At most, we should study the problem more to reduce all the scientific uncertainties. It was really because of that last concern, since we couldn't resolve these two views, we did go to a rather intensive research effort.

Although over 70 bills were introduced through 1988, the administration supported none of those bills, and I do not believe that any of them actually made it to the floor of either house for a vote.

However, as Dr. Bernabo pointed out, we did embark on one of the most intensive and expensive research efforts ever, $500 million at that time, but it pales in comparison to the climate work.

The research was driven by two basic questions, how bad was the problem, which was part of the debate in 1980, and what should we do about it.

More specifically, there were questions like, how many acidic lakes and streams are there in the United States? How many forests have been damaged? This was the scope question which was in dispute. We had anecdotal data, but we didn't have a systematic determination of this.

Is it getting worse? If so, how fast is it getting worse? This is the rate of change question.

What is the relationship between emissions and deposition? What is the relationship between deposition and environmental effects? These were the mechanism questions. They were hypotheses but we hadn't really tested them.

What is the cost of control?

Major data gathering and research efforts were undertaken and major accomplishments were achieved during the decade of the 1980s. For example, we did conduct a national lake and stream survey, the first time it had ever been done. It was a systematic, very high quality effort, which allowed us to have a much better sense of what the scope of this problem was.

We also did forest surveys and we set up monitoring systems, or added to monitoring systems that were already beginning, monitoring of deposition trends, monitoring of effects. We increased our understanding of the response of ecosystems to deposition, and we captured that in models, such as the MAGIC model. We increased our understanding of atmospheric chemistry and transport, which was captured in models like RADM.

We increased our understanding of emissions control options and likely economic decisions, which we captured in models like AUSM and CEUM, for those of you familiar with these things. AUSM was the advanced utility simulation model, which has since fallen by the wayside, and CEUM was the coal and electric utility model. These models, more than just models, were actually places where people put their knowledge together and tried to express, in the best way they could, the processes that we were trying to explain and understand. Thanks to these efforts, scientific uncertainties regarding causal links – between emissions and deposition, between deposition and acidification – were greatly reduced during the

1980s. The models reflecting our best understanding of these phenomena were linked to emissions to economic models, so that we could support credible policy analysis.

As I said earlier, there was another story line to the 1980s. In 1980, the principal instrument for achieving environmental improvement was what we have come to call command and control regulation. There was little use of economic instruments. Taxes were frowned upon, are still frowned upon, and emissions trading was still in its infancy at that time.

The bills introduced in the 1980s either mandated scrubbing or set an emissions rate to be met by every plant. Emissions trading or other flexible approaches were not given too much attention. They were mentioned, but people really didn't dwell on them too much.

In 1989, to help break the legislative logjam, the Bush administration, with the support of groups like the Environmental Defense Fund, put forth a new approach, one that relied on a national emissions cap, with trading of emissions allowances. It was designed to provide the economic savings promised by emissions trading, but with the environmental credibility lacking in previous trading efforts.

Because a national emissions cap and trading regime had never been implemented before, or even proposed, it was referred to by some as a grand policy experiment. Most of its features were unprecedented and untested. The purpose of the program that eventually emerged from Congress was to reduce SO2 emissions by 10 million tons (Figure 1). The bulk of those emission reductions would be coming from electric utilities, 8.5 million tons.

The scope of the program was to be national, except for Hawaii and Alaska, the continental contiguous states of the United States (Figure 2). It was to include every major power plant and every new power plant built. Again, this was unprecedented. We had never run a stationary source program like this before.

The program would specify what needed to be done by setting a maximum emissions cap that would require measurement of all emissions from all sources, and have automatic financial and emissions offset penalties for non-compliance.

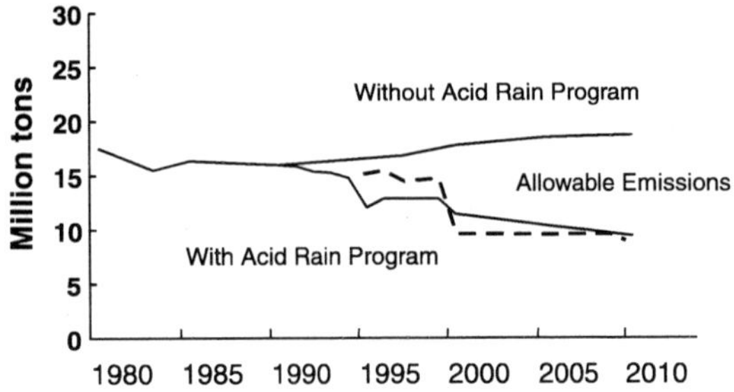

Figure 1. SO_2 reductions from power generation. Goal of SO_2 Program – Reduce SO_2 emissions by 10 million tons from 1980 level – 8.5 million tons from power generation through "cap and trade" mechanisms.

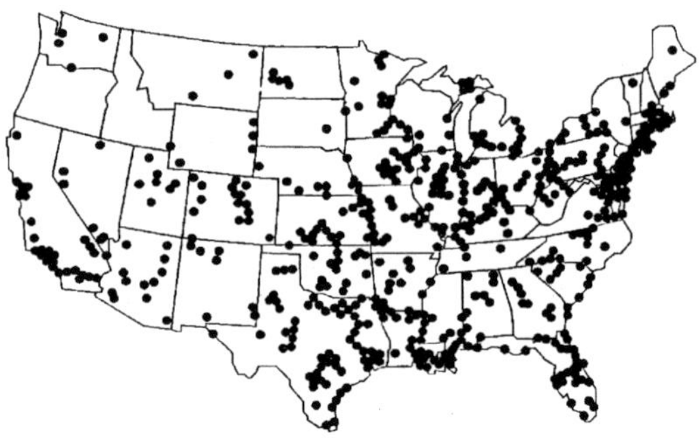

Figure 2. Scope of SO_2 program - over 2,500 sources affected.

It would also allow unprecedented flexibility for sources to apply. Unlike command and control regulation, it would not specify how or when a particular facility needed to control.

Sources could change their compliance plans or trade allowances without government approval, but the trading would not be allowed to override local standards, to protect local air quality.

So, let's look at the results quickly. Emissions declined sharply in 1995, to the surprise, I think, of everyone, including the industry and us. It was unusual, in that the reductions occurred on time. It was unusual in the significance of emissions, and they have declined since. As we move into phase II of the program, we hope that they continue to decline (Figure 3).

Most important to those of us here today is that wet sulfate deposition also declined, significantly throughout the east. All the modeling, all

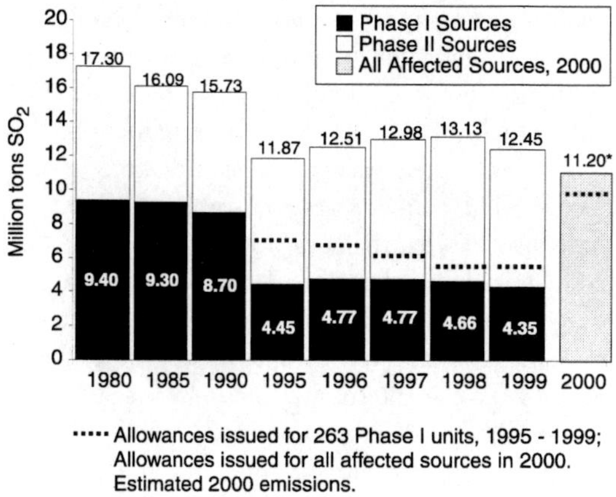

Figure 3. SO_2 emission from Title IV sources.

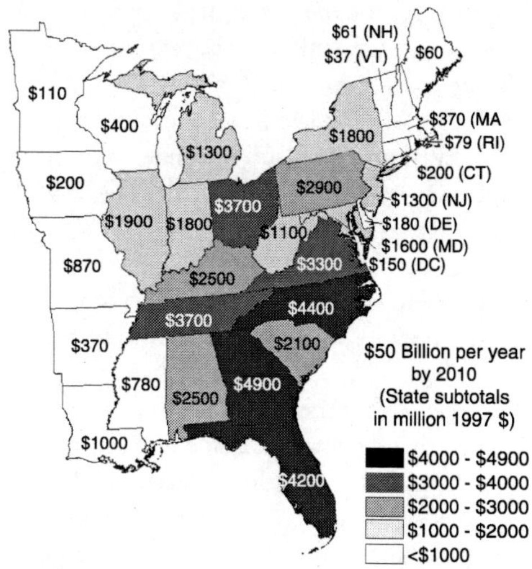

Figure 5. $50 billion in health benefits from Title IV SO_2 reductions.

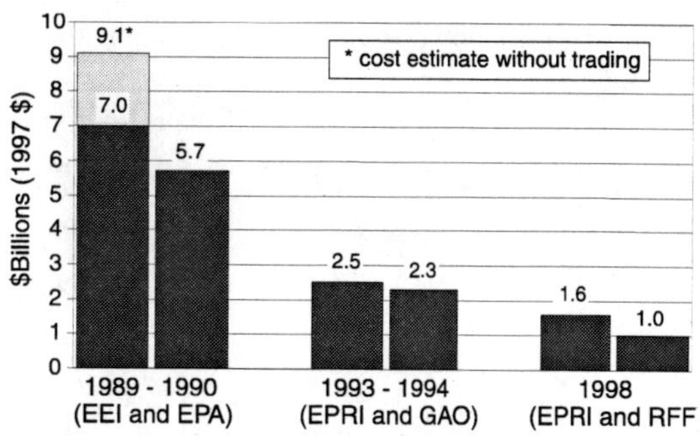

Figure 6. SO_2 cap and trade program: expected annual costs by 2010.

the source receptor work that was done proved to be correct.

We did understand the relationships and, when we reduced emissions, we did see the effects in terms of reduced deposition throughout the east (Figure 4).

This is just under the first phase of the program. There were also expected to be benefits to human health (Figure 5). Over the past decade, we have improved our ability to quantify and monetise health benefits. We see that, when the program is fully implemented over the next several years, we expect to see benefits in the range of $50 billion per year from health benefits, simply from the reductions of sulfates.

One of the biggest surprises of the program was the change in cost (Figure 6). Back in 1990, when the bill was being debated – and these numbers, by the way, for those of you who remember the numbers of the debate, have been updated to current dollars, so they are a little higher, so we can make comparisons across costs – the debate was EPA's numbers which, updated to today's dollars, would be about $5.7 billion per year, compared to industry estimates which ranged from $7 billion to $9 billion, $9 billion if the trading did not work. It was considered to be a very experimental program, and therefore there was a lot of concern about whether the benefits from the trading would actually accrue.

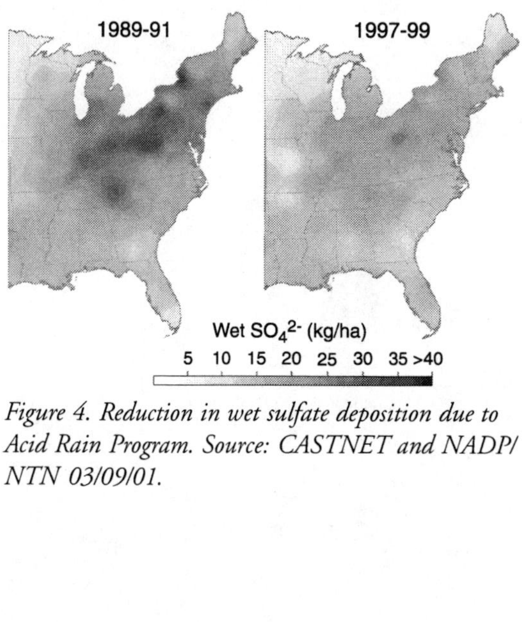

Figure 4. Reduction in wet sulfate deposition due to Acid Rain Program. Source: CASTNET and NADP/NTN 03/09/01.

A few years after the program began, it was reevaluated, both by EPRI and by GAO. Cost estimates dropped sharply. We now had real world experience as to what people were actually doing to respond to this program, and the costs were coming down. Low sulfur costs, railroad costs, scrubbing costs, various things were dropping.

A couple of years ago it was reevaluated again, both by EPRI and RFF and the costs were, again, lowered somewhat, down to the $1 billion to $1.5 billion range.

Of course, when the costs dropped, everyone asked why. How could we have been wrong in our estimates, because we know that economists are never wrong.

The fact is that there are a lot of factors combined that we hadn't considered totally when we did the analysis. I would say that the analysis done in 1990 was honest and credible with the best that we understood, but there were some things that we didn't understand.

We found that the competition across all different source options or options for control were pushing prices down more than we thought they would.

The market builds in an incentive for continuous improvement, not just meeting the standard over time, as new techniques become available, taking advantage of those new techniques. The banking provisions allowed companies to make reductions earlier, but to save those reductions. In other words, there was no disincentive to over-control. The market gives you some indication over time of what the costs actually are, so that companies have a benchmark against which to judge their compliance options. Instead of sort of guessing at what the control costs might be, they actually have an alternative to compare against, and I think they made better choices about what to do. Lastly, the way the program operates, the government is not involved in the actual trading decisions. Therefore, they could flow freely and, as costs came down, people could take advantage of lower costs. The market has improved over time. The volume of trading activity has improved and it has no negative effects on the benefits except it simply keeps the costs down (Figure 7).

We have done over 15,000 trading transactions, over 20 million allowances have been moved among accounts. This has not created any environ-

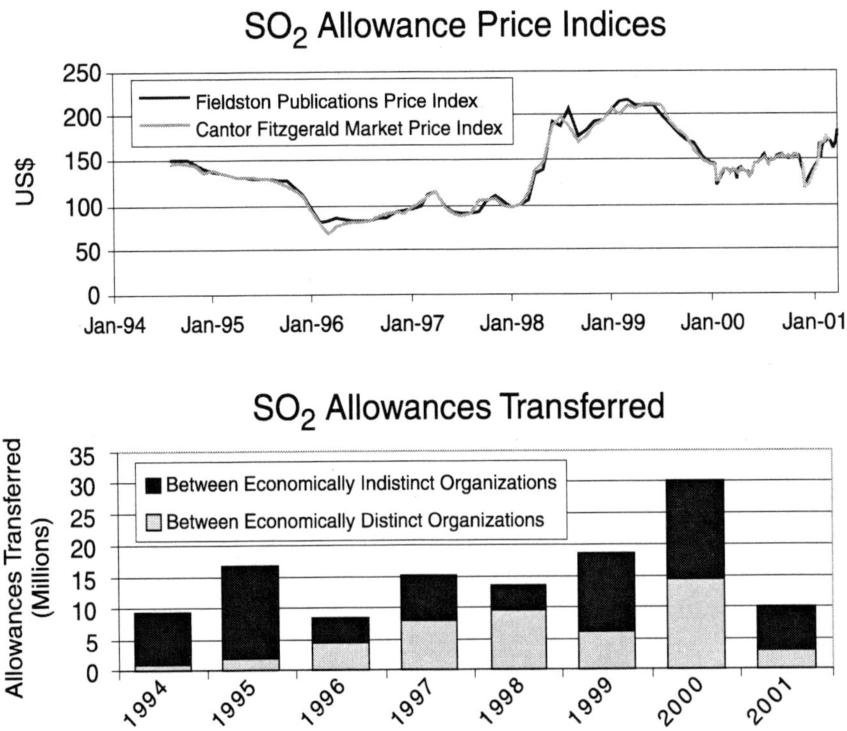

Figure 7. SO_2 allowance market. Source: Clean Air Markets Division.

mental problem, but it has significantly allowed people to reduce costs. As we look back on this policy experiment, we can summarize some of the benefits of this program. We think it has provided more certainty that a certain level of emission reductions will actually be achieved. In the face of economic growth, economic slow down, whatever the case may be, we are going to achieve those environmental benefits.

There is more regulatory certainty for sources because we have laid out a program over a number of years, and they can do their planning with respect to those targets. The compliance flexibility and the lower transaction costs have definitely helped, and there are fewer administrative resources in terms of running this kind of a program, which is also a benefit. Finally, this kind of a program allows companies to do what they naturally do, which is to try to reduce the costs. As they reduce the costs, we find that we are able to define our environmental problems and achieve benefits at a lower economic cost. Therefore, it makes it possible to look at problems and improve them at a lower cost.

I think the keys to the success of this kind of a program are probably three general areas. One is the cap itself, both environmentally, which many people look at with certainty, but also, it provides a very stable market for the trading to occur, a very predictable situation. Secondly, the accountability of the program, this was much greater than previous programs, and it needed to be. Because of the flexibility, we also needed to have greater accountability. So, we do measure all the emissions and we do show everyone all the emissions, and there is a greater degree of transparency to the program. Lastly, the simplicity of the program, which we continue to work on to keep and improve, allows for the lower transaction costs, and the lower overall cost of the programs. I think that is an important feature.

In concluding, I want to turn back to the science for a second. Since the research funds, since 1990, have all but dried up, dedicated scientists around the country, both in government and in universities and in research centers, have continued to enhance our understanding of acid rain science. While we have been running this policy experiment over the past 10 years, we have also been running a real-world scientific experiment, reducing emissions and measuring and studying the impact on our environment. However much of the essential data used for this research is provided by a network that is deteriorating, both in measuring deposition and effects. As a result of this continuing experimentation and research, we have greatly improved our understanding of the role of nitrogen, for example, and you will hear more about that, I think, over the next couple of days, both in terms of its chronic effects and its episodic effects.

We have also recognized the role of nitrogen and eutrophication of coastal waters, as well as the role of nitrogen oxides in the formation of ozone, fine particles and regional haze. Second, we have continued to improve our understanding of the role of fine particles on human health and improve our ability, as I said earlier, to monetize those benefits. Sulfates are the single largest component of this problem, and nitrates can be a significant component in parts of the west.

In conclusion, looking back over the last 20 years, I have seen the acid rain story unfold along two lines. Along one, we have proven the usefulness of a new regulatory instrument. Along the other, we have substantially improved our scientific knowledge with regard to the effects of SO_2 and NO_X. The story is not over. As I look ahead, I realize that scientific investigation is never complete. Uncertainties are never totally resolved. Policy is never perfect. I look at what we know today that we didn't know 10 years ago when we last acted on this issue. I look at the progress we have made in both the science and in our ability to manage multi-state air pollution problems. We have shown that benefits of additional reductions of SO_2 and NO_X would far exceed additional costs.

So, armed with better science and improved policy tools, I don't see why we can't move forward to solve our regional air pollution problems. Thank you.

Questions

MR. MC LEAN: We, as well as many other people, have looked at all the information as to where the trades have occurred, where the emission reductions have occurred which, of course, is the important factor. Some of those allowances are traded four or five times. It isn't really important how many times they trade. It is important where the emissions actually occur as a result of these trades. GAO has looked at this. We have looked at it. What we found is that the difference between how emissions would have occurred had we not had trading versus how emissions have occurred with the trading has shown virtually no difference.

The reason for that is there are inexpensive reductions in every state. There are inexpensive reductions at every company. When the trading actually occurs, most of it occurs within companies. Most of it occurs within states. There is limited – in a proportional sense – trading across states or between companies in different states. When trading occurs, it tends to balance. In other words, there are trades that go east to west and there are trades that go west to east.

So, we did some analysis. We have put it up on our web page that shows where the centroid of the sellers are, where the centroid of the buyers are, and they are within miles of each other, when you look at the whole country.

MS. BAUM: I just wanted to press you on one of your points. The evidence that there has been no environmental problem based on trading, brings up questions of where you have got trading going on largely – [remainder of question off microphone.]

MR. MC LEAN: In other words, the net effect of all that trading back and forth has been to leave emission reductions approximately where they were originally targeted. As Dr. Bernabo said, some of this is coincidental. Some of it is because the cheapest reductions occur in the areas with the highest emissions, and those areas have made the greatest emission reductions. In fact, when we look at some of this, for example, New York State may have actually benefited from the emission reductions, in that some of the reductions occurred a little bit more in the east than we have expected. The result is that those more immediate reductions have had slightly positive benefit. That is, if there is any effect at all, it is pretty small. I would not put too much weight on that, because they can fluctuate a few percent either way. We just haven't seen any significant geographic shift in emissions as a result of emissions trading.

MR. FRANKEL: Richard Frankel, General Accounting Office. Simple thinking or dynamic thinking suggests that there will come a point where it will get more expensive to make the next reduction. Are you starting to see that yet, or do you have a sense of where that will go? A lot of people have been overwhelmed with excitement over the fact that the cost of reductions was declining, at least basically in the market price over some of the early years.

MR. MC LEAN: My bottom line is, I would probably defer some of this to Dr. McManus, who is going to speak later from a company perspective. He probably has a better handle on what the future costs look like.

People have been projecting that the costs will increase in phase II. If we look at the trend over the last six years, I think there has been a slight increase, if you look from 1995 to today. Prices have risen from the mid-$150s to closer to $200. So, if you get rid of all the oscillations in price, I think there has been a slight rise in price. Some people expect that that will continue, maybe getting to the $250 or even $300 range.

Again, those are model predictions and I don't know exactly what will happen. Some of that depends on the economy itself. If the economy moves very strongly and we increase production of electricity, it is going to put more pressure on prices and you will probably see higher prices. If the economy slows, you are going to see less pressure and you will see less pressure on prices. That will probably be a more significant factor in the price, going forward.

AUDIENCE: A member of the NAPAP oversight review board reminded those on the board, when the first proposals were made for the cap and trade program, that the average decrease in cost, from the estimates provided before an environmental program is initiated and the final costs, vary between five and eight times. That is to say, the cost usually turns out to be one-fifth or one-eighth what it was, and your data confirms that statement. I wonder if you could comment on how far the deregulation of the railroads contributed to the decreasing costs, since the transport of western coals to the east became less expensive as the result of other regulatory policies.

MR. MC LEAN: I do think that was a significant factor. Railroad deregulation had been enacted, I think, in 1980. During the 10 years of the 1980s the prices had dropped maybe 10 percent. After 1990 they dropped, I think, another 40 percent. It was a combination of that flexibility in railroad deregulation and the demand that was being placed on western coal that combined with increased competition in western railroads and, therefore lower rates. That was definitely a factor, which then affected scrubber costs, which had to compete with lower sulfur coal prices. So, they were interactive.

MR. WOOLEY: David Wooley, Clean Air Task Force. One of your bullets talked about allowing competition from all emission control options, essentially. I just wonder if you could comment on whether or not energy efficiency or renewables really had a full opportunity to compete in the market for emission control values, and if you would recommend any changes to allow greater participation from efficiency surfaces, wind and solar sources.

MR. MC LEAN: I would sort of split that into two questions, the energy efficiency demand side and the renewables, which is really an alternative supply side option.

The energy efficiency side is much harder to measure or detect and it is much more indirect here, and I think it is much more affected by the price of electricity than anything else, particularly in moving toward a more competitive electricity market. We don't know, other than general measures, that efficiency has been improving in the United States over the last 20 years at some rate, for a combination of reasons.

On the supply side, I think what happened here is that we did have an energy conservation renewable set aside, which did give a bonus to energy conservation and renewables. It was not taken advantage of to a great degree, and the primary reason was that the cost differential between renewables and conventional power was simply too great. In any new technology coming on, the costs tend to be higher. In those areas that you want to promote a particular technology, there are a number of tools for encouraging that new technology, from tax incentives to outright subsidies, to make that technology more competitive, to allow it to deploy and then the prices to come down to be more competitive.

I think in the renewable area, we have to look at that in sort of the natural energy context, as well as environmental context. If we want to promote those technologies in those areas, we have a number of tools that we could use. In addition, for this kind of a program, they provide no additional cost. In other words, they don't need allowances. Therefore, it is a slight competitive advantage, depending on the cost of those allowances.

As one controls more pollutants and puts a price on more pollutants from other alternatives, renewables become relatively more attractive. From the way you design the program to other incentives, there are any number of ways to promote those technologies that we believe should be promoted and deployed greater in the future, and make them a little more economically competitive in the future.

SESSION I: Addressing the Acid Rain Problem - Twenty Years in Retrospect

State/Local Legislative and Regulatory Role

Richard Valentinetti – STAPPA

I am representing the State and Territorial Air Pollution Control Administrators, more commonly known as the State Air Directors.

I also have a number of other hats. One is as a state director from the state of Vermont, whose program ended up with an ambient sulfate standard in the early 1980s, trying to get the Federal Government to enforce it as part of our state implementation plan, but that didn't quite work.

Also, I have been the co-chair of the acid rain steering committee of the New England Governor's committee for years. As I talk, I am trying to represent the states. At the same time, I will tend to give some regional examples.

I want to thank the organizers for holding this conference. I think it is important. We really are seeing a lot of changes in science over the last 10 years. To get everybody together to focus is, I think, a great opportunity.

The task was to talk about what states have been doing for the last 20 years. I think that the states ended up with a patchwork of acid rain control programs throughout the country. Since the 1990s and the Clean Air Act, most of those programs have disappeared.

I took a little poetic license and went back a little further than 20 years, and really want to focus on what are the lessons that we learned as state air agencies.

In the early aspects of air pollution, we had command and control. In the 1960s and 1970s, we ended up where we were responding to catastrophes and going after the most obvious sources.

At the same time, during the 1960s and 1970s, we ended up doing a lot of controls on sulfur dioxide. We started to lower sulfur dioxide in fuels. We ended up where there was, at least in the northeast, a lot of change of fuels from coal to oil.

The regulatory programs were typically focused on controlling the largest, most obvious and most cost effective sources, unless they were old, uncontrolled and near retirement. That was 25 years ago. We grandfathered those plants, and we need to look at them again.

The other thing that we, as state air officials want to do, is get rid of these fine particulates, so you won't have these brilliant sunsets.

Acid rain is more subtle in its effects (Figure 1). As Dr. McLean said, early efforts focused not on the severity and causes of problems, but how much was it going to cost. In this city, in Washington, D.C., it seemed that the debate turned all in one week from, prove it is a problem, to how much is it going to cost to solve the problem. It was in a week that both National Geographic and Sports Illustrated ran stories on acid rain and showed fish belly up. You can talk about people dying, but seeing fish bellied up was more of an incentive in dealing with the issue.

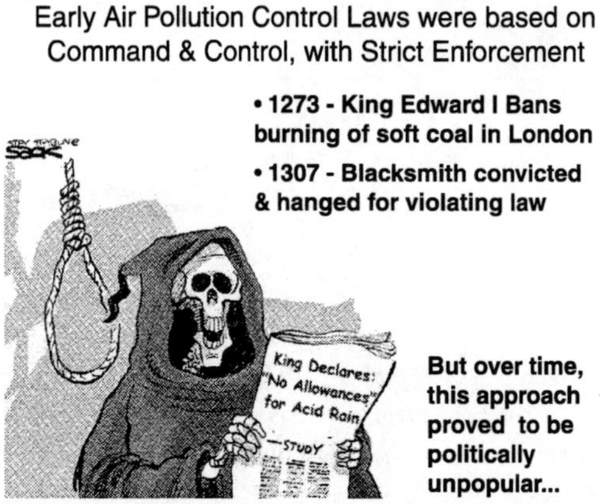

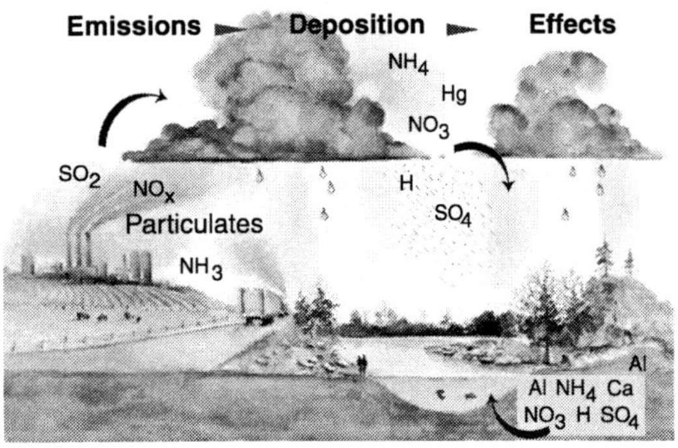

Figure 1. Acid rain is a more subtle catastrophe with more complex sources and effects.

I always remember one phrase which is, *public demand is the only thing which changes public policy.* There was a real shift in the debate during that one week.

You can see what the estimates were back in 1989 (Figure 2). It is, in fact, the actual cost.

There is good news in sulfur (Figure 3 and 4). Sulfur deposition rates have declined since 1990 and the Clean Air Act, and we have seen chemical improvements in some lakes. On the other hand, there are a number of lakes in which we haven't seen chemical improvement and we have not seen biological recovery.

The 10 years have also shown how important the issue of nitrogen is. Nitrogen was not the focus of the Clean Air Act, although there are some requirements for reductions, but we have seen no improvements in nitrogen deposition, and it is a growing concern (Figure 5 and 6). What we are finding out from the research is that there is a

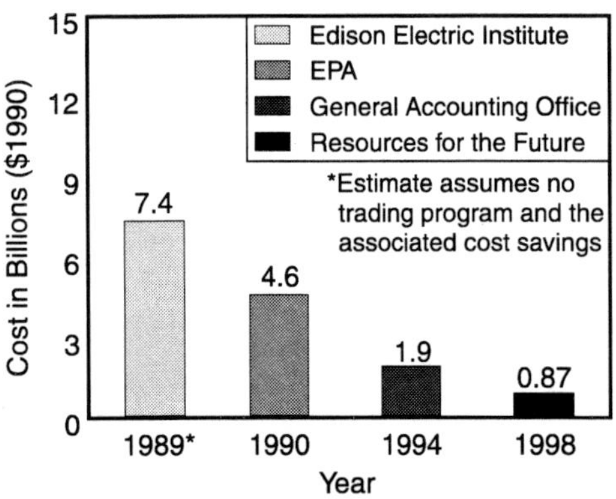

Figure 2. Estimated costs of Clean Air Act Acid Rain Controls.

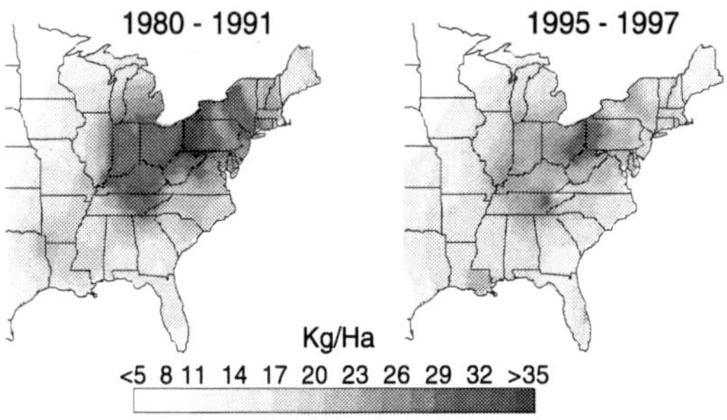

Figure 3. Wet sulfate deposition.

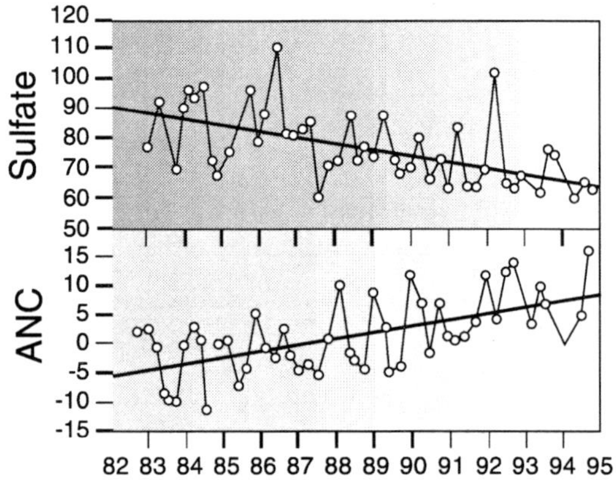

Figure 4. Trends in SO_4 and ANC at Bourn Pond, VT.

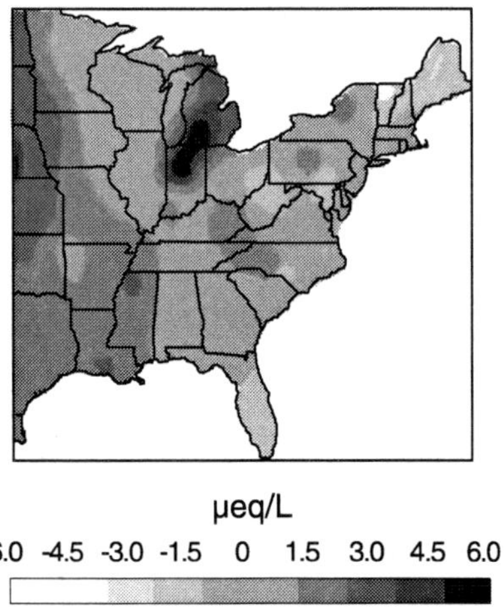

Figure 5. Changes in NO_3 concentrations in precipitation 1989-91 to 1995-97. No Improvement has been seen in nitrate deposition – there is growing concern that N can be a cause of episodic acidification of sensitive surface waters, chronic acidification in areas with N-saturated soils, and eutrophication of coastal estuaries.

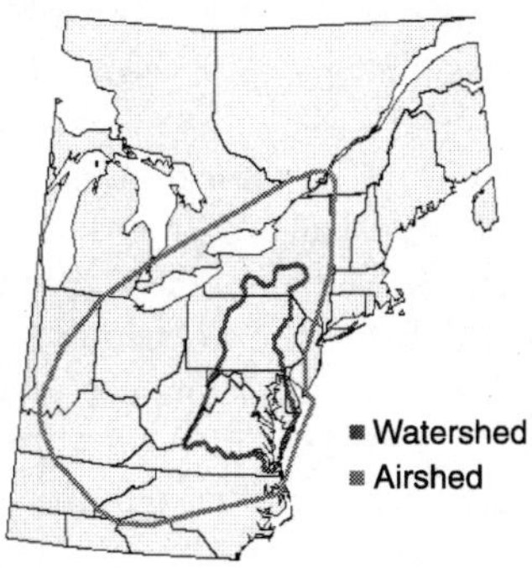

Figure 6. Chesapeake Bay watershed and airshed for nitrogen deposition.

build up of nitrogen in the snow pack, and with the controls that are forecast this will continue and we have a shock in the spring run off.

We are also seeing chronic acidification in areas with nitrogen-saturated soils. I know that Guy Fenech from Environment Canada will be talking of dealing with some of the forest issues, but this is, in fact, having a major impact on coastal areas.

The forest issue is one that we didn't recognize 20 years ago (Figure 7). Even though we are having decreases in deposition rates, the sulfuric and nitric acid has a tendency to accelerate the loss of critical nutrients from sensitive forest soils. This is preventing the nutrients from reaching the tree. The results are lower production rates.

Really, our soils are a trust fund in terms of forest productivity. We don't know how much of the principal has been used. It is important that we look at it that way, especially in the forest-rich areas. Besides, we are going to need those for CO_2 offsets in the future.

Although we have had reductions in sulfates, we still have major problems when it comes to visibility in the northeast.

What are some of the lessons that we learned from the experiment? We have learned a lot over the last 20 years (Figure 8).

One is that the sulfur concentrations and depositions respond directly and are linear. The more we reduce the sulfur, the more we can reduce the impact in terms of acid impacts. We have also learned the nitrogen reacts the same way. Since we have no controls, we are not experiencing any reductions. We need large additional reductions in both sulfur and nitrogen oxides. They are needed to assure that we do have recovery.

In 1998, the governors and premiers signed an agreement, an acid rain action plan that called for an additional 50 percent reduction in SO_2 emission and a 30 percent reduction in NO_X emissions over the commitments that had been made in the Clean Air Act, the SIP Calls and whatever.

Even though it was more of a guesstimate as to where it should be going, they felt that it was at least directionally correct. We are finding that the science is justifying those numbers.

The other thing is that there are ancillary benefits in a lot of other areas, such as regional

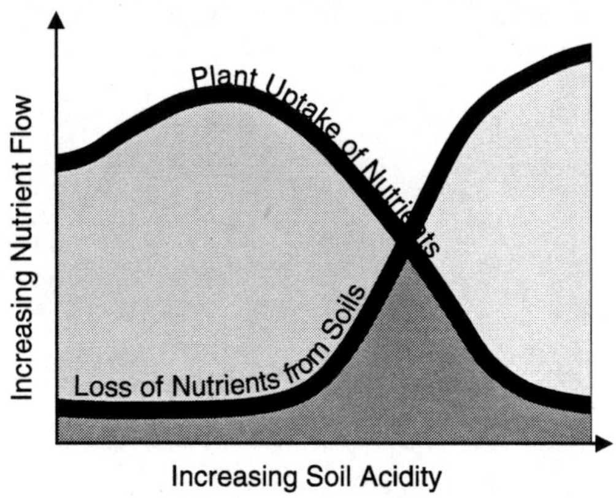

Figure 7. Even as deposition rates decrease, continued deposition of sulfuric & nitric acids accelerates loss of critical nutrients (Ca, Mg, K) from sensitive forest soils in the northeastern US and Canada.

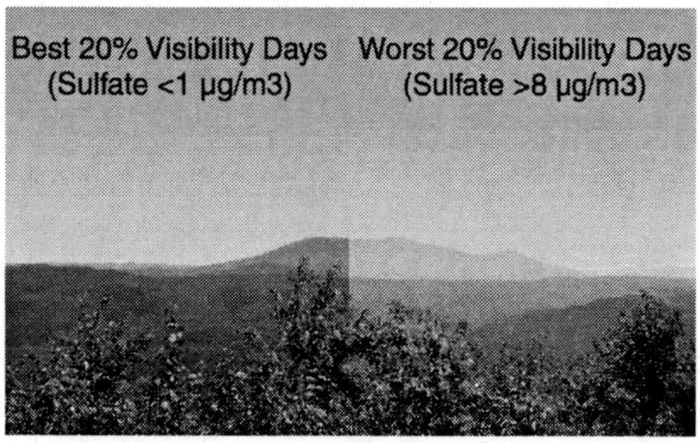

Figure 8. Effects of sulfate on visibility. While sulfate aerosol levels have improved, sulfate still accounts for 2/3 of the visibility impairment in the Northeast, including Class I areas like Vermont's Lye Brook Wilderness Area.

haze, the sulfates, the acid aerosols, the fine particulates and ozone. There has been widespread benefit from the market system. Although I am not crazy about the market system in a lot of ways, it has pushed forward in dealing with control issues. Although I am as skeptical as other people as far as where these reductions are occurring, I think that in a full evaluation, this is a program that has worked and worked well.

One of the options is that, since we have a $4 billion cost savings, in terms of projections to actual costs, that it is like those emissions were on sale. I don't know about you, but we should be able to take that money and put it back into the program. I think it is something that we should focus on, let's spend that money that was on sale.

We need to include reductions of both sulfur and nitrogen oxides and continue to seek the most cost effective emission reductions. At the same time – and I realize this is an acid rain conference – at the same time, we have to look at the other pollutants. We need to come up with a multi-pollutant bill to deal with these issues.

I am going to invent a word here. The Clean Air Act has been too model-pollutant-centric.

We end up where we start , dealing with one issue . Focus on that, put the controls on that, and then move on to the next. I don't think that this helps the environment and I don't think it helps industry in terms of planning for the future, and the power industry needs a long planning horizon.

We really need to take a holistic approach in looking at this issue, and try to come up with the control systems that do the best benefit in terms of reducing the pollutants, all of them at some level.

Talking about states and acid rain programs 10 years ago, we are going through the same thing right now as far as state programs dealing with multiple pollutants. A lot of them start in the northeast, mainly because we want to have clean hands and we have a smaller number of power plants than they have in other parts of the country.

There are a number of programs that have been announced and are in the process of going through regulations in Massachusetts, New York, New Hampshire and Connecticut.

Massachusetts' recent announcement indicated that they were using a four-pollutant approach, looking at controlling carbon dioxide from their power plants.

Trading programs. If we have a patchwork of state programs out there with trading or have different values on what the trade is worth, we are never going to have a trading program that works.

I shouldn't be saying this talking for the states because, of course, all governors love flexibility. In this case, if we are going to make a trading program work, we need a national trading program to deal with these issues.

In conclusion, we do understand what the problems are, the sources and the most effective way to deal with them.

We need to work with industry to come up with solutions. If all else fails, we will have to – and we will do it state by state--go back to command and control. Thank you.

Compliance Experience of Regulated Entities

John McManus – American Electric Power

I grew up in upstate New York, went to college in upstate New York. While I was in college, I climbed a fair number of the high peaks in the Adirondacks. When I came to graduate school in 1976 and I read some of the early papers by Gene Likens and Ellis Cowling on this issue of acid rain, it looked like an interesting topic to do some research on, so that is what I worked on.

I learned a few things in graduate school, one being that trying to collect siltation samples without contaminating them is a tricky thing to do when you don't have a budget, which I didn't as a graduate student; that I am not very good as a laboratory chemist; and that acid rain is a fairly complex issue.

Twenty-five years later, we are still here talking about it, which suggests something else to me, that I may still have some employability with my company, because I know a little bit about this issue.

Looking at it from an industry perspective, some of what I will say you have already heard from Dr. McLean, because we are talking about the same program.

My company's experience on this is not a whole lot different than the industry in general. Some of the things that I say apply to a lot of companies in terms of things that we did in the 1980s, and the compliance measures that were taken in the 1990s. So, it is not just American Electric Power [AEP] that you will hear about here.

A little background on AEP, for those who aren't familiar, we operate in 11 states as a result of a merger that was completed last year with Central and Southwest Utilities. We have coal fired power plants and natural gas power plants, primarily in the mid west and in the middle south.

I will just give a few quick statistics which are somewhat relevant to this. We generate a fair amount of electricity. We have about 38,000 megawatts of capacity, of which 25,000 is coal, which is directly relevant to this issue.

Most of the coal capacity is in the mid west, about 20,000 megawatts of it, with the rest in Texas, Oklahoma and Louisiana. We have 4.8 million customers. We have a transmission distribution system, and 2000 revenues of about $13.7 billion. So, we are a fairly large utility, which is not a real surprise.

Getting more direct to this issue, our perspective on acid rain, AEP started to become engaged in the issue in the late 1970s, as it started to become both a scientific and a policy issue. That was not just a coincidence. That is when I started working with AEP and told my management, you should pay attention to this.

In the 1980s, we sponsored a fair amount of research on acid rain, as did the industry in general, through the Electric Power Research Institute. One particular project that we did was called the Utility Acid Precipitation Study Program, which was pretty much the state of the art precipitation program at the time.

It used many of the same techniques to collect and analyze precipitation as did the National Acid Deposition Program, but the UAPS program was actually a daily monitoring program and not a weekly program.

UAPS collected a fairly extensive database, primarily in the eastern part of the country, of daily precipitation chemistry, which allowed for a fair amount of analysis on what really drives the chemistry, more so than a weekly sampling program did.

AEP was active in a program called Living Lakes, which was looking at the concept of actually liming acidified lakes in the Adirondacks to get some immediate benefits, recognizing that an emission reduction program would take quite a bit of time before you would actually see a response in the ecosystem.

We closely followed the NAPAP program in the 1980s, and were actively engaged in the development of Title IV and eventually the regulation.

Once the allowance program was introduced and utilities started to get a better understanding of the concept of allowances and the value that they had, it became a real free-for-all to get as many allowances as possible.

Interestingly, I think the original bill that was introduced from the first Bush administration only affected sources above 1.2 pounds per million BTU, and the cleaner sources I don't believe are in the allowance program. Many of those sources were in the western part of the country, but it was still intended as a national program.

As those western companies realized that they were going to be left out of the allowance program and that new power plants would have to have allowances, everyone decided that they wanted to be part of the program. That was an interesting response, because usually companies like mine looked for ways to get out of these kinds of programs.

The introduction of the market approach and the allowance concept really did change the dynamic of this considerably. Once Title IV was passed and the regulations began, AEP, as well as a number of other utilities, were able to work with Dr. McLean's group on their acid rain advisory committee.

Looking at how we actually have complied with Title IV, the initial focus was on a self-sufficient compliance plan, and this was true of almost every single utility in the country, with the exception of one that I can think of.

Because the market concept was new, no one knew how active it would be, whether allowances would be readily available to use for compliance. Every company looked at ways to comply just within their system. So, the market itself did not play a significant role as a compliance tool, at least during most of Phase I.

For a company like AEP, which had at the time about 45 coal-fired units scattered across five states, the ability to use allowances for our system provided a tremendous amount of flexibility. Just being able to move allowances and develop a system-wide compliance plan for our own units, regardless of the rest of the market, resulted in a significant cost savings.

It did lead to some complexities. At the time, all of our states and all of our plants were regulated by state rate commissions, and each state had a different perspective on how they thought allowances should be handled and how allowances should be accounted for, as they moved across state lines.

The result of that is, we have, at AEP, a very complex approach to how we have to account for allowances. Dr. McLean may not know this so much, but I know there are some people who work in his division that I think hate AEP, because of the number of allowance transfers that we have to send to them when we are doing compliance or any allowance trades.

The allowance system is a great approach. It does have some underlying issues that, looking to future programs, hopefully could be handled in different ways.

The actual compliance steps that we took in Phase I were a combination of pollution controls, different fuels and using allowances. We put scrubbers on one of our largest plants in Ohio at a cost of a little over $600 million, did some fuel switches to lower sulfur coal. We converted three units to have the capability to burn natural gas, which we tested for about one month and have never used since then, especially when gas prices went to $10 this past winter. It is not a cost effective compliance measure.

When we look at Phase II, compliance, at least initially, will be through fuels, looking at different mixes of low and medium sulfur coal, and drawing down allowances from the bank that was built up during Phase I.

We don't anticipate putting scrubbers on, at least in the near term, and I don't think there have been many scrubbers at all installed in the industry as a Phase II compliance measure yet.

NO_X is part of the program. As has been pointed out, it was not the major focus of Title IV, and it was directed at combustion-type controls, not retrofit controls, or post-combustion controls.

We have installed low NO_X burners or essentially the equivalent of that on every one of our boilers to comply with the NO_X provision.

The result of this, you can see the reduction in SO_2 that occurred between 1994 and 1995. That was primarily a result of starting up the scrubbers on our Gavin plant (Slide 1).

You then see that SO_2 emissions on our system went up for a couple of years, and then it is starting to gradually come down.

That increase, I think, is a direct result of the allowance market. If you recall the allowance price graph that Dr. McLean put up, allowance prices

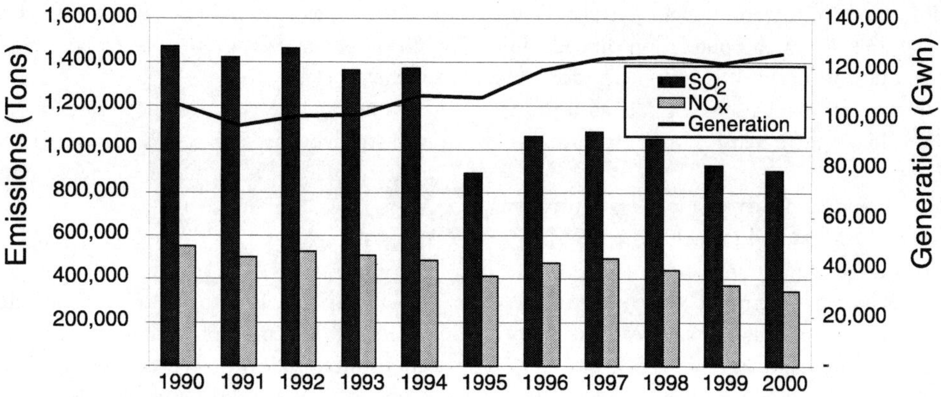

Slide 1. AEP East Emission Trends 1990-2000.

dropped significantly after 1995, hit a low in 1996, and then have gradually come back.

The coal market and the allowance market are directly interrelated. As allowance prices dropped, it became cost effective to look at higher sulfur coals, which even with the cost of sulfur embedded in them, became economical choices. I don't know if this type of result is reflected in the broader industry numbers, but we saw some direct relationships with the market price in our system.

NO_X reductions are much more modest. Again, the target for the program was more modest and at AEP, at least, we had very few units that were affected by the Phase I NO_X program. Now that we are in Phase II, all of our units have NO_X controls.

Now, where do we go from here? When you look at the Title IV program as well as some others, it is important to remember that there are still significant reductions in the pipeline.

There is a large bank of SO_2 allowances that was built up industry-wide during Phase I. That bank will gradually be drawn down.

The Title IV Phase II cap, if you look at it on an emission rate basis, people tend to think of it nominally as a low sulfur, 1.2 pound per million BTU program.

Because it is a cap, and because there has been a significant amount of growth since the mid-1980s, which was the baseline for the program, it effectively is, we think, about a 1.8 to 1.9 pound per million BTU program.

It is more stringent than people think it is, once the bank of allowances is used up. When you look at what fuels you could use for coal for compliance, to stay within that emission rate, you would effectively have to convert every coal unit to Powder River Basin coal, which isn't practical and isn't economical, and has a whole host of political implications for some of the states that we operate in.

We don't expect that to happen. What it says, though, is that there will have to be additional scrubbers installed under the Title IV program to stay within this cap, sometime in the next few years.

Looking at NO_X, the Title IV target was somewhat modest, but we do have another program starting now in the east, the NO_X SIP CALL.

We started up just this past week, SGR Systems, at our Gavin Plant. In at least the first few days, they are achieving 90 percent reduction in the NO_X emission rate. We, as well as many other utilities, will be installing additional NO_X controls. So, there is a lot of NO_X reduction that will occur in the next two to three years.

Just to point out, again, a value of a market-based approach, especially when you have built some incentives into it, the reason we started up our Gavin SGR systems on April 30 is that, starting May 1, we can earn early reduction credits as an incentive.

There are other regulatory programs that Dr. McLean's agency is working on that will directly affect sulfur, nitrogen and mercury emissions in the next few years, at least in this decade.

When you look at acid rain as an issue, do new

research findings justify additional controls? The Hubbard Brook Research Foundation put out this report a month or so ago. It is very nicely done. As I went through it, I had a sense of deja vu. I didn't see a whole lot in there that I thought was really new.

It was a good summary of a lot of information that I felt was understood in 1990 when NAPAP issued its final report.

So, it raises a question as to whether we are going over a lot of things that we already knew, or how much is really new, and that is one of the purposes of this conference and there should be some new information presented. The idea that it will take decades to see recovery of some of the aquatic systems in the northeast is something that was understood by NAPAP.

I guess I would argue that Congress designed the program with that understanding. Congress made a policy decision in 1990. Is it appropriate to revisit that policy decision now and change the program? That is an important question that will be looked at, and it raises the question of the potential for legislative action.

There is a lot of discussion on three pollutant bills, and three pollutant plus carbon bills. What are the realistic expectations that Congress will do something to change the Clean Air Act and address some of these problems in a more coherent coordinated fashion?

It comes down to what we think is the main question. Do we expect another 10 years of debate like we had in the 1980s? I certainly don't expect that. The real question, when we look at the existing fleet of coal-fired power plants in the country and what we do with them, is how much reduction do we want and how fast do we want it?

I think that is where the debate will be in the future, I am not sure how long it will take, a year, a couple of years.

We would hope that ultimately what is decided relies on market mechanisms. We think that the allowance program has been a tremendous success. It provides a lot of flexibility. It is obvious that it reduces costs and it is easy to administer.

Working with Dr. McLean's division on this program, although there are some issues where we have had some differences, it has been a very easy program, from our perspective, to administer. I won't say it has been a dream, but it has been much easier than some of the other environmental programs that we have.

So, market mechanisms allow, from an administrative perspective, a fair amount of simplicity as well. With that, I will take questions.

QUESTIONS

AUDIENCE: Isn't the trick to get the old clunkers off line and replace them with new, cleaner sources? How do we see that best acted?

MR. MC MANUS: I guess I take issue with old clunkers. We have some of our units that were built in the 1950s that have some of the best heat rates of any units on our system. For coal fired units, they are very efficient.

AEP in general, as a fleet, has one of the best average heat rates of any company in the country. We have some very good units that are older units that are very economical to run. So, I wouldn't consider them clunkers.

Should we be replacing them quickly or over time is the question. How fast do you do this?

There is a tremendous debate going on right now on energy. The energy systems in this country in the last year or so have gone in a direction that people didn't expect. Natural gas prices, gasoline prices, the electricity market in California, the coal market in the east, all have seen significant shifts.

The role that rail deregulation played in keeping allowance prices low is an important role, because it forced low sulfur coal providers in the east to keep prices down. That is changing now. Coal is in short supply, whether low or high sulfur, and we are seeing that reflected in allowance prices.

There is a major energy policy issue here. How quickly do we want to convert to a more efficient energy system? How much do we rely on coal? Is that something you can do in a short period of time? That is what we think is a key part of the debate. The idea of old clunkers and grandfathered power plants, we tend to disagree with that perspective. Especially when you look at Title IV, every coal plant in the country is now regulated for sulfur dioxide and nitrogen dioxides.

Granted, they are not as clean as if you built a brand-new power plant, but you can't turn over the fleet of power plants that provided over half the electricity that this country uses in a short period of time. It is just not practical.

AUDIENCE: Focusing on NO_X, I think it is a bit disingenuous to say that the Congress decided in 1990 not to pay a whole lot of attention to nitrogen. There was a widespread view that nitrogen wasn't so much of a problem because, after all, it was a useful fertilizer.

We have seen that that really isn't the case any more, both in the Chesapeake Bay, the job we did last year on the Adirondacks and other places. Even the SIPCALL and Section 126 stuff is not putting a cap on nitrogen. If 50 percent on sulfur is of questionable sufficiency, in the views of at least some, is 10 percent on nitrogen even a serious bite?

Then also on nitrogen, people probably know about the ozone formation study that was published in Science this past week, which raised the unfortunate problem that small flues are much higher producers of ozone than large flues, which sort of protect the NO_X from being converted into ozone.

The article didn't bother to mention that it protects it from being turned into ozone by turning it more into nitrate, which is nitrate deposition.

It looks like not only may there be a substantially greater need for NO_X control than has been done so far, but also that it may be more difficult.

MR. MC MANUS: Just a couple of comments on that. I am not sure that was a question as much as some useful insights.

I think Congress did place less emphasis on nitrogen in 1990. It was almost sort of an afterthought, because the debate in the 1980s was on sulfur.

The title IV NO_X program is a modest approach. It is combustion controls. It has reduced AEP's nitrous oxide emissions by about 25 percent, which is not as much as we get with SO_2.

The NO_X SIPCALL program is a fairly aggressive program. Granted, it is only in the summertime for five months. It does cap effectively emissions in those five months. For AEP, we will have to reduce our emissions in that five-year period by another 75 percent, on top of what we have done for Title IV. So, on an annual basis, that is still a fairly significant reduction in NO_X emissions.

When you look at the power plant sources of NO_X, in comparison to other sources, before the NO_X SIPCALL program, power plants contributed probably between 20 and 25 percent of total NO_X emissions in the United States.

With the reductions we have seen in the east for the SIPCALL, that is going to go below 20 percent – I am not sure how far below.

Transportation sources, which are now over 50 percent, will be, as a percentage, even higher.

If we really want to focus on nitrogen oxides, it is going to be important to look at more than just power plants.

It is not that you can't do more with power plants, year round versus just in the summer, but it may not get the benefit you think, when it is a smaller contributor to total NO_X than it was in the past.

MR. BENNETT: Ed Bennett with New York State. Just one question on the enforcement of the major modification provisions of the act and how this would play out with the Title IV provisions, could you comment on that?

MR. MC MANUS: I am not sure of the question.

MR. BENNETT: It is getting to the older units and the major modifications occurring with them, making them live forever. Like you say, they do operate, they operate well and they are 50 years old.

MR. MC MANUS: Specifically as it relates to the litigation brought by the Federal Government, New York State and some others, because it is active litigation, I am not going to say a whole lot about that.

I guess I would take issue with major modification versus routine maintenance. Certainly what we have done has been consistent with the regulatory program.

We fully have understood that program, and having worked in our environmental department for my 24 years with AEP, I know, as we considered what we did at our power plants, as we considered new plants, we did it all within the context of that regulation.

That will play out eventually in the next year or two, and we will see where that takes us.

SESSION II: Research and Analysis - What the Reports Say

Bernard Melewski - Adirondack Council

MR. MELEWSKI: I am the legislative director and counsel for a not-for-profit organization in New York called the Adirondack Council.

Just a little bit about us. All we do is work to protect the Adirondack Park in New York. For the vast majority of you who are not from New York, the Adirondack Park is about six million acres. It is about 20 percent of the total land mass in New York State.

It is the largest park of any kind in the lower 48 states. It contains the largest accumulation of wilderness east of the Mississippi and it is systematically being destroyed by acid rain, despite the fact that we have had, for over 100 years, a provision in our state constitution that deems state land within our park, which is about three million acres now, as forever wild, to be forever protected.

So, we have a serious problem in the Adirondacks, and the Adirondack Council has called upon our Congressional Delegation to try to get Congress to pay attention to the issue and see if we can address it in the future.

One of the important tools for us has been the plethora of reports that have come out of a variety of sources since the 1990 Clean Air Act amendments.

Two of them, as we will discuss, with some forethought, were required by Congress, those from the U.S. Environmental Protection Agency, and also from the National Acid Precipitation Assessment Program.

We have three speakers here who are going to touch on some of these reports. I just wanted to give you a brief idea of the quite large number of reports that have come out.

In 1994, the General Accounting Office (GAO) reviewed air pollution and allowance trading. It concluded that a cap and trade allowance program offers an opportunity to reduce emissions at less cost.

In 1995, the U.S. EPA released the "Acid deposition standard feasibility study" report to Congress.

In 1997, the U.S. Geological Survey, produced "Acid Rain in our nation's capital. " It is actually a walking guide to the effects on buildings and monuments here in Washington, D.C. from acid rain.

In 1998, the National Acid Precipitation Assessment Program, "The Biennial Report to Congress. "

In 1998 the Adirondack Council released "Acid Rain – a Continuing National Tragedy. "

In 1998, Trout Unlimited published "Acid Rain, Current and Projected Status in the Coldwater Fish Communities of the Southeastern United States. "

In 1999, the U.S. Geological Survey, "Soil Calcium Deposition Linked to Acid Rain in the Eastern United States."

Again, in 1999 and released in 2000, the Ecological Society of America, "Acid Deposition, the Ecological Response. "

In the year 2000, GAO again, "Acid Rain, Emissions, Trends and Effects in the Eastern United States. " Also, EPA did an analysis of the acid deposition in the Acid Rain Control Act, which was then sponsored by Senator Moynihan, Senate bill number 172.

Most recently the Hubbard Brook Research Foundation, "Acid Rain Revisited, Advances in Scientific Understanding since passage of the 1970 and 1990 Clean Air Act Amendments. "

Many of these documents are available on the literature table or, if you ask any of our speakers, you can get access to these documents.

Acid Rain Revisited

Charles T. Driscoll – Syracuse University

Elevated emissions of sulfur dioxide, nitrogen oxides and ammonia have resulted in acidic deposition and subsequent effects on terrestrial and aquatic resources in the northeastern U.S. and elsewhere in North America, Europe and Asia. These effects include changes in soil chemistry, stress to forest vegetation, acidification of surface waters and disturbance to aquatic biota. In this overview, a summary will be given : 1) of spatial and temporal patterns in emissions of sulfur dioxide and nitrogen oxides and atmospheric sulfur and nitrogen deposition; 2) effects of acidic deposition on soil, forest vegetation and surface waters; and 3) the response of terrestrial and aquatic ecosystems to possible future controls in emissions of air pollutants.

Emissions of sulfur dioxide in the U.S. increased markedly after the Industrial Revolution, peaking in early 1973 at 28.8 million metric tons (USEPA 2000). These emissions were largely (~60%) associated with electric utilities. In contrast, emissions of nitrogen oxides increased more gradually through the early decades of this century, increased more prominently after World War II, and peaked at 21.8 million metric tons in 1990. Nitrogen oxide emissions have remained relatively uniform since about 1980. Following the 1970 Amendments to the Clean Air Act, there have been marked reductions in emissions of sulfur dioxide. Emissions have decreased 32% from 1973 to values of 19.6 million metric tons in 1998 largely due to controls on electric utilities. These controls have resulted in decreases in concentrations of sulfate and hydrogen ion in atmospheric deposition in the eastern U.S. since the early 1970s (Likens et al. 1996). There is a strong relationship between sulfur dioxide emissions in the emission source area for the northeastern U.S. and atmospheric sulfur deposition for sites in the Northeast (Driscoll et al. 2001). Moreover with Title IV of the 1990 Amendments of the Clean Air Act there have been additional reductions in sulfur dioxide emissions and atmospheric deposition of sulfate and hydrogen ion since 1995 and these reductions are anticipated to continue until 2010.

Elevated emissions of sulfur dioxide, nitrogen oxides and ammonia have resulted in high atmospheric deposition of sulfur and nitrogen in the northeastern U.S.. This disturbance has had deleterious effects on forest ecosystems, including soil, vegetation and surface waters. The response of soil to acidic deposition includes: 1) accumulation of elevated pools of sulfur and nitrogen, 2) depletion of nutrient cations (i.e., calcium, magnesium) from soil exchange sites, and 3) the mobilization of aluminum from soil, resulting in high concentrations of inorganic forms of aluminum in waters draining soil (Driscoll et al. 2001). Clearly acidic deposition has accelerated the acidification of base-poor forest soils.

There is considerable concern about the potential effects of elevated atmospheric deposition on trees. However, it is difficult to make a direct linkage between acidic deposition and forest health because trees are exposed to many diverse stresses. Nevertheless, recent studies have suggested that exposure of high elevation red spruce forests to elevated concentrations of hydrogen ion in atmospheric deposition reduces cold tolerance sufficient to cause freezing injury (DeHayes et al. 1999). Linkages for other tree species are less well established but acidic deposition clearly accelerates the leaching of nutrient cations from foliage and soil, causes the interference of root function by increased availability of aluminum and competitive inhibition of calcium, magnesium and phosphorus assimilation (Cronan and Grigal 1995). Much current research is directed at establishing a linkage between acidic deposition and decline of sugar maple (Horsley et al. 2000).

Elevated atmospheric inputs of sulfate, nitrate and ammonium have had clear effects on surface waters in acid-sensitive regions of the Northeast. Based on the U.S. Environmental Protection Agency Environmental Monitoring and Assessment Program (EMAP) program, of 1,812 lakes in the Adirondack region of New York (>1 ha) 10% are chronically acidic (acid neutralizing capacity < 0 μeq/L) and 31% potentially experience episodic acidification (acid neutralizing capacity < 50 μeq/

L; Driscoll et al. 2001). In 6,834 lakes (>1ha) in New England 5% are chronically acidic while 10% are susceptible to episodic acidification. A total of 1,875 lakes in the Adirondacks and New England are sensitive to acidic deposition (acid neutralizing capacity < 50 μeq/L). For a large fraction of these lakes (83%) the composition of anions is dominated by sulfate plus nitrate (> 50% of the total anion composition) and therefore have been acidified by acidic deposition. The anion composition of the remainder of these lakes (17%) is dominated by organic anions. While atmospheric deposition contributes to the acidity of these lakes, they are likely naturally acidic due to high inputs of organic acids.

Episodic acidification occurs when a surface water experiences a short-term decrease in acid neutralizing capacity that is generally associated with hydrologic events (i.e., precipitation events, snowmelt). Several mechanisms contribute to episodic acidification, including dilution of basic cation concentrations and increases in concentrations of sulfate, nitrate and organic anions (Wigington et al. 1996). In surface waters with low acid neutralizing capacity values (<50 μeq/L) in the Northeast, short-term increases in concentrations of nitrate are the dominant mechanism of episodic acidification (Schaefer et al. 1990). These increases in nitrate coincide not only with decreases in acid neutralizing capacity but also marked increases in inorganic monomeric aluminum.

Decreases in pH and increases in concentrations of inorganic monomeric aluminum have diminished species diversity and the abundance of plankton, invertebrates and fish in acid-impacted surface waters of the Northeast. In the Adirondacks, a significant positive relationship exists between pH and acid neutralizing capacity of lakes and the number of fish species present in those lakes (Gallagher and Baker 1990). A survey of 1,469 Adirondack lakes conducted in 1984 through 1987 show that 24% of the lakes in this region (i.e., 346) do not support fish. These lakes had consistently lower pH and acid neutralizing capacity and higher concentrations of aluminum than lakes that contained one or more species of fish. Even acid tolerant fish species such as brook trout have been eliminated from some waters in the Northeast.

Acid episodes are particularly harmful to aquatic life because abrupt changes in water chemistry allow fish few areas of refuge. High concentrations of inorganic monomeric aluminum are directly toxic to fish and are the primary cause of mortality during acidic episodes. High acidity and concentrations of inorganic monomeric aluminum disrupt the salt and water balance of fish, causing red blood cells to rupture and blood viscosity to increase. Studies show that viscous blood strains the fish's heart, resulting in a lethal heart attack (McAvoy and Bulger 1995).

Recovery from acidic deposition involves decreases in emissions resulting from regulatory controls, which in turn leads to reductions in acidic deposition and allows for chemical recovery. Terrestrial and aquatic biota have clearly been impacted by acidic deposition. Critical chemical conditions for effects on terrestrial biota include a molar ratio of exchangeable calcium to exchangeable aluminum in soil of < 1.0 and soil percent base saturation of < 20% (Cronan and Grigal 1995). Critical chemical conditions for effects on aquatic biota include surface water pH values < 6.0-6.5, values of acid neutralizing capacity < 50 μeq/L, and concentrations of inorganic monomeric aluminum > 1-2 μmol/L (Driscoll et al. 2001). When emission controls and subsequent decreases in acidic deposition are sufficient to ameliorate adverse chemical conditions such that these thresholds are not exceeded, biological recovery will be enhanced. Biological recovery is likely to occur in stages; since organisms vary in sensitivity to acidic deposition they will not recover at the same rate. Current understanding of species response to improvements in chemical conditions is incomplete, but research suggests that stream macro-invertebrates may recover relatively rapidly (i.e., within 3 years), while lake zooplankton may take more than a decade to fully re-establish (Gunn and Mills 1998). Fish populations in streams and lakes should recover in 5-10 years following recovery of macro-invertebrates and zooplankton which serve as food sources. It is possible that, with improved chemical conditions and the return of other components of the aquatic food web, the stocking of streams and lakes could help accelerate the recovery of fish. The recovery of terrestrial ecosystems is even more difficult to predict than aquatic recovery. Given the life span of trees and delays in the response of soil to decreases in acidic deposition it is reasonable to suggest that decades will be required for affected trees on sensitive sites to recover once chemical conditions in the soil are restored.

The time required for chemical recovery varies widely among ecosystems in the Northeast. This variability is a function of: 1) historical rate of atmospheric sulfur and nitrogen deposition; 2) the rate and magnitude of decreases in acidic deposition; 3) the extent to which exchangeable basic cations have been depleted from soil, 4) the extent to which sulfur and nitrogen have accumulated in soil and the rate at which they are released following decreases in atmospheric deposition; 5) the weathering rate of soil minerals and the associated supply of basic cations to the ecosystem; and 6) the rate of atmospheric deposition of basic cations.

Since widespread monitoring of acid-sensitive surface waters was initiated in New York and New England in the early 1980s, changes have been evident in their chemical composition. There have been uniform decreases in surface water concentrations of sulfate in the Adirondacks, Catskills and New England in response to decreases in sulfur dioxide emissions and atmospheric deposition of sulfate (Stoddard et al. 1999). There have been no systematic trends in concentrations of nitrate. In New England these changes have coincided with modest increases in the acid neutralizing capacity of surface waters. In contrast there has been no significant recovery in the acid neutralizing capacity of Adirondack and Catskill surface waters. The reason for these regional differences in rates of recovery appears to be due to the extent of depletion of exchangeable nutrient cations in soil (i.e., calcium, magnesium). Higher inputs of acidic deposition in New York may have facilitated greater leaching losses of exchangeable nutrient cations than in New England thereby delaying the recovery of surface waters in response to decreases in acidic deposition.

Scientists and engineers have developed models that depict the physical, chemical and biological processes within forest watersheds. Acidic deposition models can be used as research and management tools to investigate factors responsible for the historical acidification of soil and water, as well as ecosystem response to anticipated future changes in acidic deposition. The model PnET-BGC was used to assess past and potential future changes in soil and stream chemistry in response to changes in acidic deposition at watershed 6 at the Hubbard Brook Experimental Forest, New Hampshire (Gbondo-Tugbawa et al. 2001; Gbondo-Tugbawa and Driscoll 2002; Table 1). Hindcasts suggest that prior to 1850 and the advent of acidic deposition, the soil base saturation at Hubbard Brook was 22%, stream sulfate concentration was 10 µeq/L, acid neutralizing capacity was 40 µeq/L, pH was 6.3, and concentration of monomeric aluminum was 1.5 µmol/L. By 1970, near the peak of acidic deposition, the soil percent base saturation decreased to 12%, stream sulfate concentrations increased to 60 µeq/L, acid neutralizing capacity decreased to 5 µeq/L, pH decreased to 4.8 and monomeric aluminum concentrations increased to 12 µmol/L. Subsequently model predictions of future changes in soil and stream chemistry were made to assess ecosystem response to emission controls associated with the 1990 Amendments of the Clean Air Act and additional reductions in emissions of sulfur dioxide emissions for electric utilities (Table 1). This analysis suggests emission controls have started recovery of acid-impacted forest ecosystems in the Northeast from acidic deposition. However, despite aggressive controls on emissions from electric utilities the rate of chemical recovery of acid-impacted soils and surface waters, like those at the Hubbard Brook Experimental Forest, will be slow.

Acidic deposition is a pervasive problem that has had greater effects on soils, trees and surface waters than had been previously projected. Although the 1970 and 1990 Amendments of the Clean Air Act have had positive effects, emissions remain high compared to background conditions. Given the accumulation of sulfur and nitrogen and the loss of available nutrient cations in soil, many forested areas of the Northeast have become more sensitive to inputs of acidic deposition. These effects will delay the rates of chemical recovery following decreases in atmospheric deposition.

Nevertheless model calculations suggest that deeper emissions cuts and faster cuts will accelerate recovery from acidic deposition.

QUESTIONS

MS. BRADT: Patricia Bradt, Muhlenberg College. How do you anticipate a recovery of the calcium and the magnesium baselines in soil? Do you think that is going to happen? Is it going to leach in from the bedrock?

MR. DRISCOLL: It is a very slow process. These soils have been developing over 14,000, 15,000 years. They are very low in bases to start with because of the nature of the mineralogy.

To redevelop those base cations will take a very long period of time. That is a large part of the reason why we expect recovery to be so slow.

If you want, you can lime. Liming experiments have been done, but that is an artificial way. If you allow nature to take its course, you have to reduce the inputs of acids and allow it to redevelop, which will be a long time frame.

MS. BAUM: Ellen Baum with the Clean Air Task Force. Is there any point where recovery won't happen, if there isn't a certainly level of cut? Could it be so delayed or the input so high that, even with some level of reductions, we are not going to get recovery?

MR. DRISCOLL: I guess it depends on what you mean by recovery. In the Adirondacks, our model calculations suggest that some of those sites that are very sensitive and showing higher loadings than we see in parts of New England, are still losing base resources. Those calculations would suggest that we are still losing this very important stock of material.

In general, I think that, except for maybe the

Table 1. Results for projections with PnET-BGC showing past and projected future values of critical chemical of stream water and soil at watershed 6 at the Hubbard Brook Experimental Forest, New Hampshire. Background conditions were determined based on historical reconstructions of atmospheric emissions and deposition. Future prediction were made for the year 2050 under the scenarios of the 1990 Amendments of the Clean Air Act (CAAA) and the 1990 CAAA plus an additional 40% reduction in utility emissions of sulfur dioxide and the 1990 CAAA plus an additional 80% reduction in utility emissions of sulfur dioxide.

	Background conditions			**Emission scenarios**		
Chemical indicator	1850	threshold	1970	2050 with the 1990 CAAA	2050 with 40% addition utility SO_2 reduction	2050 with 80% addition utility SO_2 reduction
Wet sulfate deposition (gS/m^2-yr)	0.23	N/A	1.7	0.9	0.7	0.5
Stream sulfate (µmol/L)	10	N/A	62	34	29	24
Stream acid neutralizing capacity (µeq/L)	43	>50	-5	-1	1	3
Stream pH	6.3	>6.0	4.8	5.3	5.5	5.7
Stream aluminum (µmol/L)	1.5	<2.0	12	6.4	5.6	4.3
Soil % base saturation	22	>20	13	12	13	14

most sensitive sites, we have by and large turned the corner and we are starting to move into a direction of recovery driven by these lower sulfate concentrations.

The big question is how long is that going to take. At the current levels for sensitive sites, we are talking many, many decades before we would see significant recoveries in chemical indicators that we would expect to lead to improvement in biological resources.

References:

Cronan CS, Grigal DF. 1995. Use of calcium/ aluminum ratios as indicators of stress in forest ecosystems. Journal of Environmental Quality 24: 209-226.

DeHayes DH, Schaberg PG, Hawley GJ, Strimbeck GR. 1999. Acid rain impacts calcium nutrition and forest health. BioScience 49:789-800.

Driscoll CT, Lawrence GB, Bulger AJ, Butler TJ, Cronan, CS, Eagar C, Lambert, KF, Likens GE, Stoddard JL, Weathers KC. 2001. Acidic deposition in the Northeastern United States: Sources and inputs, ecosystem effects, and management strategies. BioScience 51:180-198.

Gallagher J, Baker, J. 1990. Current status of fish communities in Adirondack Lakes. Pages 3-11 to 3-48 in: Adirondack Lakes Survey: An Interpretive Analysis of Fish Communities and Water Chemistry, 1984-1987. Ray Brook (NY): Adirondacks Lakes Survey Corporation.

Gbondo-Tugbawa SS, Driscoll CT. 2002. Evaluation of the effects of future controls on sulfur dioxide and nitrogen oxide emissions on the acid-base status of a northern forest ecosystem. Atmospheric Environment 36:1631-1643.

Gbondo-Tugbawa SS, Driscoll CT, Aber JD, Likens GE. 2001. Evalution of an integrated biogeochemical model (PnET-BGC) at a northern hardwood forest ecosystem. Water Resources Research 37:1057-1070.

Gunn JM, Mills KH. 1998. The potential for restoration of acid-damaged lake trout lakes. Restoration Ecology 6: 390-397.

Horsley SB, Long RP, Bailey SW, Hallet RA, Hall TJ. 1999. Factors associated with the decline of sugar maple on the Allegheny Plateau. Canadian Journal of Forest Research 30:1365-1378.

Likens GE, Driscoll CT, Buso DC. 1996. Long-term effects of acid rain: response and recovery of a forest ecosystem. Science 272: 244-246.

MacAvoy SE, Bulger AJ. 1995. Survival of brook trout (*Salvelinus fontinalis*) embryos and fry in streams of different acid sensitivity in Shenandoah National Park, USA. Water Air and Soil Pollution 85: 439-444.

Schaefer DA, Driscoll CT, van Dreason R, Yatsko CP. 1990. The episodic acidification of Adirondack lakes during snowmelt. Water Resources Research 26:1639-1647.

Stoddard JL, Jeffries DS, Lukewille A, Clair TA, Dillon PJ, et al. 1999. Regional trends in aquatic recovery from acidification in North America and Europe. Nature 401: 575-578.

[USEPA] U.S. Environmental Protection Agency. 2000. National Air Pollutant Emission Trends, 1900-1998. Washington, DC: U.S. Environmental Protection Agency, Report EPA – 454/R-00-002.

Wigington Jr PJ, Baker JP, DeWalle DR, Kretser WA, Murdoch PS, Simonin HA, Van Sickle J, McDowell MK, Peck DV, Barchet WR. 1996. Episodic acidification of small streams in the northeastern United States: Episodic Response Project. Ecological Applications 6: 374-388.

National Acid Precipitation Assessment Program Analysis

Michael Uhart – National Acid Precipitation Assessment Program (NAPAP)

I am here to provide you a little bit of background, not to share any new information. You will be hearing the new information from people who are much more capable of providing it than I am.

I am here to tell you what has happened in the past as far as NAPAP is concerned, and to provide some perspective on what you might hear from the remaining speakers today and tomorrow.

Basically, I have broken this talk down into two pieces. One is NAPAP in the 1980s. It was a focused monitoring and research program. Altogether, agency contributions were about $500 million.

There was, as you have heard, quite a bit of NGO and utility and state involvement. The result, not only from the assessment, but also from the annual reports, the hearings, as Dr. Bernabo mentioned before, was Title IV of the 1990 Clean Air Act Amendments.

In Title IX of those Amendments, NAPAP was reauthorized to continue, but its role was changed.

NAPAP still must monitor all that is necessary to produce assessments. Basically our role now is to assess the effectiveness of Title IV and the costs and the benefits, which is good, because it is not very often that somebody gets a chance to actually tell Congress how good a job they did in enacting legislation. That is essentially what NAPAP does.

NAPAP's mandate was to coordinate federal acid rain research and monitoring, evaluate the status and trends of certain variables I will talk about, and to conduct periodic assessments.

Those assessments basically come out every two years. We are asked to essentially answer two questions, and I say "we" meaning the NAPAP agencies which, besides NOAA, also include the Departments of Energy, Agriculture, Interior, EPA and NASA.

We are asked to answer two questions, what are the costs, benefits and effectiveness of Title IV, and what reduction and deposition rates – notice this is deposition rates and not emissions – are needed to prevent adverse ecological effects. One major challenge to us is to define what is an adverse ecological effect that is measurable.

Our report to Congress was delivered to Congress in 1998. It is available on the web. It is also available on CD ROM.

Of the statistics I have for 2001, we have had over 10,000 requests from 2,000 different users in these first few months.

The scope of the assessment was emissions, amine concentrations and deposition up through 1995. Costs and benefits were often in sections of the assessment.

The assessment focused on what we have learned since 1990. Basically, we had the 1990 integrated assessment and the benefits of those 10 years in the 1980s to work from.

Assessments are incremental. We don't reinvent the wheel. It is a lot easier to see what has happened since the last time we did an assessment.

Assessments provide incremental improvements in scientific knowledge. For that reason, assessments shouldn't be considered to be separate items. I think it is important to connect periodic assessments. Otherwise, you really don't have an improvement and a general consensus of knowledge over a period of time.

We also emphasized that the first year was bad timing. We were asked to do the first assessment in 1996. Yet 1995 was the first year of compliance as far as sulfur dioxide was concerned. As far as looking at the effectiveness of Title IV, it was rather difficult, because we only had one year to work with.

We broke down the assessments into several specific questions. What are the economic benefits related to effects areas? What is the effect of Title IV?

We used several measures of effectiveness. Basically, we looked at state variables or what might be considered state variables.

The effectiveness variables that we looked at were emissions, amine concentrations, depositions and, to the extent possible, actually the response of sensitive receptors, and you will see the list of sensitive receptors that we included in the assessment.

Also, if you are going to understand what the is going on, you have to look at the relationship between those sensitive receptors, how do you get from emissions to deposition, how do you get from deposition to responses of sensitive receptors.

Those effectiveness measures I mentioned – emissions, deposition, et cetera – here is where my pitch comes in. Those come from monitoring.

They don't come from anything else. They come from consistent monitoring and, since this is a national program, it has to be some sort of nationally coordinated or cooperative kind of monitoring.

Without that kind of monitoring, we would not have an assessment. Even the poorest assessment, you couldn't do, unless you had monitoring. If we didn't have any continuing emissions monitors and we didn't even have the old method of coming out with emissions, how would we know what happened with emissions. So, monitoring is absolutely necessary to any kind of assessment.

If you are also going to take a look to satisfy policy makers as to what is the cause and effect relationship, how do we know this is what causes that, and also to help them decide what they are going to legislate or what kind of rules they are going to implement or what kind of programs they are going to design, you have to give them some sort of scenario, what the is going to happen in the future.

For these two factors, you need models. Many of the people you are going to listen to today run these models. These models are absolutely important if we are going to understand the relationships that are going on, and be able to predict the relationships in some future scenarios.

Nationally, monitoring is jeopardized. You might hear it here or there but the National Atmospheric Deposition Program (NADP) is under constant threat of being reduced.

Aquatic monitoring networks are being reduced. Some of these monitoring networks that are essential for future assessments, not only for acid rain but for also other environmental issues, are being jeopardized.

A lot of those efforts are, for the most part, coordinated through the federal government. As far as modeling is concerned, the federal government has a significant part, EPA doing much of that, NOAA also doing parts, and all the other agencies involved.

Also, a lot of it is done by institutions and people like Dr. Driscoll.

What NAPAP does is to put all this together and to come up with some sort of unbiased consensus as far as the executive branch of government is concerned.

Now, in the last assessment, just some very general findings that we came up with. We had good SO_2 reductions and NO_X reductions were very limited. Both of those outcomes were expected.

Following that, you would expect sulfate deposition decreased, but little change in nitrate deposition.

A lot of these things were pretty much known to everybody. NAPAP assessments aren't designed to come up with the newest things. The reason that we do the assessments is to bring a lot of different people together, a lot of information together, and to come up with some sort of consensus, so you don't have the "I say this" and "I say that" kind of business.

There are limited observable ecosystem responses to Title IV. A lot of this has to do with confounding factors, the lack of monitoring information and just lack of information on anything other than on a very local basis.

As was mentioned, even though in 1990 it was known that nitrogen does play a significant part, I think over the years we have been getting more and more appreciation for exactly what the role of nitrogen is.

Compliance costs – you have heard this. The compliance costs are less than what we thought they were going to be. Emissions trading has been effective. Brian McLean touched on this, so I won't waste your time.

Aquatic ecosystems – you will see updates on most of this stuff over the next day and a half or so. Many lakes and streams have experienced decreases in sulfate concentrations.

There is evidence of some recovery in the New England lakes. Mind you, this is what NAPAP came up with in its 1996-1997 assessment. So, things might have changed since then.

It was said throughout the report in many sections that additional reductions of sulfur and nitrogen deposition would be necessary for recovery of sensitive Adirondack lakes, as well as other ecosystems.

Other ecosystems – including forest ecosystems, we reiterated the fact that high elevation spruce fir forests are the most sensitive and are seeing quite adverse consequences directly related to acidic deposition.

Most forest ecosystems in the east, south and west are not currently adversely affected, and the key word there is currently.

There were statements in the assessment saying, basically, if you continue deposition, even though they are not currently affected, they will become affected, and if they are not reduced especially in areas where deposition is already high, then those forests could see some very adverse consequences, mainly due to chronic multiple-decade exposure.

Obviously, high deposition rates have an impact on episodic acidification also. There is a gradual leaching of soil nutrients. You have heard this and you will continue to hear it in the next day and a half. At least as far as sulfur is concerned, its role in these soil nutrients, reductions in sulfur deposition, at least in the short run, will see some improvements in sensitive forest.

Visibility was a difficult problem because a lot of the modeling that we had hoped that we could take advantage of wasn't there.

Luckily, visibility is a pretty much straight physics kind of question. Atmospheric science is very well known as far as visibility is concerned. So, pretty much general models can be used. There is a very good relationship between ambient concentrations and visibility.

However, sulfate concentrations were 15 percent lower, and this is just in the eastern United States.

We looked at particular areas, as far as visibility was concerned, but we didn't have exactly the modeling to show what exact sources impacted those areas.

Even though the concentrations were 15 percent lower, we didn't see the atmospheric science reduction of eight percent that would have been expected. That could have been for a variety of reasons.

Materials and cultural resources – the important factor here is that cultural resources are at greater risk, or the monetary value of that risk is higher than purely operational sources. These assets include buildings, bridges, cars, those kinds of resources.

Then we got into the physics, a little bit of it, exactly what the causes were of some of these problems. That was also referred to in a book by the National Park Service that was mentioned earlier.

One of the things that NAPAP did was bring human health into the forefront a little bit, and basically make connections. This is where the most troubling part in our clearance process was, how exactly we worded the relationship between fine particles and human health, since they are based on epidemiological studies.

Those studies don't prove a cause and effect relationship, but they are pretty good at saying what most people believe is actually happening.

There was a lot of talk back and forth of exactly how much we could say as far as fine particles, sulfates and nitrates included, which are a large fraction of total fine particles.

It turned out that human health effects are a significant part of the benefits of reducing acidic deposition, primarily its precursors.

I mention agriculture. I don't know why; I guess it is that everybody expects to see agriculture there.

Basically, nitrogen and sulfur are nutrients, but the biggest factor for agriculture is ozone damage not damage directly from NO_X or SO_2.

The economic benefits – Dr. Burtraw will be talking more about the economic benefits later. The human health and visibility were two things that we were able to evaluate. I think Dr. Burtraw did a very good of evaluating those.

They are very significant. They are much greater than what the compliance costs are, and basically, the economic benefits in the future are pretty well unknown at this time. It depends on a lot of economic factors. I think Dr. Burtraw can tell us exactly what is going to happen to the economy over the next five or 10 years.

The linkages we did mention. I think the linkages are important when we talk about control of multiple pollutants.

We have tropospheric ozone – fine particles and health, sulfates and radiated forcing as a climate issue, energy production and deposition of nutrients and heavy metals.

Those are all linkages that have to be considered. They are not being strictly considered by NAPAP, but I think they should be considered. It is not in our mandate to look at those things.

Also, as I mentioned before, this is an incremental assessment. We are going to be doing another one. The drafting is in progress.

We will undergo peer and agency review and clearance. I have no idea how long that will take. It took at least three times longer last time than everybody told me it would take, but it is a necessary process.

The whole idea is that the assessment will be delivered to Congress hopefully late in this calendar year. This is an executive department report to Congress. So, basically this is an agency document, although we had several authors who were not in the agencies and the peer reviewers are not in the Federal Government. They are all external peer reviewers.

I want to make one last point. Assessments are primary. I mentioned that before. I think periodic assessments are necessary so that you can improve the science or at least improve the understanding of the science by the people who are making the decisions. With that, I will end.

Questions

MR. KAHL: Steve Kahl, University of Maine. You commented on the New England situation and whether there might be an update. Indeed, the conclusion that there is recovery in the northern New England lakes is no longer correct. That might have been true through the early 1990s, but it is not true today.

I cannot speak for Vermont because they are represented, but for New England – New Hampshire, Massachusetts – we cannot say we are in a recovery phase.

MR. UHART: I know it is not a question, but I am going to take your statement and run with it. You know, you have Maine, Vermont. That is one of the problems when you do a national assessment, is how to put those thing together.

As I mentioned before our last assessment, we were put at a disadvantage in coming out with an assessment when acid rain controls under Title IV were only in existence for one year. That is what happens. That is another reason why you have to do assessments every four years. You can't just do a snapshot.

MS. BRADT: Patricia Bradt, Muhlenberg College in Allentown. We haven't seen any recovery in eastern Pennsylvania either, to reinforce what was said for Maine.

MR. PENNINGTON: Peter Pennington, Shepwood(?) Environmental Institute. You said that monitoring is under threat, and that your aquatic monitoring in particular is decreasing. Could you give some more information on that, and are you likely to be using NGOs to bring in that information?

MR. UHART: As far as atmospheric deposition monitoring, our NADP program, there are many non-federal members of that network, who are the mainstays of the network. After all, we are asked to talk about deposition and not necessarily emissions.

There are a lot of partners in that network. It is a very long-standing network. There are budgetary threats from the federal side and there are also continuing budgetary threats from the states that are involved, because some of their budgets are going down, too. So, that is a problem.

MS. LAMBERT: This is Kathy Lambert from the Hubbard Brook Research Foundation.

Just to give a few specifics, some of us have worked to put together a two-page description of monitoring needs, that outlines five major monitoring programs related to acid rain and their funding needs.

We will make some copies and put that out on the table. As many of you know, we are in the middle of the budget process, so this is a good time to get familiar with these needs and contact your members of Congress.

So, the need is quite substantial, both in terms of annual operating costs and modernizing equipment which is, in many cases, World War II vintage equipment. I will put these out there, if someone is interested.

MR. UHART: I will follow up on that just a little bit. The Committee on Environment and Natural Resources also put out a document on the need for continued monitoring.

In regard to reviews, besides the external peer review, there are two primary reviews that this next assessment will come under. One is from the CENR itself. NAPAP is made up of six agencies. Those six agencies are principals in the CENR. So, rather than go through the agency review, it was felt a couple years back that it would probably be more efficient to go through a CENR review, and those agencies would be included anyway.

Following that is a review by the National Science and Technology Council (Office of Management and Budget).

Obviously, a lot of the statements that are made here are going to have funding implications. That is where OMB jumps in and that is where probably the most interesting part of the clearance process will occur.

Beyond Title IV: Perspectives on Additional Reductions

Rona Birnbaum – Clean Air Markets Division, U.S. EPA

I am going to touch on a couple of areas, in terms of Title IV and perspectives looking beyond Title IV. Then I am going to talk about some of the work that we have been doing to look at, evaluate, and project potential future scenarios and the implications of those. I am going to talk about using the existing mechanism and its continued and appropriate applicability for any future reductions.

Back in 1995, the EPA released the Acid Deposition Standard Feasibility Study report to Congress, where we for the first time introduced the notion that it looks like, based on improved models, we may not be able to see a recovery that we had hoped to see when the 1990 amendments were passed. This issue, this concept, hasn't just been created in the last year or two years.

Likewise, EPA issues other reports. The Great Waters report to Congress is another report that comes out every two years or three years, I believe. It is there, too, that we see among the issues associated with atmospheric depositions, the issue of nitrogen deposition and its impact on coastal eutrophication.

There are other numerous EPA documents that reinforce the need to continue to look at atmospheric deposition and project the impacts.

Other reports came out. The Ecological Society of America (ESA) report back in 1999, looking at the ecological response, documented many of the things that will be discussed here, basically that in some regions we are seeing recovery and in other regions we are not seeing recovery, and reductions under the Title IV program may not be sufficient.

Likewise, looking at the nitrogen deposition in coastal areas, ESA also released a report in 1997 that raised concerns whether further reductions in nitrogen might be necessary to address the coastal issue.

There is a General Accounting Office report that was recently mentioned, that came out just last year. It focused primarily on some of the most sensitive ecosystems, such as those in the Adirondacks, also pointing to the fact that the expectation for recovery may not be as apparent as we thought it would be, and emphasizing nitrate levels in particular. Finally, you will hear about Hubbard Brook work that just was released.

So, we have here sort of an evolution of where we were and where we are right now. Basically, in 1990, the expectation was that a 50 percent reduction in SO_2 emissions from the utility sector was going to do something rather substantial, that the effect would be rather rapid in terms of surface water chemistry anyway, and it also documented, of course, that without doing anything, things would get worse.

In 1995, again, EPA and NAPAP started reflecting on the notion that actually our improved models are beginning to tell us that reductions may not be sufficient, but there are still some issues to be looked at. Now, by 2001, the results of long-term monitoring and the results of efforts like the Hubbard Brook work, as well as numerous other research studies that have been published are really solidifying the case that further reductions may, indeed, be necessary.

You heard from Brian McLean earlier today regarding the emissions reductions that we have seen under the acid rain program.

Basically, we have indeed seen significant emission reductions (Slide 1). We have not only seen them on the grand scale, but we have seen the largest reductions occur, in fact, in those highest emitting states. In the state of Ohio in particular, there was about a 40 percent reduction.

The documentation shows that recovery is, indeed, not happening yet in some of the most sensitive systems (Slide 2). It still reflects some of the older data from Maine. It also talks to the issue of the necessity for long-term surface water monitoring and, indeed, the agency is continuously looking at that issue. The office of research and development, that currently funds the surface water monitoring, is also involved.

We have looked at the issue of trading and the potential for hot spots. Obviously, this is a topic that was discussed and debated back in the 1980s. It comes up now and again during the 1990s and, indeed, most recently as well. Obviously, tracking

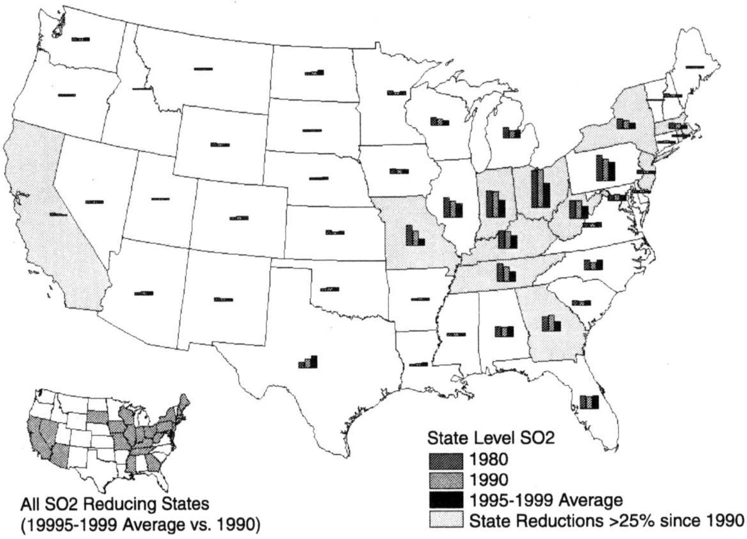

Slide 1. Total utility SO_2: 1980, 1990, 1995-1999 average.

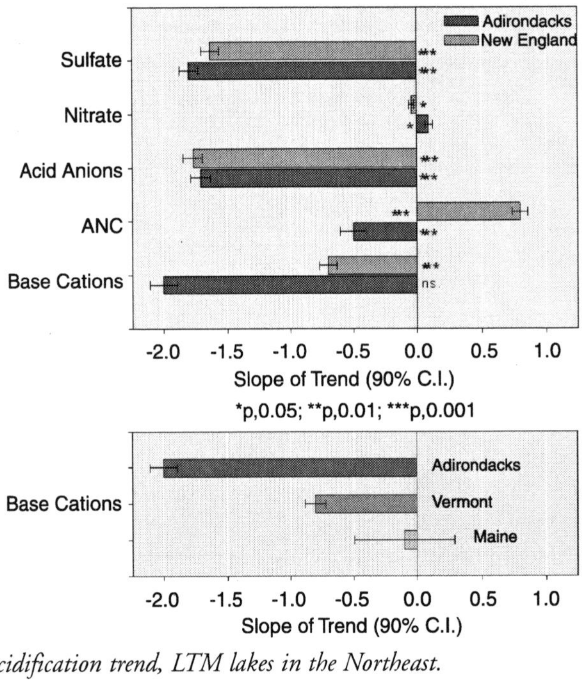

Slide 2. Regional acidification trend, LTM lakes in the Northeast.

this issue is critically important to us. If there are any hot spots, we want to know that and we want to understand why.

We have not seen any regional emission shifts based on trading; the overwhelming majority of allowances were retired in the same state in which they were allocated. There have not been any significant in-flows of emissions; indeed, the greatest emission reductions have occurred in those high submitting states. Finally, most of the acquiring activity has been taking place in the Midwest, the same region that has seen the highest reductions (Slide 3).

This is basically the data that goes into those summary results, where we looked at the geographic mean center, sort of the epicenter, of

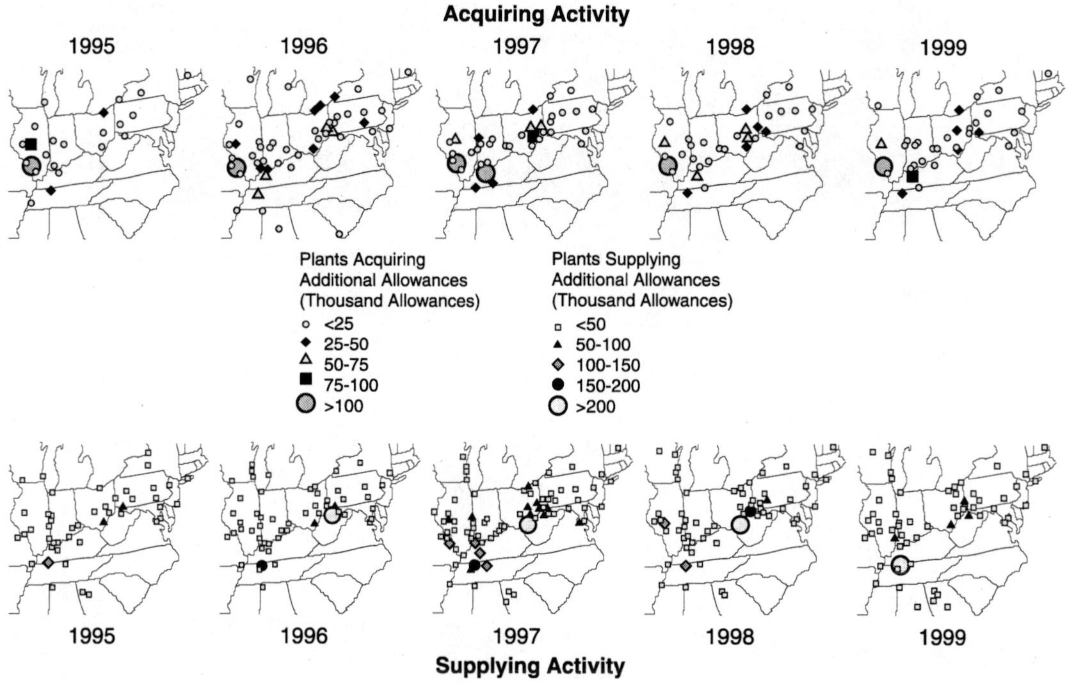

Slide 3. Geographic mean centers of SO_2 allowance trading activity 1995-1999; Plants acquiring allowances for compliance; beyond each year's allocation; plants supplying these allowances.

acquiring versus supplying. From 1995 through 1999 acquiring activity has remained fairly constant and, supplying activity has remained constant. They don't change a whole lot.

Let's move into looking beyond Title IV. What we have started to do is look at modeling projections, evaluating what would be potential impacts of further controls.

We have found significant benefits to human health and large-scale reductions in fine particles. We have looked at visibility improvements and we have seen some broad sulfur and nitrogen deposition reductions in sensitive ecosystems.

We have also looked back at what was done early on in NAPAP, looking at the source receptor maps. Indeed, they continue to show that transport is still important and broad regional large-scale implications are apparent. This is actually something that I begged our modelers to let me use today. It is not completed yet. Basically, if you look at just four of the acid sensitive regions in this country and you look at the air sheds that contribute NO_X to the nitrogen deposition problem, the policy implication, of course, is that, while the goal would be to protect a single ecosystem, it would more likely protect a series of ecosystems, not to mention improving visibility and other regional air pollution issues. You are looking at continued, broad, regional implications here.

Maps have been done for NO_X air sheds for estuaries, in this case, Hudson River, Chesapeake, Pimlico and Altamaha.

Again, the message here is this sort of dual utility, it documents the importance of nitrogen deposition to those particular estuaries, but on the broad level, the policy implication is that, if the design of any policy is to protect numerous sensitive areas and numerous environmental issues, you are looking at broad, regional issues.

In some of our analyses, we have gone beyond Title IV, what I am calling Title IV plus. An additional 50 percent reduction beyond Title IV and about a 60 percent reduction in NO_X beyond Title IV shows some substantial improvement (Slide 4 and 5). Again, what is important here is not just that the reductions occur, but that the impact of the reductions are broad and not only affect sensitive ecosystems in the high elevation

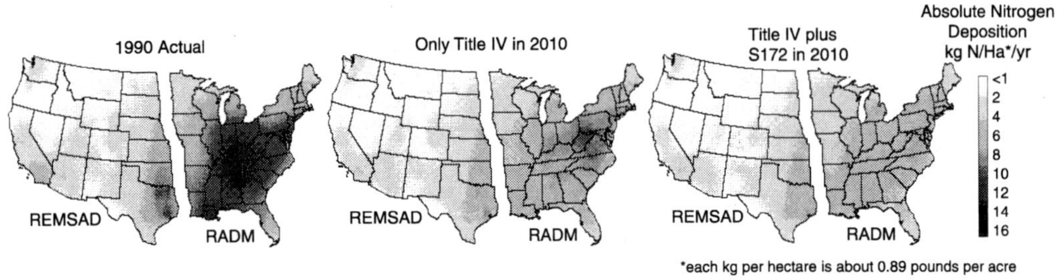

Slide 4. Absolute nitrogen deposition.

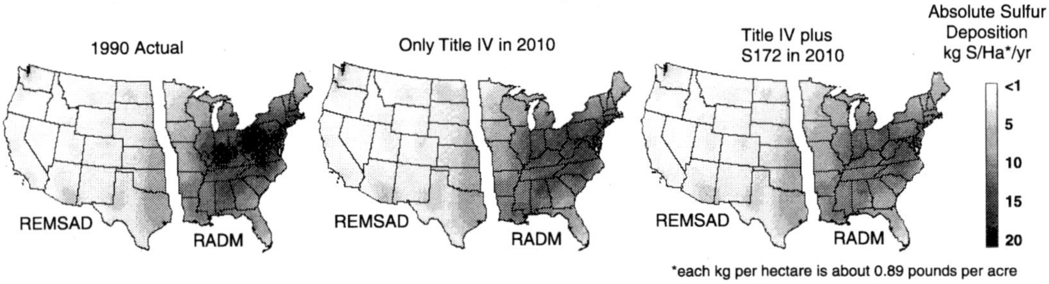

Slide 5. Absolute sulfur deposition.

complex terrain but also the lower elevation coastal sensitive areas.

We are looking at improvements in annual PM, in visibility, and other areas (Slide 6). Again, there are broad implications in terms of changes that would come from further reductions.

We also have been looking at both the benefits as well as the costs of further reductions. What is interesting to note is total benefits broken out by state. The highest benefits come in the highest emitting states. States in the Ohio valley are, in fact, achieving the greatest health benefits from the current program and will benefit from further reductions. In summary, as you have heard already and will continue to hear, there seems to be mounting evidence pointing in the direction that further reductions do seem to be necessary to get to the recovery stage envisioned by Title IV. There are substantial benefits, particularly when we look at PM and visibility, not to mention reduced sulfur deposition and nitrogen deposition.

As long as we are still dealing with an air

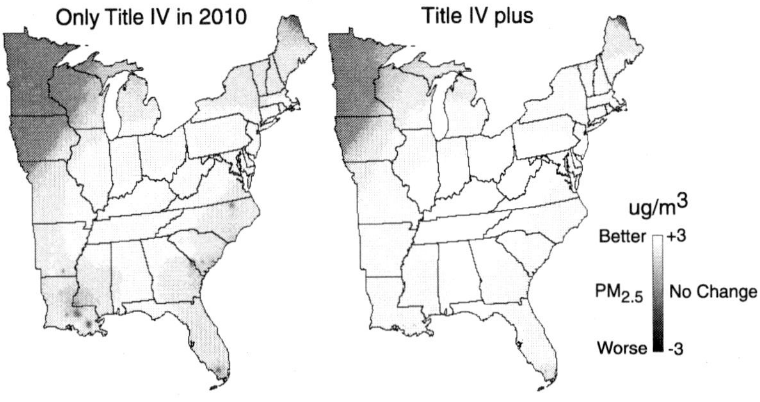

Slide 6. Improvements in annual $PM_{2.5}$ provided by Title IV and further emissions reductions (from 1990 Base).

pollution problem that has long range implications and is regional in nature, it seems to be a good matter of faith to reconsider the goal, look at the cap, and continue to use a mechanism that seems to be addressing the problem in a fairly effective way. Thank you.

QUESTIONS

MR. STACEY: Paul Stacey with the Connecticut Department of Environmental Protection. I think we have heard a common theme here of the need to be more comprehensive in our outlook particularly for nitrogen reductions.

You made a great case of why we have to look at multimedia approaches, air and water, to set these targets.

Yet, as I look at these various programs, they are really not providing a linkage to other programs that are also nitrogen reductions. Also, EPA's air/water interface plan really didn't gel in terms of setting direction.

What is the final end point? Do we need to manage for extra protection for ozone, for acid lakes? I wonder if you could respond to this by maybe giving the EPA's position on when this is going to gel, when we are going to get together and comprehensively come up with a solution as to what the final end point might be for nitrogen reduction to solve all these issues.

MS. BIRNBAUM: I think we are moving in that direction in terms of the EPA program and offices working in a joint manner, in ways that we have never worked before, the air program working with the water program, working with NOAA's estuary program, working with states.

We see the same thing happening in the states where they perhaps also suffer from the same sort of pipeline kind of approach with air people and water people.

I think both at the federal level and the state level perhaps as well, folks are starting to come together.

We are preparing reports and doing assessments that are more integrated in nature. I think you will hear from many people at EPA that there are reasons to reduce sulfur, and there are reasons to reduce nitrogen, a whole array of reasons.

While that is happening, I hope that we also then move into how can we look at strategies and emission reduction goals, collectively.

I think what is happening right now is that necessary first step. I think we are moving in the direction that you mentioned.

MR. WOODWORTH: Neil Woodworth, Adirondack Mountain Club. I was concerned earlier in this presentation when there was a statement that NAPAP is indicating that eastern forests outside New England were not showing quite the same level of damage.

I think the most recent studies indicate that forest soil changes are pretty serious and that, as more time goes on, we are going to find out that there will be a sharp increase in impact.

Have any of you changed your modeling or the method of assessment for forest soil chemistry changes?

MS. BIRNBAUM: Yes. One of the challenges, of course, is to keep up with the science and, as the science evolves, it is fundamentally important to make sure that the models that we use at EPA and anywhere in the government are really in step with where the science is.

We have made significant improvements and changes to both the surface water models as well as to the models that we have been using for analyses that look at forest soils.

That has been changing and it is primarily because one of our goals is to really keep up with the researchers and to ensure that new scientific evidence is conveyed into these estimates that we do at EPA.

We are certainly trying hard to make that happen. If you ever see that that is not happening, you should tell us.

Remarks of Senator Charles E. Schumer

I would like to begin by thanking the Center for Environmental Information, The Adirondack Council, and the many other event sponsors who have made this day possible. I would also like to thank Elizabeth Thorndike for that very kind introduction. It is a great honor for me to speak to you here in our nation's capital, to discuss this NATIONAL problem.

Perhaps it is no surprise that as a U.S. Senator representing New York State, I am interested in halting the destructive effects of acid rain. After all, my own state is "ground zero" in the fight against it. From Long Island Sound to the lakes and streams of the incomparable Adirondack Mountains, the damage that acid rain has caused is only too apparent.

Today is a hopeful day, because I feel that we are – at long last – coming to view this problem not in regional terms, but as a national challenge that requires a coordinated and cooperative response. In order to solve the acid rain problem, we will have to push past the tired regional antagonisms.

Part of my enthusiasm for this conference today stems from a meeting that I held with Bernard Melewski of the Adirondack Council almost one year ago. The subject then was how we could go about passing legislation to reduce sulfur and nitrogen oxide emissions. He asked me what I needed. I told him simply that we would need a national support network. We would need people across the country first to recognize the extent to which their own communities were being affected by this pollution. Then we would need to get them committed to helping us pass acid rain legislation, which is of such urgency.

Well, I understand that nearly 80 percent of our audience here today resides OUTSIDE the Empire State. That sounds like the beginning of a national support network! Thank you, Bernard.

There is no longer any credible doubt that acid rain has long-term deleterious effects upon our natural environment. Study after study has confirmed this fact. Two reports mandated by Congress in 1990 – the EPA's 1995 report and that of the National Acid Precipitation Assessment Program – clearly demonstrate that deeper cuts in sulfur dioxide and nitrogen oxide levels are absolutely necessary.

Work by groups as dissimilar as the Ecological Society of America and the General Accounting Office here in Washington have reaffirmed these conclusions.

Just recently, the Hubbard Research Foundation reported that the Clean Air Act Amendments of 1990 have accomplished little in the way of tackling the acid rain problem, and that the problem is indeed more severe than many had previously suspected.

While this is a well-researched and analyzed subject, it is by no means an academic one. It is a personal one, for whoever has been lucky enough to visit and enjoy our national and state forests.

Not too long ago, I visited New York's Adirondack Park. As a lifelong Brooklyn resident, I am continually amazed at the scope and size of this wonder. We New Yorkers are very fortunate to live in a state where over 100 years ago our elected officials had the foresight to amend the state's constitution to set this land aside for future generations to enjoy.

The Adirondack Park is the largest park in the lower 48 states, and it is the greatest assemblage of wilderness east of the Mississippi River. But the more recent history of the Adirondacks is not nearly as positive.

Acid rain has destroyed nearly 80 percent of the high-elevation red spruce forests there. Over 20 percent of the lakes have been deemed "dead," and close to 55 percent are considered highly acidic. It is estimated the number of "dead lakes" there will double over the next 40 years.

For me, the destruction of this national treasure alone is significant enough to pass federal legislation. As I said earlier, however, we need to educate all Americans about the effects of acid rain in their own communities, if we are going to be successful. In our efforts, the Adirondacks should not be thought of as just New York State's tragedy, but as a cautionary tale to those who would do nothing about acid rain.

In the Southern Appalachians, for instance,

scientists believe that thousands of miles of streams may soon be void of trout. Currently, 59 percent of the trout streams throughout the Southern Appalachians are considered highly vulnerable to acidification.

Studies have indicated that Maryland receives some of the highest concentrations of sulfate and nitrate deposition in the United States. Acidification is having a detrimental effect on fish populations throughout the state, such as striped bass, white perch, yellow perch and American shad. Research indicates that the sources of the acid rain reach as far away as Indiana and even Canada.

Even Western states are now feeling its effects. The National Science and Technology Council (NSTC) found that episodic acidification may be occurring to a significant degree in many high-elevation western lakes and streams in the Sierra Nevada, Cascade and Rocky mountain's which are highly sensitive to acid rain.

As I mentioned earlier, we can no longer resort to the traditional regional antagonisms with regard to acid rain. While New York's position is precarious, it is no longer unique.

Upwind Midwestern States and downwind Northeastern ones must recognize the national significance of this problem. And we must exercise care and caution in our approaches to enlist the President's support, because this issue is too important to wait.

In this Congress, Senator Clinton and I have sponsored two bills that would combat acid rain throughout the nation by reducing power-plant emissions, as power plants are the primary source of acid rain causing pollutants. The first bill, the Acid Rain Control Act, would lower the current caps on sulfur dioxide and create a cap and trade program for nitrogen oxide emissions. It would also authorize the EPA to regulate mercury emissions from utilities. This bill was originally authored by former Senator Daniel Patrick Moynihan, whose leadership on these issues requires no elaboration.

Our second bill, the Clean Power Act, would similarly cut power plant emissions of sulfur dioxide, nitrogen oxide, and mercury, while also reducing carbon dioxide emissions. This bipartisan bill also has the support of Senators from several different states like Jim Jeffords of Vermont, Susan Collins of Maine, and Joe Lieberman of Connecticut.

I don't need to tell you the extent to which these smokestack pollutants are killing our forests and lakes. But I would like to underscore the fact that they are causing asthma in our children, emphysema and other lung ailments among Americans of all ages, and destroying the facades of some of our great civic buildings. If you require any proof of this, please make the journey over to the U.S. Capitol at the conclusion of today's program.

We have the bills to address these problems, the next question is where we go from here.

Let us resolve to use this great occasion – this milestone conference – to redouble our efforts to broaden support for these measures. We need to reach out to the Administration. While the President has wavered on his support for carbon reductions and we need to change his mind on that score, he has, at least, recognized the importance of reducing sulfur and nitrogen levels.

It may not be as big an opening as we might have hoped, but the door is open and we cannot let this opportunity pass.

We also need to enlist the support of Members of Congress who don't yet understand that acid rain has serious implications for their own constituents.

We need to make it clear that the Virginia trout are worth protecting. That the front range of the Rockies may be in danger. That the futures of Chesapeake Bay and New England Maples are not certain. Finally, we need to convince legislators that the historical monuments in Washington, DC, are under fire.

I promise to work with Senator Clinton, Congressmen Boehlert, Quinn, McNulty, McHugh, and Sweeney to build support on Capitol Hill for these sensible and long overdue measures. I am hopeful about our Senate prospects.

In previous years, I would have thought that the best bet would be jamming these bills into a larger measure by exercising a senator's prerogative to filibuster – a sort of scorched earth approach. Now I am optimistic that we can do this through more normal channels.

Environment & Public Works Committee

Chairman Bob Smith and Ranking Member Harry Reid have pledged their support to moving these bills through the committee, and I believe they will help. Then we will begin the systematic lobbying of each and every of the U.S. Senate's 100 members.

New York's Attorney General Eliot Spitzer, who introduced landmark litigation to force polluting utility companies to comply with existing law, will surely continue to do his part.

We have a good chance, but we will need your continued help, and need a great deal of it.

I thank you for your generous welcome. I am inspired by your commitment, and I look forward to working with your organizations to get these bills passed in the 107th Congress. Thank you.

Luncheon Speaker

SESSION III: Acid Rain Impacts - State of the Science

Mary Barber – Ecological Society Of America

The presentations in the following section builds on the earlier discussions to focus on our scientific knowledge related to the effects and impacts of acid rain. In particular the authors discuss the last 10 to 20 years of monitoring and research. Our current monitoring system and knowledge of emissions and depositions are presented by Dr. Van Bowersox. Dr. Greg Lawrence discusses our improved understanding of the effects of acid rain on forested systems and especially the links between soil processes and changes in surface waters. Freshwater systems are addressed by Dr. Arthur Bulger who highlights fish response to acid conditions, the physiological mechanism of response and the possibility of recovery. Dr. Bob Howarth goes on to discuss coastal systems and in particular the over-enrichment resulting from nitrogen reaching these systems from terrestrial and atmospheric sources. Human health effects are discussed by Dr. Lippman who emphasizes respiratory effects and our growing knowledge about particulates. Dr. Mary Striegel concludes this session with an update on research on treatments and mitigation of acid rain effects on stone buildings.

It is apparent that our understanding of the effects of acid rain has continued to grow over the last decade. As a scientific community we must remember that our results not only need to be analyzed and then integrated with results from other systems and disciplines, but that these results also need to be translated and disseminated to a broader audience of less technically trained individuals. Many of the authors of the following papers work with organizations that are well aware of this translation need. As policy and decision makers, we must acknowledge that funds need to be available for this translation phase of scientific study as well as for the research and analysis stages. In this way, scientific knowledge will more likely inform policy debate.

Sources and Receptors: Monitoring the Data

Van Bowersox – Illinois State Water Survey

In my talk, I will present data from the National Atmospheric Deposition Program (NADP) to show how the NADP enables scientists, educators, decision-makers, and the public to track pollutant sources and environmental exposure to chemical deposition. Several speakers already have presented maps from the NADP National Trends Network, which for more than 20 years has collected precipitation for chemical analysis.

The NADP is a cooperative research support program. It receives funding from nine federal government agencies and from State Agricultural Experiment Stations, universities, state and local governments, Indian tribes, and non-governmental organizations. Well over 100 organizations support the NADP. The inside back cover of the informational brochure, *Nitrogen in the Nation's Rain*, lists NADP "cooperators," whose continued commitment makes this program so successful. NADP has three precipitation chemistry networks, the National Trends Network and Atmospheric Integrated Research Monitoring Network (which measure acid precipitation species, nutrients, and base cations), and the Mercury Deposition Network (which measures total mercury in precipitation).

NADP data are available on-line at the Internet address, http://nadp.sws.uiuc.edu, approximately eight months after sample collection. Last year, more than 100,000 maps and data files were downloaded from this site, which received nearly a million "hits."

Much of this talk will address long-term trends in precipitation chemistry in the United States. Professor Jim Lynch and Jeff Grimm of Penn State University and I published an article describing methods we used for trend analysis in the journal *Atmospheric Environment* in 1995 (vol. 29, pp 1231-1246). Briefly, the approach is to calculate bimonthly weighted-average concentrations, reducing the number of data points from weekly concentration measurements to six average concentrations per year. This reduces the scatter or variance of the data. We then take the logarithm of these average values to make the distribution more nearly normal, i.e., more like a bell curve. Next, we apply a linear least squares model, which gives a best-fit straight line showing the change in concentration over time. This model includes terms that effectively remove the seasonal variations in the data. A paper in the journal *Water, Air and Soil Pollution* (vol. 85, no. 2, pp 595-601) shows that the results of this model compared favorably with other types of trend models applied to a common data set. Finally, two new papers appeared in 2000, one in *Environmental Science & Technology* (vol. 34, no. 6, pp 940-949) addressing concentration trends, and one in *Atmospheric Environment* (vol. 34, pp 1665-1680) addressing deposition trends.

Before presenting this latest update on NADP data, I'd like to acknowledge Bob Larson and Roger Claybrooke, both in the NADP Program Office, for their assistance in preparing this presentation.

Let's begin by re-examining the National Acid Precipitation Assessment Program (NAPAP) conclusions published in NAPAP's State-of-Science (VI) report on deposition monitoring methods and results. I was one of the primary authors of that report, published in 1990. One conclusion was that "the weight of evidence suggested that all analytes (except H^+) exhibited a decreasing (not significant) trend in deposition and concentration between 1979 and 1987 for the eastern half of the United States." In short, what we found was that sulfate and nitrate had decreased at nearly every station in the eastern United States but acidity had not. And, we determined that the reason for this was that decreases in acid species – sulfate and nitrate – had been offset by decreases in base cations.

Let's look, now, at what we've learned since the conclusion of NAPAP I. I will present data from the NADP National Trends Network (NTN), which has been in operation for more than 20 years at some stations. Refer to Figure 1, a map that shows the location and length of record of NADP/NTN sites in the contiguous United

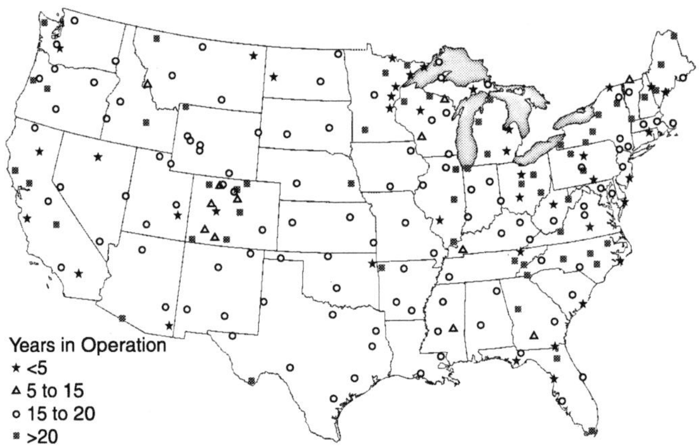

Figure 1. National Atmospheric Deposition Program National Trends Network (NTN).

States. Approximately 170 stations have 15 or more years of data; in this presentation I will summarize data from 150 of these stations.

Figure 2 presents sulfate concentration maps for a 15-year period, beginning in 1985 and ending in 1999. Each map shows three-year average concentrations centered on the year indicated at the top of the map. Showing three-year averages helps smooth the transition from one year to the next. Scan the maps in Figure 2, beginning with 1985 and ending with 1999. Red depicts high concentrations and green depicts low

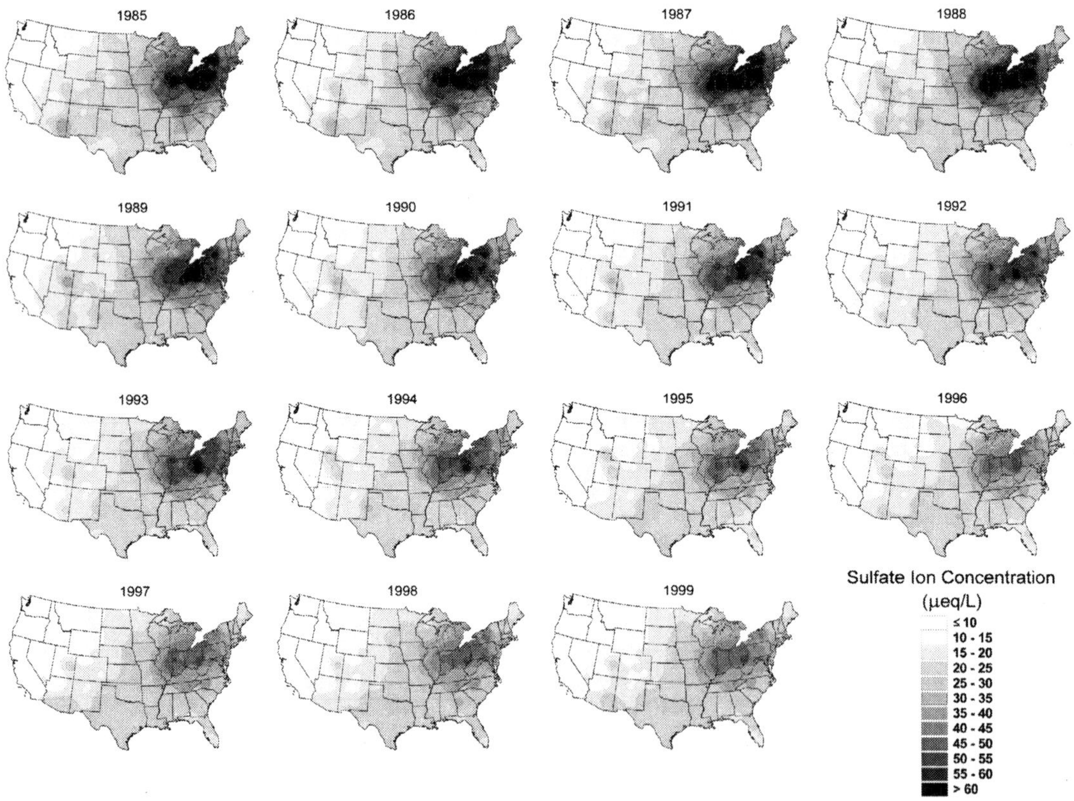

Figure 2. Sulfate ion concentration 1985-1999.

concentrations (see the legend). Highest concentrations are greater than 60 micro-equivalents per liter. Un-neutralized, sulfate at these levels would cause rainfall to have a pH of ~ 4.2. At 10 micro-equivalents per liter, the darkest green color, un-neutralized sulfate would cause rainfall to have a pH of 5, below which scientists generally call rain, "acid rain." So, sulfate alone could cause rainfall to be acidic almost everywhere in the United States.

This is what we conclude from the maps in Figure 2:

1. Highest sulfate concentrations occur in the northeastern United States, from Iowa and Illinois in the Midwest and Tennessee and North Carolina in the South through Vermont and New Hampshire in New England.
2. Sulfate concentrations decrease from values above 60 micro-equivalents per liter in 1985 (Illinois to New York) to 45 – 55 micro-equivalents per liter in 1999.

Figure 3 is based on the trends work by Lynch, Grimm, and Bowersox, discussed earlier. Pictured are two concentration maps. The left panel shows annual sulfate concentrations averaged over a 12-year period, 1983 to 1994, and the right panel shows annual sulfate concentrations averaged over a 5-year period, 1995 to 1999. Decreases in sulfate concentrations between the early and late periods are evident.

Figure 4, also based on the Lynch, Grimm, and Bowersox trends work, shows sulfate concentration differences between the observed and model – estimated 1995-99 sulfate concentrations. Blue denotes sulfate decreases, yellow/red increases, and grey no change. Recall that the 1990 Clean Air Act Amendments (CAAA-90) required reductions in sulfur dioxide emissions effective 1 January 1995. The CAAA-90 targeted 110 coal-fired electric utility plants in 21 states, 17 east of the Mississippi River. The white plus signs in Figure 4 mark the locations of these targeted utilities. Notice that sulfate decreases are largest in and downwind of the Ohio River Valley, the location of many of the targeted utilities.

Figure 5 summarizes the changes observed in sulfate deposition at NTN sites since January 1995, when the CAAA-90 limited sulfur dioxide (SO_2) emissions at targeted plants. In the Northeast (green area), where 68 percent of the total U.S. SO_2 emissions occur, sulfate deposition decreased at 77 percent of the NTN sites. At many sites there was a marked (step-function) drop in sulfate deposition in 1995. In the Southeast (pink area), where 25 percent of the total U.S. SO_2 emissions occur, sulfate deposition decreased at 68 percent of the NTN sites. In the West (yellow area), sulfate

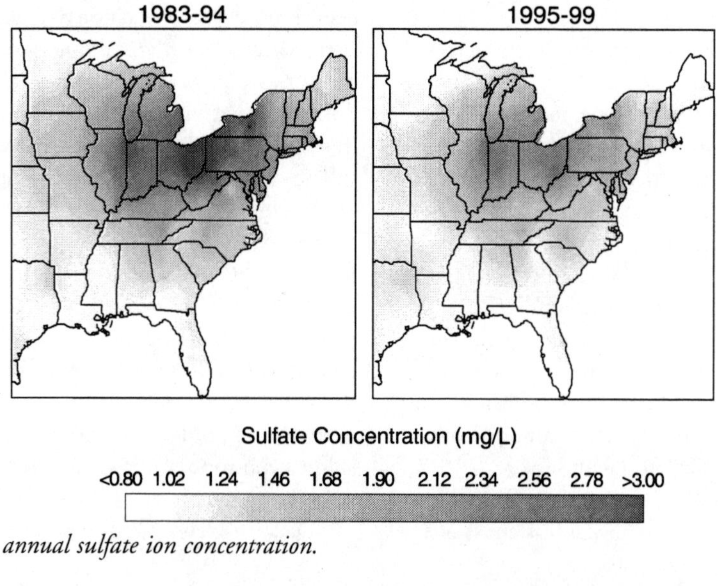

Figure 3. Mean annual sulfate ion concentration.

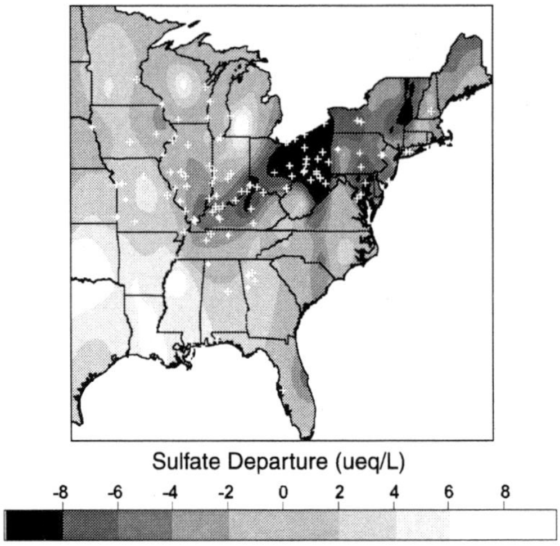

Figure 4. Departures of mean annual sulfate ion concentration during 1995-99 from values predicted by the 1983-94 trend.

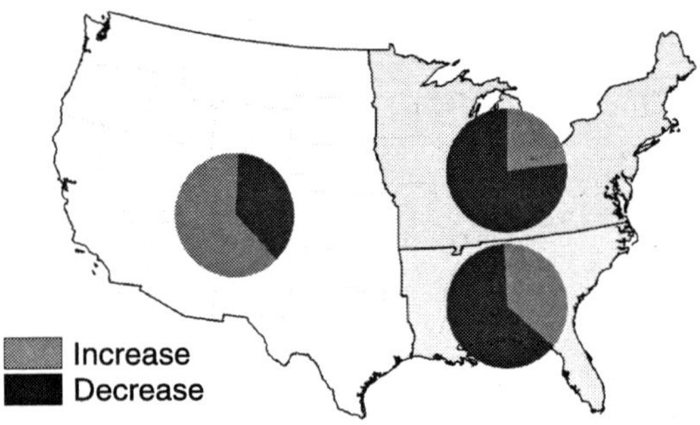

Figure 5. Frequency of sites with increasing or decreasing sulfate deposition : 1983-1994 versus 1985-1999.

deposition actually increased at most (63 percent) NTN sites.

It is worth noting that 85 percent of U.S. SO_2 emissions result from fuel combustion, 65 percent from burning fuels to generate electricity. Annual electric utility SO_2 emissions dropped more than 2.8 million tons, following implementation of emission limits set by Title IV (acid deposition controls) of the CAAA-90. In eastern Canada, Ontario and Quebec Provinces also reduced SO_2 emissions at about the same time as reductions occurred in the United States. SO_2 emissions from these two Provinces are about the same as from West Virginia and New York combined.

Next, we will look at three states, Ohio, New York, and Vermont, to see how sulfur emissions and deposition changed following CAAA-90 SO_2 reductions. Following the same format as in Figures 3, 4, and 5, I will compare emissions and deposition in 1995-99 ("post" CAAA-90 reductions) with 1983-1994 ("pre" CAAA-90 reductions).

Figure 6 shows SO_2 emissions and wet sulfate deposition in Ohio in the "pre" and "post" periods. Both emissions and deposition are shown in units of thousands of tons of sulfur, so the sulfur mass emitted can be compared directly with the sulfur mass deposited in precipitation. This does not mean that the sulfur deposited in Ohio was necessarily emitted in Ohio. Further, the sulfur deposition does not include dry deposition, which occurs when sulfur-containing gases and particles deposit to vegetation, soil and other surfaces during dry weather. The deposition estimates in this figure are calculated entirely from NTN sulfate concentration and precipitation data.

In Figure 6, the grey bars show Ohio's annual sulfur emissions, which are well over 1 million tons in 1983-1994 and are 28 percent lower in 1995-1999. Blue bars show wet sulfur depositions, which are about 10 percent of Ohio's sulfur emissions and drop 13 percent from the "pre" to the "post" period. Unless dry sulfur deposition is more than an order of magnitude larger than wet sulfur deposition, Ohio emits more sulfur than it receives.

Data for New York appear in Figure 7. From

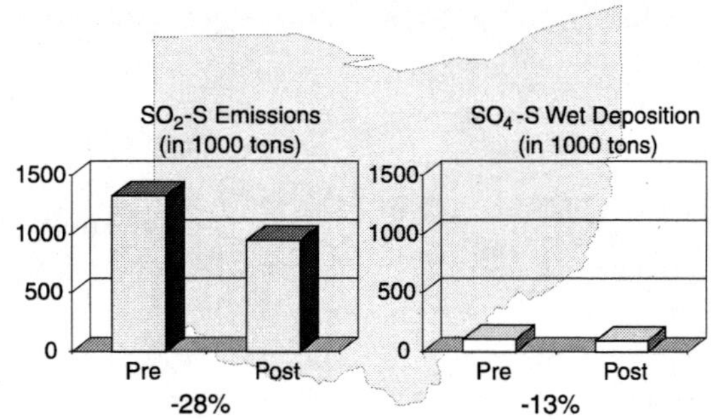

Figure 6. SO_2 emissions and wet sulfate deposition in Ohio.

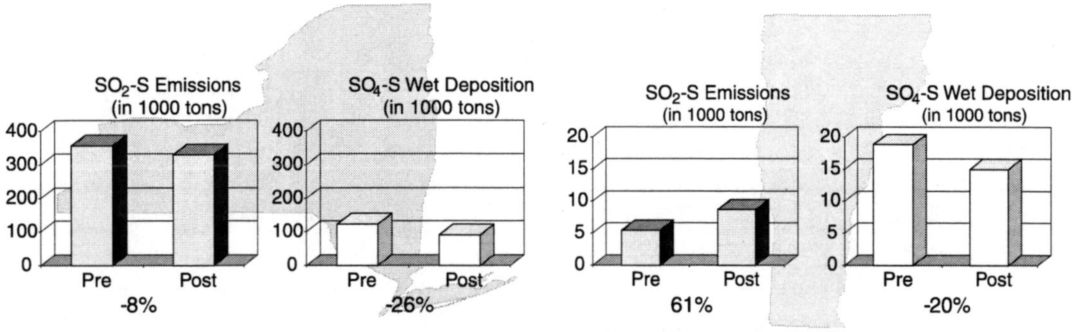

Figure 7. SO_2 emissions and wet sulfate deposition in New York and Vermont.

the "pre" to "post" period, SO_2 emissions decreased 8 percent, while wet deposition decreased 26 percent. Sulfur emissions and deposition are more nearly balanced in New York than in Ohio. Nevertheless, if wet and dry sulfur deposition are roughly equal in magnitude, New York's sulfur emissions exceed total sulfur deposition. Ohio's sulfur emissions (Figure 6) are three to four times higher than New York's sulfur emissions.

It is apparent that sulfur emissions in Ohio and New York greatly exceed the sulfur deposited by precipitation. The opposite is true in Vermont, where wet deposition is 2 to 3½ times greater than emissions. Vermont's emissions and wet deposition are shown in Figure 7. Although Vermont's sulfur emissions increased 61 percent between the "pre" and "post" periods, it is important to note that Vermont's emissions are less than one percent of Ohio's and less than three percent of New York's. Despite the emissions increase in Vermont, wet deposition decreased 20 percent.

Emissions and wet deposition data from these three states (Ohio, New York, and Vermont) help improve our knowledge of how changes in emissions affect wet deposition. These data point to the importance of models, such as the Regional Acid Deposition Model, that simulate the relationships among emissions, transport, transformation, and deposition of atmospheric sulfur. Also important is knowing the dry deposition of sulfur on a regional basis, which is the goal of the U.S. Environmental Protection Agency's Clean Air Status and Trends Network and the National Oceanic and Atmospheric Administration's Atmospheric Integrated Research and Monitoring (dry deposition) Network.

Summarizing what we've learned about sulfur deposition since the conclusion of NAPAP I:

1. In the East, sulfate deposition decreased by ~12 percent on average. Largest decreases – 15 to 25 percent – occurred in and downwind of major source areas – Illinois, Indiana, Ohio, West Virginia, Pennsylvania, New York, Maryland. Sulfate did not decrease everywhere in the East, although decreases were observed at 73 percent of NTN sites and about half were statistically

significant. Sulfur deposited by precipitation is only about 1/4th of the sulfur emitted.
2. In the West, sulfate deposition was level to increasing, with increases observed at 63 percent of NTN sites.

Next, we will examine nitrate in precipitation. The largest U.S. sources of nitrogen oxides, which are transformed into the nitrate we observe in precipitation, are vehicular exhaust and emissions from utility and industrial fuel combustion. These sources comprise nearly 90 percent of nitrogen oxide emissions.

Figure 8 presents nitrate concentration maps for the same 15-year period as in Figure 2 for sulfate. In Figure 8, the highest concentrations (red-most color) are around 33 micro-equivalents per liter, which is about half the maximum sulfate concentration. Highest concentrations occur in states from Illinois and Michigan to New York. If un-neutralized at these concentrations (24 – 33 micro-equivalents per liter), nitrate would result in a pH of ~4.5.

Scan the maps in Figure 8, beginning with 1985 and ending with 1999. Unlike sulfate, nitrate does not appear to increase or decrease in any general sense. Instead, the concentrations exhibit year-to-year variations but no consistent trends. Since nitrate concentrations exhibit no obvious trend while sulfate concentrations have decreased, nitrate is growing in importance as a cause of acidic deposition, particularly in the eastern United States. Aside from being an acidifying agent, it is important to remember that nitrate is a nutrient, which can stimulate plant growth. In coastal estuaries such as the Chesapeake Bay, researchers have linked atmospheric nitrogen deposition to eutrophic conditions in which excessive growth of aquatic plants leads to low oxygen concentrations that can be harmful or lethal to fish and other aquatic animals.

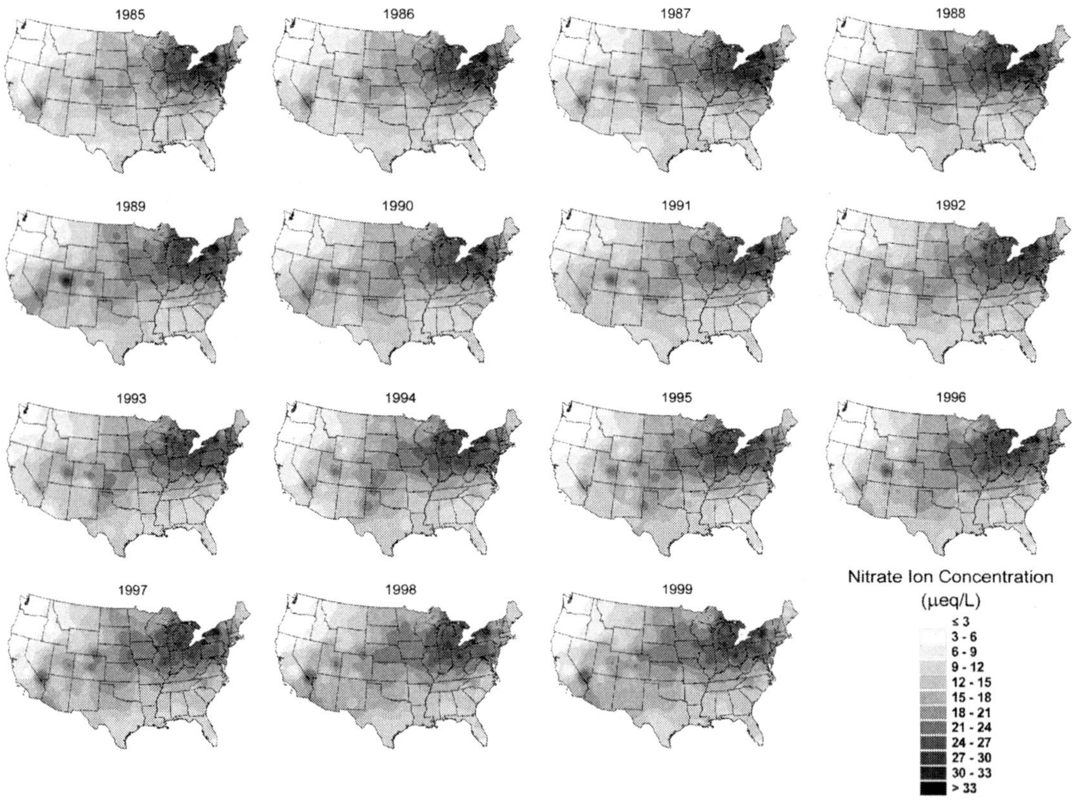

Figure 8. Nitrate ion concentrations 1985-1999.

Figure 9 presents the 15-year record of another nutrient in precipitation, ammonium. Ammonium is the chemical form that ammonia takes in precipitation. Ammonia enters the atmosphere as a gas from animal waste, fertilizer applications, natural microbial decomposition in the oceans and soils, and fuel combustion. Precipitation scavenges gaseous ammonia from the air as well as particles which contain ammonia that has combined with other pollutants, such as sulfate and nitrate. Unlike sulfate and nitrate, which make precipitation acidic, ammonium neutralizes these acids in precipitation. The scale in Figure 9 is the same as it is for nitrate concentrations in Figure 8, which means that when the colors in these two figures match, the acidic contribution of nitrate is effectively balanced (i.e., neutralized) by ammonium. In general, nitrate and ammonium are in balance in Illinois; nitrate exceeds ammonium east of Illinois; and ammonium exceeds nitrate west of Illinois to the Rocky Mountains. From the Rockies to the West Coast, nitrate generally exceeds ammonium except in ammonium hot spots, such as in Utah and northern California.

Scan the maps in Figure 9, beginning with 1985 and ending with 1999. Over this 15-year period, the area of high concentrations in eastern South Dakota and Nebraska and western Minnesota and Iowa expands to an area stretching from North Dakota to the Texas panhandle and from Nebraska to Illinois. Concentrations in parts of this area increased by as much as 50 percent. This is a source area containing confined animal feeding operations and active row crop production; however, ammonium concentration increases are disproportionately greater than changes in these agricultural activities. Clearly, we need to improve our knowledge of ammonia emissions and cycling to understand what we've observed in this area.

Figure 9 also shows that some NTN sites

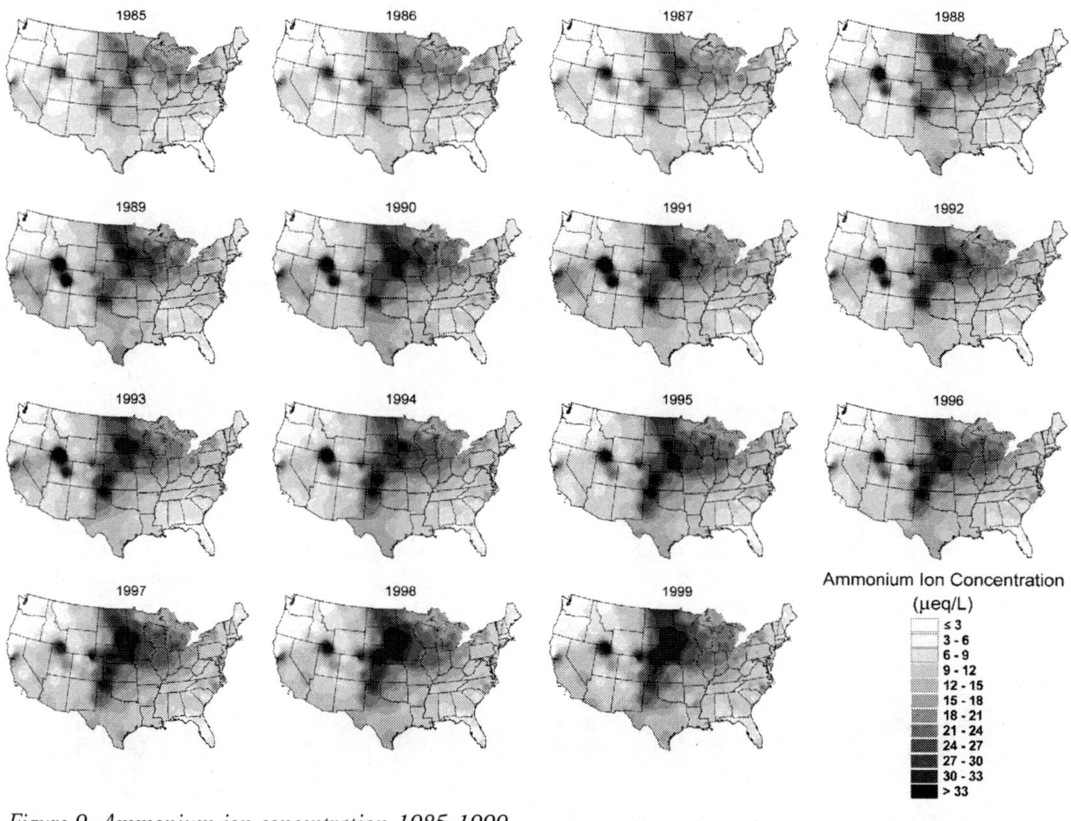

Figure 9. Ammonium ion concentration 1985-1999.

apparently are influenced by local ammonia sources. Persistent hot spots occur in California, Utah, and New York, and an expanding hot spot appeared in eastern North Carolina in the mid-1990s. Ammonium concentrations at one or perhaps two (hot spot) sites are substantially higher than at surrounding NTN sites. Finally, it is important to note that ammonium acts to neutralize precipitation acidity, but once deposited in an ecosystem, it is a proton donor, meaning it behaves as an acid. Ammonium deposition is an acidifying phenomenon.

On the same scale as Figures 9 and 10, Figure 10 shows the 15-year sequence of base cation concentration maps. Base cations include the combined concentrations of four constituents: calcium, magnesium, sodium, and potassium. These are alkali and alkaline earth elements that derive primarily from wind erosion of soils, dust from unpaved roads, cultivation, and construction and mineral processing activities. Road salt is also a principle source of sodium and seasalt is a principle source of sodium and magnesium. Base cation concentrations in Figure 10 appear to exhibit inter-annual fluctuations with no clear evidence of regional increases or decreases over the 15-year (1985-1999) period. The downward trend reported in NAPAP State-of-Science Report VI, that "decreases in acid species – sulfate and nitrate – had been offset by decreases in base cations," apparently ended in the late 1980s. A careful trend analysis is needed to demonstrate whether regional changes in base cation concentrations have ended.

The final question I will address is what happened with hydrogen concentrations. In other words did precipitation acidity increase or decrease? Figure 11 displays mean annual 1983-1994 hydrogen ion concentrations ("pre" CAAA-90 sulfur emissions reductions) alongside mean annual 1995-99 concentrations ("post" CAAA-90 sulfur emissions reductions). Clearly, hydrogen ion concentrations decreased from the "pre" to the "post" periods, which is evident in the hydrogen ion difference map in Figure 12. There is a high

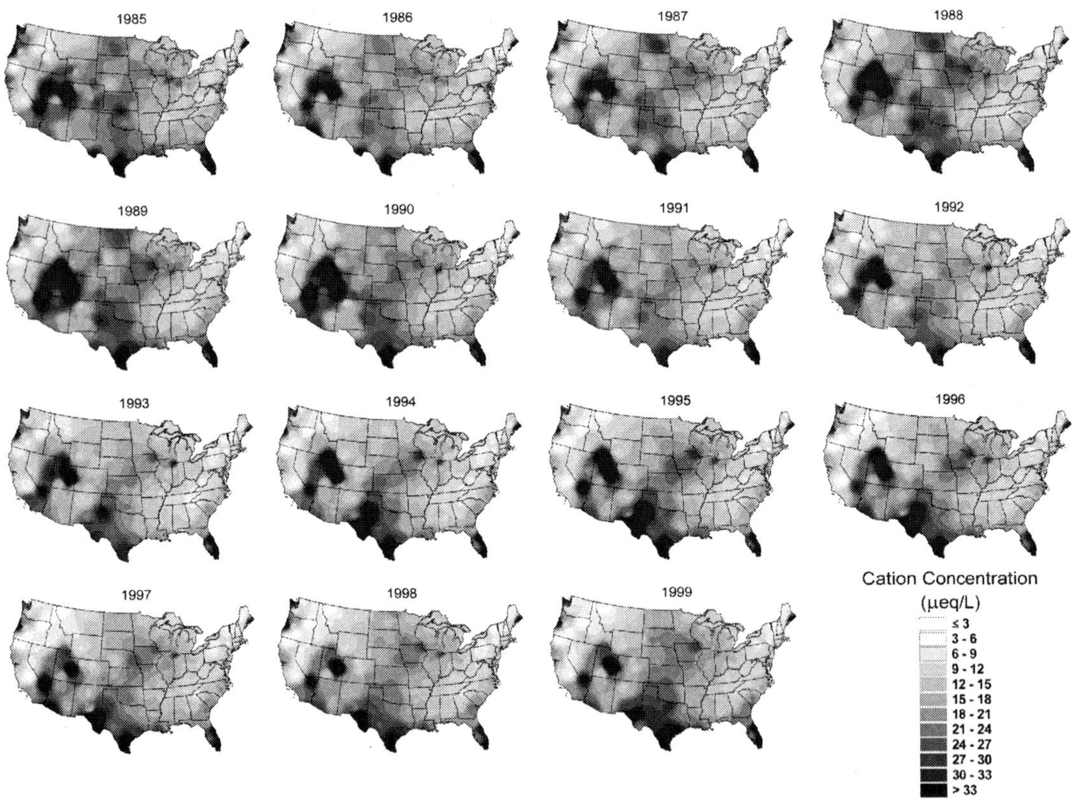

Figure 10. Cation concentration 1985-1999.

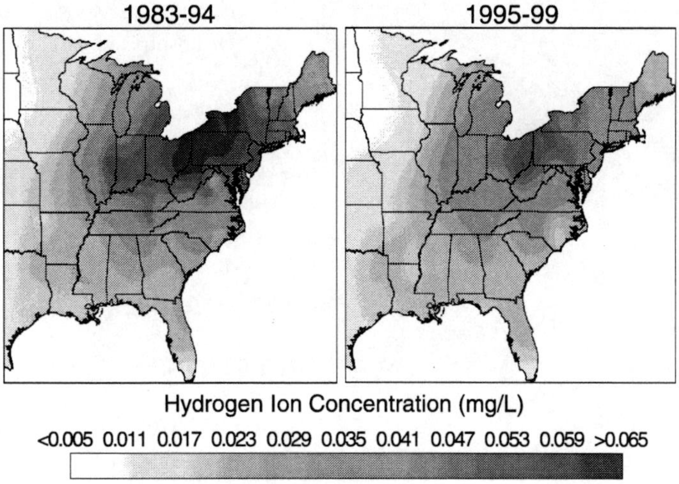

Figure 11. Mean annual hydrogen ion concentration.

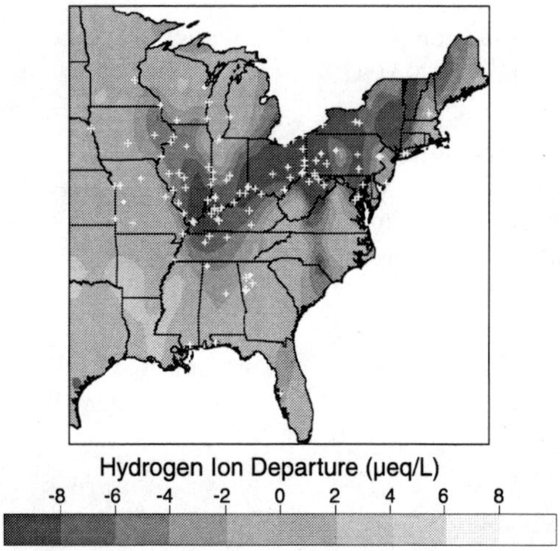

Figure 12. Departures of mean annual hydrogen ion concentration during 1995-99 from values predicted by the 1983-94 trend.

degree of spatial coherence between the sulfate changes in Figure 4 and the hydrogen ion changes in Figure 12. Both maps display changes in the same concentration units. Direct comparison shows almost identical sulfate and hydrogen ion decreases from Ohio to the east and north through New England to Maine.

Most evident is the footprint of hydrogen ion decreases in and downwind of the Ohio River Valley, where sulfur emission decreases were largest. The plus signs in Figure 12 mark the locations of electric utility plants targeted for sulfur emissions reductions in Phase I of the CAAA-90. Except for a small area mostly in Louisiana, Mississippi, and Arkansas (see yellow areas in Figure 12), hydrogen ion was level to decreasing throughout the eastern United States.

Decreases generally occurred in a step-wise manner consistent with the "pre" (1983-1994) to the "post" (1995-1999) Phase I sulfur emissions reductions. Figure 13 summarizes the changes observed in hydrogen ion deposition at NTN sites since January 1995. In the Northeast (green area), hydrogen ion deposition decreased at 90 percent of the NTN sites. In the Southeast (pink area), hydrogen ion deposition decreased at 84 percent of

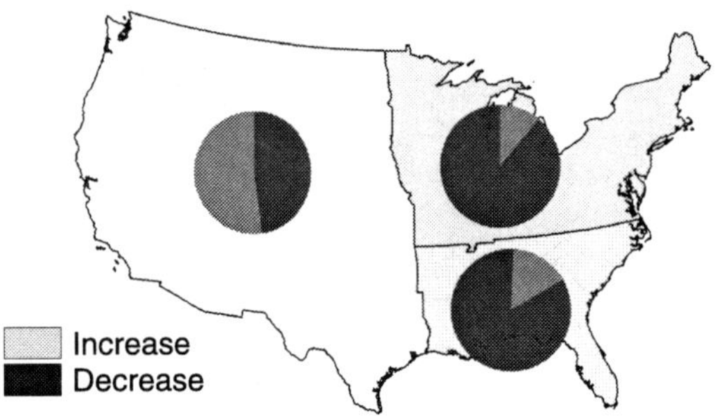

Figure 13. Frequency of sites with increasing or decreasing hydrogen ion deposition: 1983-1994 versus 1995-1999.

the NTN sites. And in the West (yellow area), hydrogen ion deposition decreased at about the same number of NTN sites as it increased (48 percent to 52 percent, respectively).

Summarizing what we've learned about hydrogen ion (acidity) deposition since NAPAP I:

1. In the East, hydrogen ion deposition decreased at 88 percent of the NTN sites (63 percent were statistically significant, much higher than in the NAPAP I conclusions). Largest decreases matched the sulfate decreases and generally occurred in and downwind of the Ohio River Valley from Ohio through New England.
2. In the West, hydrogen ion deposition remained essentially unchanged.

In conclusion:

1. Acidic deposition in the northeastern United States decreased since January 1995. The largest decreases – 10 to 25 percent – occurred in and downwind of the Ohio River Valley.
2. Over the last 15 years, nitrate concentrations have been level to increasing. Relative to sulfate, nitrate has grown in importance as a cause of acidic deposition.
3. Ammonium concentrations have increased, primarily in the Midwestern United States but also in some hot spots, such as eastern North Carolina.
4. Long-term monitoring allows us to evaluate ecosystem exposure, changes in natural and anthropogenic sources, and source receptor relationships, and is an invaluable resource for informing policy decisions.

Acid rain, are the problems solved? Figure 14 shows the NADP/ NTN average annual pH maps for 1985 and 2000, both displaying substantial areas that receive precipitation with pH levels below 5.0. Compare the maps and answer the question. Thank you.

Questions

AUDIENCE: I wonder if you could expand upon the increase in ammonium and some of the sources of that.

MR. BOWERSOX: It is an interesting question and it is one that several scientists at the Illinois State Water Survey are examining right now. Joe Scudlark of the University of Delaware is in the audience and he, too, is examining this question. Ammonia sources include animal waste, fertilizer, automobile exhaust, human waste, power plants, senescing crops, and natural sources such as microbial degradation of nitrogen in soils and oceans.

MS. BAUM: On that ammonium question, I am not used to thinking about ammonium as a cation. So, I am thinking that the cations are

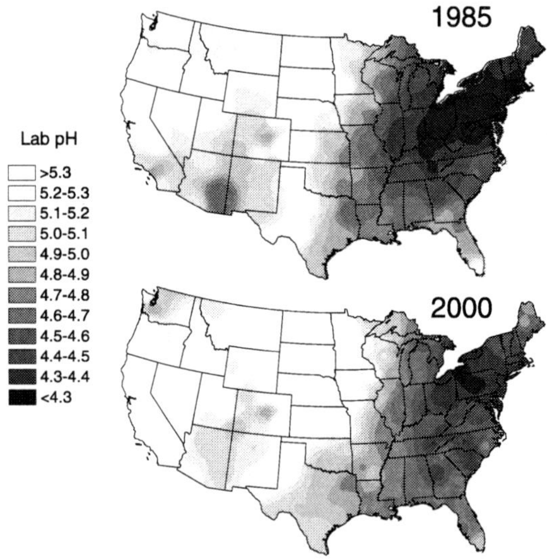

Figure 14. NADP/ NTN average annual pH maps for 1985 and 2000.

actually seeing a reversal. If you took ammonium out of there, would you come up with that same conclusion?

MR. BOWERSOX: No, cation increases are being driven by ammonium.

MS. BAUM: So, the ammonium could account for the cation increase.

MR. BOWERSOX: In precipitation, that is right.

MR. COWLING: I just wanted to pick up on the comment you just made. I would urge us all to start thinking about ammonium. It is essential to understand that ammonium is an acidifying deposition. The Europeans use net acidifying deposition. We use in America, acidic deposition. Senator Schumer spoke about nitrogen oxides and sulfur oxides. Ammonium is acidifying in the environment and we need to be thinking about total nitrogen deposition in this country. It is total nitrogen deposition that contributes to eutrophication. It is not just nitrogen oxides. It is not just ammonium, and it is not just those inorganic forms of nitrogen either.

In the most recent rather careful measurements in my state, one third of the net deposition of nitrogen is occurring in organic forms. You can take these two numbers, the sum of the nitrates and ammonium and add to that half again that much. That is what we are dealing with in terms of nitrogen deposition to ecosystems at the present time.

MR. BOWERSOX: Those are excellent points. I might add that Ellis Cowling and Jim Galloway are organizing a 2001 conference on nitrogen that will occur October 14-18 at the Bolger Center. You can find out more about this meeting from the NADP web page, where we have a link to that conference announcement. The whole nitrogen issue is the focus of this meeting.

MR. KAHL: Steve Kahl, University of Maine. Were the trends you showed for ammonium corrected for the handling problems?

MR. BOWERSOX: I don't think our handling problems were any worse in the 1980s than in the 1990s, so it's unlikely the temporal changes depicted in the maps were affected by handling. We do have a bias, certainly, in our ammonium data. Ammonium levels are affected by this bias. Some researchers have found NTN measurements to be up to 15 percent lower than ammonium in preserved samples at the same sites. We have looked at ammonium biases at some side-by-side NTN and AIRMoN stations and find no bias at some stations and at others a bias of 15 percent. Our calculations of ammonium deposition are biased low accordingly. Thank you for your question.

Blood, Poison and Death: Effects of Acid Deposition on Fish

Arthur J. Bulger Jr. – University of Virginia

This might be the title of a murder mystery. In fact, there isn't any mystery. The negative effects of acidification on fish are clear and unambiguous, and the ecological consequences are clearest for fish. It is simply a matter of toxicology.

First I would like to set the stage. In the eastern third of the United States, most of the values for volume-weighted annual average pH of rainwater start with a four (Figure 1). This is about ten times as acid as estimated for rainwater pH in pre-industrial times. This means that "pure" rainwater in the eastern United States is acutely lethal to all native fish species. Not even a goldfish could survive in unbuffered rainwater. This is the central issue I will speak about.

"Acid Rain" implies the deposition of acid materials in wet deposition (rain, snow, fog, cloud) as well as in the dry deposition of dust (particles) and gases; scientists use "acid deposition" for this process, because it explicitly includes dry as well as wet deposition; indeed, deposition of acid materials in dry form is often equal to the amount deposited in wet form. Acid deposition is responsible for the loss of hundreds of fish populations in Europe and North America.

The effects of acid deposition on fish involve complex processes at different spatial scales, from atmospheric transport to cell membrane transport. Nevertheless, the interaction of these processes can be summarized in five major points, as follows.

Acid deposition effects on fish.

1) The source of the acid: fossil fuels.

The burning of fossil fuels releases sulfur and nitrogen oxides into the atmosphere, where they are converted to sulfuric and nitric acids. These acidic materials may be transported long distances in the atmosphere before they are deposited in wet or dry form on landscapes.

2) Landscapes differ in sensitivity, based on geology.

Whether or not acidic deposition produces negative effects on the animals living in streams

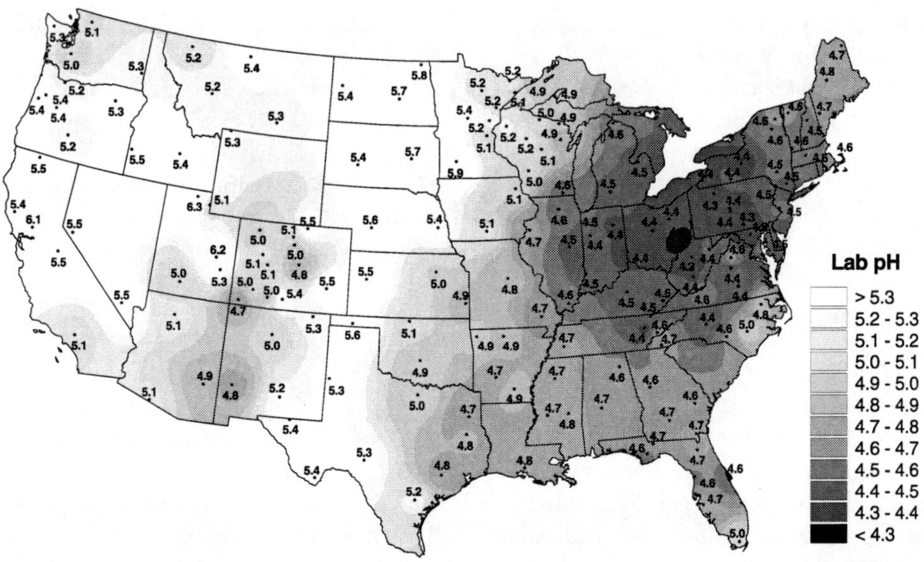

Figure 1. Hydrogen ion concentration as pH from measurements made at the Central Analytical Laboratory, 1999 (National Atmospheric Deposition Program, National Trends Network).

and lakes depends largely on the bedrock geology of their catchments. In landscapes underlain by limestone (carbonate bedrock), which provides substantial buffering of acidity, negative effects due to acidification are neither expected nor seen in water bodies. Basaltic, granitic, and siliciclastic (such as sandstone) bedrock types represent a series of decreasing levels of buffering capacity, such that modest amounts of acidic deposition produce conspicuous negative effects in sandstone catchments. Since buffering capacity ultimately depends on the weathering of acid-neutralizing material from the bedrock, hard bedrock types produce less buffering capacity for streams than soft bedrock types. Mountains by their very nature are more resistant to weathering than surrounding lowlands (that's why the mountains are still there), so mountain streams and lakes are usually the most sensitive to acidification due to the lower buffering capacity of their catchments. In contrast, large valley streams and lakes are often the recipients of upstream weathering products, and are often less sensitive to acidification as the result of their greater buffering capacity.

3) The role of aluminum: metabolic poison.

Aluminum is the most abundant metal on the earth's surface, and the third most abundant element. It is non-toxic and insoluble under acid-neutral conditions, but very toxic to fish and other aquatic species under acidic conditions; unfortunately, the solubility of aluminum increases exponentially as pH falls below 5.6; its maximum toxicity occurs at about pH 5.0. The deposition of acids results in the release of aluminum from soils, which is carried in solution to streams and lakes. Both the aluminum and the hydrogen ion (derived from sulfuric and nitric acids) are toxic to fish, but in most streams and lakes the aluminum is the primary lethal agent; fish can survive more acidic conditions (i.e., lower pH) in the laboratory in the absence of aluminum.

4) Site of toxic action: fish gill

The site of the toxic action of both hydrogen ion and aluminum is the fish gill. The gill is a complex organ responsible for oxygen and carbon dioxide exchange, as well as maintaining the proper salt and water balance the fish's body. It is this later function which is always compromised by acid and aluminum stress; respiration is also compromised at higher concentrations of aluminum.

Freshwater fish maintain salt (sodium chloride) in their blood at concentrations similar to those in humans and most other vertebrates. The proper functioning of most body cells, and especially blood cells in this context, depends on keeping salt concentrations in body fluids within rather narrow limits. Since salt concentrations in the blood are much higher than the water in which they swim, fish constantly lose a small amount of sodium and chloride from the blood by passive diffusion across the thin skin of the gills. The lost sodium and chloride are replaced by an energy-requiring process (active transport) using biochemical "pumps" in the gill membranes which transport sodium and chloride from low concentration in the external stream water to higher concentration in the blood.

Aluminum and hydrogen ion poison the biochemical pumps which transport sodium and chloride into the body; they also weaken the junctions between gill cells, making them leak more sodium and chloride than they otherwise would. The rapid loss without replacement of sodium and chloride produces a cascade of negative physiological effects in the body.

It is a common misconception that stream acidification causes acidification of fish blood, with negative effects on oxygen transport; this does not occur at observed levels of acidification in nature which produce fish death.

5) The cause of death: circulatory collapse secondary to electrolyte imbalance.

The key indicators of incipient mortality under acute acid and aluminum stress are the concentrations of sodium and chloride in the blood plasma. When either or both (sodium and/or chloride) fall more than 30% below normal, death occurs within hours.

The proximal cause of death is ionic dilution of the blood plasma. This causes blood and body fluid disturbances which ultimately kill the fish through circulatory collapse.

Under normal conditions, plasma ionic (electrolyte) concentrations and body cell ionic concentration are in equilibrium. Under acute acid stress, ions are lost more rapidly from the blood plasma than from blood and muscle cells; as a result, there is an osmotically-driven shift of water to the cells from the plasma; blood plasma volume may drop as much as 30%; at the same time, the red cells swell due to the osmotically-driven shift of water from the plasma; the result is a doubling of blood viscosity. The heart is unable to circulate this much thicker blood at a rate sufficient to supply oxygen to body tissues, including the heart itself, so the fish dies of circulatory collapse secondary to ionic imbalance.

In short, the heart cannot supply its own tissues with sufficient oxygen to sustain cardiac activity. So, the fish has a lethal heart attack. Molecules released in Ohio give trout in Virginia heart attacks.

There are wide differences in the sensitivities of fish and other vertebrates to acidity. Acidification eliminates the most sensitive species first. The result is a "dose/response" relationship between acidity and the number of fish species, which looks like this (Figure 2):

This is the number of fish species in lakes with various pH classes in the Adirondacks. These are the lakes that still host fish. About a quarter of all the lakes are fishless in the Adirondacks; at least 20 percent are fishless because they are too acid to support any fish.

We see the same phenomenon in Virginia with a smaller sample size (Figure 3). As Senator Schumer mentioned today, acid rain has moved into Virginia.

Virginia and the Southern Appalachians have had a delayed response to acidification because the soils are thicker. Those soils retain sulfur. When that sulfur retention capacity is exceeded, all the sulfur that comes into the watershed exits to the stream or lake. The southeast is expected to have a slower recovery as well, because after the sulfur deposition is lowered, it will take some time to release the retained sulfur.

Another issue of importance in considering acidification in the northeast vs. the southeast is that lakes are numerous in the northeast and much less so in the southeast. Streams respond quickly to acid deposition, producing large acid episodes; episodes are usually of smaller magnitude in lakes because of dilution in the larger volume of lakes.

Here are some fish modeling results for Virginia (Figure 4). About 30 percent of the probable historic trout streams in Virginia have become too acidic to support any fish (summing categories "chronically acidic", 6% and "episodically acidic", 24%). They don't support brook trout any more. Brook trout is the region's most

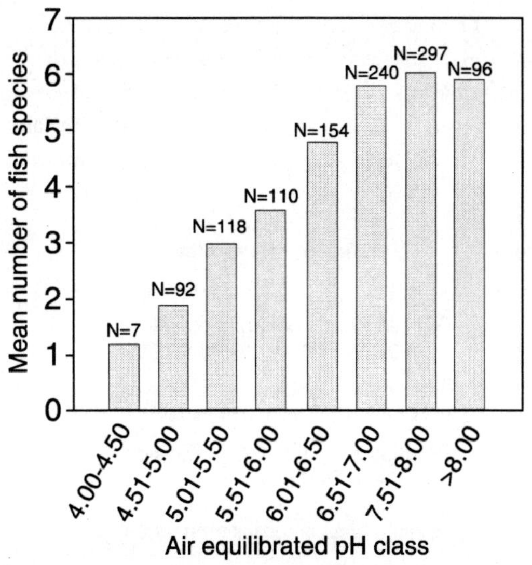

Figure 2. Acidity and number of fish species in Adirondacks waters (NY).

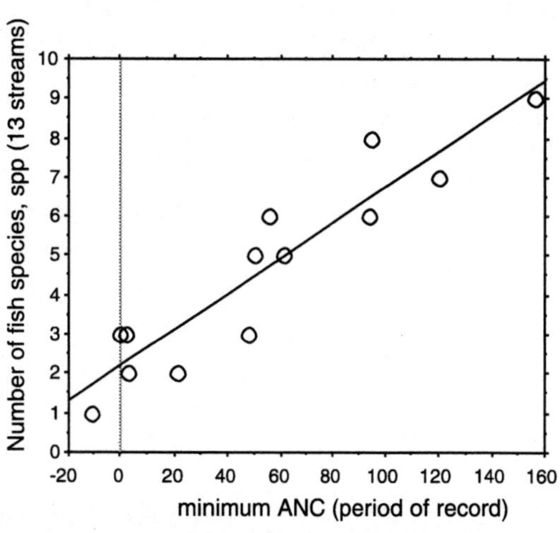

Figure 3. Thirteen streams, with Meadow Run (lowest ANC) showing only one species, reflecting recent loss of Blacknose Dace.

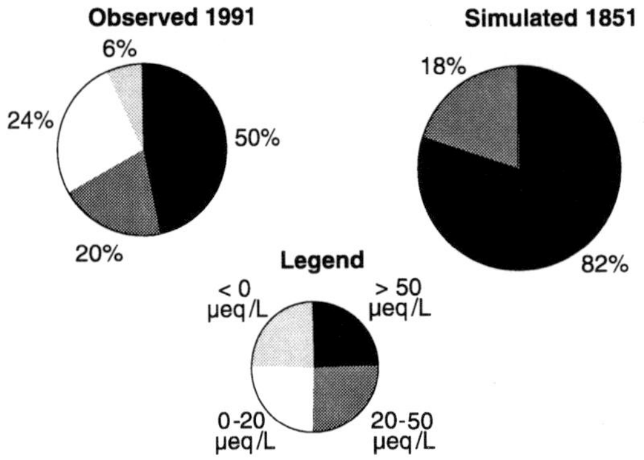

Figure 4. Percentage of streams in alkalinity classes.

acid tolerant species. When brook trout are gone, there are no fish left. By contrast, we estimated that there were no such streams among this population of about 300 Virginia trout streams in pre-industrial times.

This figure (Figure 5) defines the relationships among these four alkalinity classes and likely responses of brook trout populations in Virginia.

The "Southern Appalachian Man and the Biosphere" program (US Forest Service) estimated that, in the Southern Appalachians, there are about 30,000 miles of potential trout streams. Virginia has about 40 percent of the Southeast's trout stream miles. If one is willing to extrapolate from the 30% of Virginia's trout streams which are too acidic for any fish, then roughly 9000-10000 miles of the trout stream mile in the Southern Appalachians may have been lost as suitable fish habitat due to acidification. This must be regarded as a guess, since it is likely that some streams lost their fish populations before any records were made; streams which have become too acid now for fish may not have had fish in the past for other reasons.

This figure (Figure 6) shows further modeling results for Virginia trout streams. The pie diagram at the right is the distribution of streams in the four alkalinity classes, at the base or reference year of 1991. Three scenarios were modeled: continued acid deposition at 1991 levels, a 40% reduction relative to 1991, and a 70% reduction in acid deposition relative to 1991.

Even seventy percent reductions in sulfur deposition, relative to 1991, aren't good enough. There will still be more streams which become acidic and fishless in Virginia, and very probably in other parts of the southern Appalachians as well.

We conclude that reductions in sulfur greater than 70 percent will be needed just to stop further losses, and very substantial reductions will be needed for recovery.

Conclusions for the northeast suggest that reductions in sulfur deposition so far have not produced recovery from acidification, and acid rain is still a widespread problem. Substantial recovery of acidic streams like Hubbard Brook (NH) will occur in 25 years, if power plant SO_2 reductions are lowered by 80 percent. Lakes and streams in sensitive landscapes are still acidic.

The 1990 Clean Air Act amendments will not achieve chemical recovery that will permit biological recovery.

The reason we still have fish in the eastern third of the country is that most landscapes can buffer at least some of the acidity in the rain. However, it has become clear in the last ten years that the buffering capacity of some landscapes has

Brook Trout Category	ANC Class	ANC Range $meqL^{-1}$	Brook Trout Response
Suitable	Not acidic	> 50	Reproducing brook trout populations expected where habitat suitable
Indeterminate	Indeterminate	20-50	Extremely sensitive to acidification; brook trout response variable
Marginal	Episodically acidic	0-20	Sublethal and/or lethal effects on brook trout possible
Unsuitable	Chronically acidic	< 0	Lethal effects on brook trout probable

Figure 5. Brook trout categories, ANC classes, and ANC thresholds for brook trout responses to acidification in forested, headwater catchments in western VA (Bulger et al., 2000).

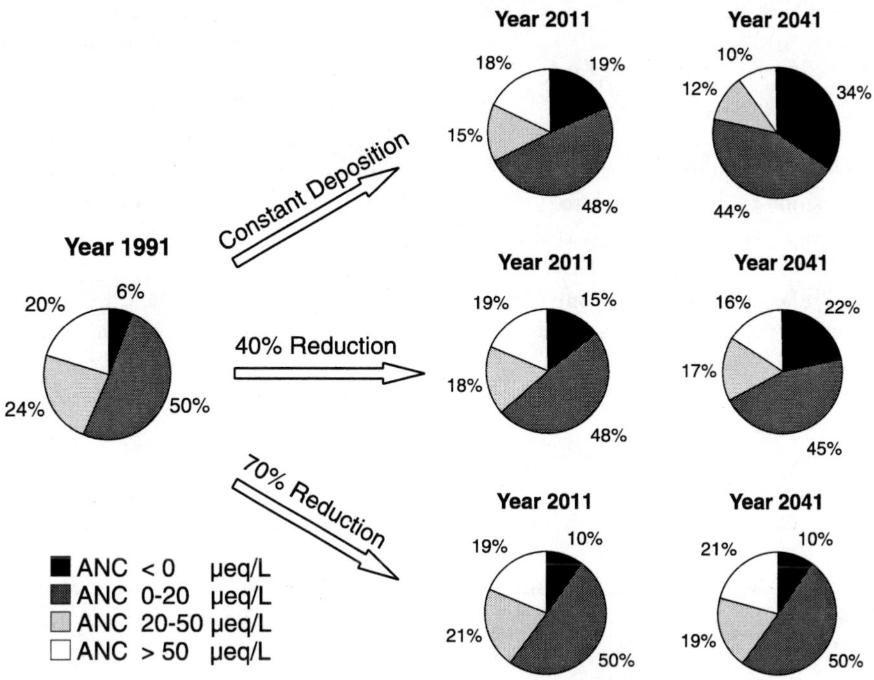

Figure 6. Percentage of streams in alkalinity classes for additional Virginia trout streams.

been substantially reduced, and so acidification will continue without further reductions in emissions.

Can we imagine a recovery? Chemical recovery of streams and lakes is dependent on the rate at which acid deposition is lowered, and the rate of local weathering of bedrock, and is likely to be highly variable.

However, once chemical recovery has occurred and stream conditions reach a pH of about 6, biological recovery is likely to occur on relatively short time scales. In streams, large invertebrates may recover in as little as three years. They are very good colonizers: many are the young stages of insect species that fly as adults.

In lakes, functional zooplankton communities may recover in about 10 years. However, from post-liming studies and recovery from experimental acidification studies, there are no examples where the exact zooplankton communities were restored, but all the zooplankton community functions appeared to be in place after about 10 years.

Fish comminutes might recover to some degree in perhaps 10 to 20 years. Once conditions are suitable for fish, they could be reintroduced rather than waiting for natural colonization events. In terms of the recovery of species composition, the last species in the stream may be the first one back. The most sensitive species, the first to go, will be the last to come back.

Questions

AUDIENCE: I am sorry if I missed this. You mentioned an 80 percent reduction in sulfate deposition and I would like to know for what baseline and also what has to happen to nitrogen depositions.

MR. BULGER: All the models that I am talking about were based on sulfur deposition, especially in the southeast. I didn't know what to do with nitrogen deposition.

AUDIENCE: The base year for the 80 percent reduction in the northeast, do you know what the base year is for the sulfate reductions in the northeast?

MR. BULGER: Substantial recovery of streams like Hubbard Brook, which is a pretty acidic stream, will occur in 25 years if power stations drop SO_2 emissions by a decrease of 80 percent relative to 1990.

Acknowledgments

I would like to thank the Center for Environmental Information for the invitation to speak here today. I also thank the US National Park Service, Shenandoah National Park, and the Hubbard Brook Research Foundation for their support and especially for the opportunity to collaborate with many fine scientists, many of whom are authors of the following references.

References

Driscoll, C. T., G. B. Lawrence, A. J. Bulger, T. J. Butler, C. C. Cronan, C. Eager, K. L. Lambert, G. E. Likens, J. L. Stoddard, and K. C. Weathers. 2001. Acidic Deposition in the Northeastern US: sources and inputs, ecosystem effects, and management strategies. Bioscience 51:180-198.

Driscoll, C. T., G. B. Lawrence, A. J. Bulger, T. J. Butler, C. C. Cronan, C. Eager, K. L. Lambert, G. E. Likens, J. L. Stoddard, and K. C. Weathers. 2001. Acid Rain Revisited: advances in scientific understanding since the passage of the 1970 and 1990 Clean Air Act Amendments. Hubbard Brook Research Foundation. Science Links Publication. Vol. 1, no. 1.

Bulger, A. J., B. J. Cosby and J. R. Webb. 2000. Current, reconstructed past and projected future status of brook trout *(Salvelinus fontinalis)* streams in Virginia. Can. J. Fish. Aq. Sci. 57: 1515-1523.

Bulger, A. J., B. J. Cosby and J. R. Webb. 1998. Acid Rain: Current and projected status of coldwater fish communities in the Southeastern U.S. in the context of continued acid deposition. Trout Unlimited, Arlington Virginia. 28 p.

Coastal Ecosystems

Robert Howarth – Environmental Defense

I have been asked to talk about the coastal effects associated with acid deposition. Last week, several of us met under Ellis Cowling's and Jim Galloway's direction in Woods Hole, and this is a direct outgrowth of that. The point I want to make here is partly to put the coastal nutrient pollution issue in context, and also to show what Jim Galloway calls the cascading effects of nitrogen in the landscape (Figure 1).

Nitrogen has a variety of potential environmental effects, both on human health and the functioning of ecosystems. Unlike many pollutants, nitrogen does not have a single effect, rather it has this huge range of effects as it moves through the environment.

The nitrogen can come either from fossil fuel combustion, NO_X, that we have heard mostly about, or ammonium, which is the growing problem we have heard a little bit about this afternoon.

The NO_X clearly contributes to ozone and smog and then goes on to contribute to fine particles and haze problems.

The ammonium waste also contributes to fine particles and haze problems. Both of those, when deposited onto forests, can have a variety of effects, some potentially beneficial such as increased growth, but all of the negative effects that Charlie Driscoll talks about.

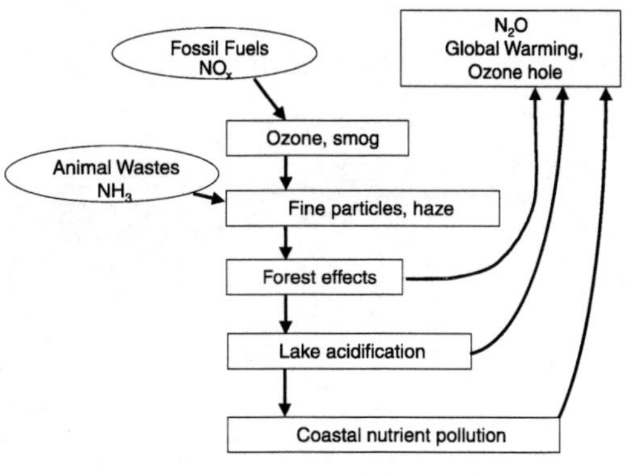

Figure 1. The cascading effect of nitrogen on human health and ecosystem functioning as a single atom moves through the environment.

As that leaves the forest or the upstream landscape, it contributes to lake acidification. Then the same nitrogen, as it reaches the coast, is contributing to the coastal nutrient pollution I will be talking about.

Now, there are losses along the way to the atmosphere as N_2O, which is a long lived gas with a residence time of 130 years in the atmosphere. That, too, has deleterious effects, in that it is a strong greenhouse gas, and it is a major and growing contributor to ozone depletion in the stratosphere, creating the ozone hole.

Today I will talk principally about what the consequences of nitrogen are in the coastal zone, and what the sources of this nitrogen pollution to the coast are.

The linkage to atmospheric deposition and acid rain is that we now believe that a very large percentage of the nitrogen pollution in the coastal zone originates through atmospheric transfer. Globally, an estimated two-thirds of the nitrogen that is created from human activity for agriculture and through fossil-fuel combustion is transferred through the atmosphere and some point. Some of this falls directly onto the surface of the oceans, while the rest is deposited onto the terrestrial landscape. A significant amount of this deposition onto land is then exported in streams and rivers to coastal marine ecosystems (Figure 2).

Terrestrial systems (N deposition Tg/yr)	
NOy	25
NHx	40
	65

Oceans (N deposition Tg/yr)	
NOy	20
NHx	20
	40

Figure 2. Estimates for deposition of both NO_y and ammonia and ammonium globally, onto the terrestrial landscape and onto the oceans. NO_y is 55% from combustion sources. Agriculture contributes 45% of NO_y and 85% of NH_x. 67% of human-created N cycles through atmosphere. Data are courtesy of Jim Galloway.

Our best estimates are, in the United States, somewhat more than one third of the nitrogen reaching the coast comes through the atmosphere at some point in time.

My talk is focused on nitrogen and not sulfur. Seawater systems have tremendous amounts of sulfur in them and adding sulfur makes no difference.

On the other hand, seawater systems are limited in their production by nitrogen. If we add more nitrogen, there is a very, very strong effect of increased algal and plant growth.

The concern in marine ecosystems is not with acid effects. Unlike the previous talk, I am not talking about the toxicity of acid rain, because seawater is very well buffered and handles the acidity fine. The concern is with the nitrogen itself.

Now, nitrogen is a nutrient, and nutrients are defined as things that nourish, so we think of them as a positive. The problem is simply having too much of a good thing. In the United States, and in fact much of the world, the coastal oceans now have too much of a good thing.

The consequences of this excessive nitrogen fertilization are summarized in this report, a report from the National Academy of Sciences that came out last summer. It is called Clean Coastal Waters: Understanding and Reducing the Effects of Coastal Nutrient Pollution. The report is the result of a three-year-long committee at the National Academy that I chaired, and we looked at all aspects of coastal nutrient pollution (Figure 3).

Clean Coastal Waters: Understanding and Reducing the Effects of Nutrient Pollution

NRC 2000

Figure 3. A report from the National Academy of Sciences' Committee on Causes and Management of Coastal Eutrophication. Online: www.nap.edu/catalog/9812.html.

One of our major conclusions is that the atmospheric source of nitrogen to coastal rivers and bays, particularly the atmospheric deposition of nitrogen onto the terrestrial landscape with subsequent export to coastal rivers and bays, is very much underappreciated and under-regulated.

What happens when we add nitrogen to coastal waters? Well, nitrogen is usually the limiting nutrient in coastal marine ecosystems. You add nitrogen, you get more algal and plant growth. In moderation, this is not necessarily bad, and can in fact result in greater fish harvests. However, excessive nitrogen inputs lead to a series of deleterious ecological effects which are harmful to fish and other marine life.

One of these changes is a decrease in dissolved oxygen. Low levels of oxygen, which we call hypoxic conditions, are unable to support fish and other higher forms of life. We are seeing increased incidences of hypoxia and even anoxia – the complete absence of dissolved oxygen in seawater – globally in many parts of the coastal oceans, including waters in the United States. This figure (Figure 4) shows the largest example we have in the United States, an area off the plume of the Mississippi River in the Gulf of Mexico, the coastal waters of Louisiana, an area that has grown to be as large as 20,000 square kilometers in some recent summers. That is an area the size of the state of New Jersey with oxygen levels too low to support fish and other higher forms of life.

There is some evidence that these "dead zones," as they are called, have existed for quite some time in many regions. However, we are seeing more of these low-oxygen areas, and many such as the dead zone in the northern Gulf of

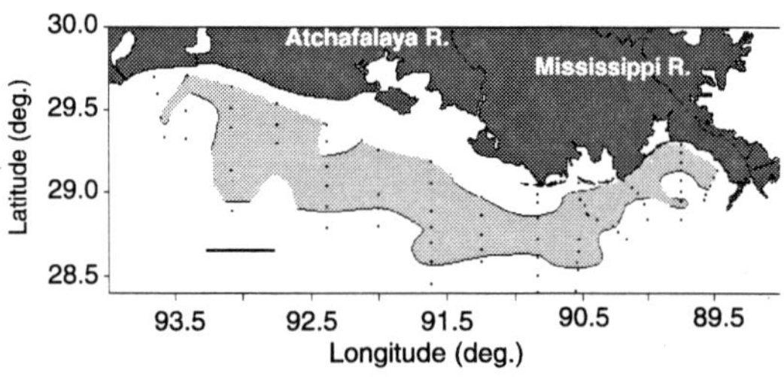

Figure 4. The extent of the hypoxic zone in the northern Gulf of Mexico in July of 1999. Bottom dissolved oxygen less than 2.0 mg/L. Figure is courtesy of Nancy Rabalais.

Mexico are clearly growing in size. This is the result of increased nitrogen flows to these waters.

I want to reiterate that this is not simply a problem of coastal Louisiana. These dead zones, areas of excessively low oxygen, are now commonplace in the coastal oceans of the world. Here in the United States, they are prominent features in Chesapeake Bay, Long Island Sound, and several other systems.

Adding nitrogen to coastal waters also increases the incidence and duration of harmful algal blooms. Things like brown tides and red tides and physteria blooms can all be encouraged by adding nitrogen to coastal waters. Some such blooms are natural events. Harmful algal blooms have always occurred, but they are increasing greatly in their severity and length of duration, and that appears to be at least in part a consequence of putting more nitrogen into coastal waters.

Nitrogen also has a major influence on biodiversity. When we fertilize coastal rivers and bays with nitrogen, we lose biodiversity in a very, very obvious way.

One aspect of that is degrading sea grass habitat. This slide (Figure 5) shows a relatively pristine subtropical sea grass bed. If you have a good eye, that is an area of about one meter by one meter. If you have a good eye and you are a good botanist, you might be able to pick out something like 10 to 15 different species of plants, just visually, in this area of approximately 1 square meter.

If you add nitrogen to a system like that, the habitat quality is degraded (Figure 6). The productivity of the seagrasses increases, at least to a point, but the habitat value for nursery grounds for fish and shellfish is much less. You also lose biodiversity, and in this degraded seagrass bed, only two species are apparent. If the seagrass meadow receives even more nitrogen pollution, the seagrasses die out, further degrading the habitat quality.

Ecological changes among phytoplankton and zooplankton are harder to observe, but nitrogen pollution also alters the ecological structure of plankton communities, and lowers their biodiversity as well.

The first survey that was ever made in the United States to try to put a systematic estimate on the extent of the problem of coastal nitrogen pollution was done by NOAA just in the last few years. They released their report in December of 1999. It is an excellent study and I recommend it to anybody who is interested in this problem.

The NOAA study concluded that, although the data quality are variable, that there are sufficient data around to estimate the spatial extent of this problem of coastal nutrient pollution.

They concluded that something like two thirds of the coastal rivers and bays of the United States are moderately to severely degraded from this problem, roughly one third moderately degraded and one third severely degraded.

This slide (Figure 7) shows the geographic extent of the problem. Degradation from nitrogen pollution occurs in some of the coastal rivers and bays of each region in the United States. Our NAS committee concluded that nitrogen is the largest pollution problem in the coastal waters of the United States.

Figure 5. A seagrass meadow receiving only low amounts of nutrient inputs.

Figure 6. A seagrass meadow receiving elevated nutrient inputs.

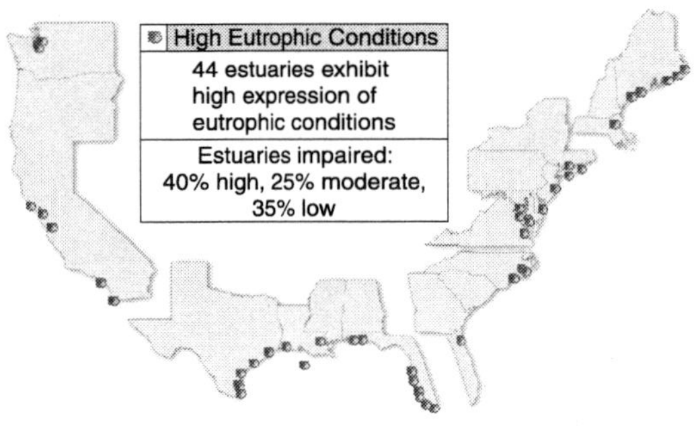

Figure 7. The geographic distribution of coastal rivers and bays in the United States that are highly degraded from nutrient pollution. Reprinted from the NOAA report of Bricker et al. (1999).

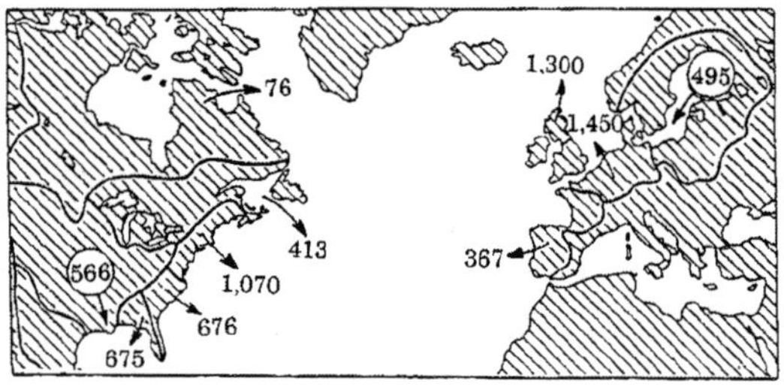

Figure 8. The flux of nitrogen to the coastal oceans from various regions surrounding the North Atlantic Ocean (riverine N export, kg/km²/yr). Reprinted from Howarth et al. (1996).

Where is the nitrogen coming from? I have already said a large part of it is from atmospheric deposition. Let me just give you a little bit of the background that has been pulled together to determine where nitrogen is coming from.

This slide (Figure 8) shows a large scale analysis to try to determine what the fluxes of nitrogen to coastal oceans are and where the nitrogen might be coming from. It shows the flux at the scale from the land to coastal oceans, at the scale of large regions, such as the entire Mississippi River Basin or the entire northeastern United States. The fluxes are per area of watershed (kilograms of nitrogen per square kilometer per year).

You see a lot of variation from as little as 76 kilograms per square kilometer per year up there in Hudson's Bay and Labrador, to a rather high number in the northeastern United States of over 1,000 kilograms per square kilometer per year, and higher fluxes yet in parts of Europe.

We can explain that variation very simply, by looking at what the inputs of nitrogen under human control are to the landscape. Those inputs are the NO_y deposition from fossil fuel combustion, inorganic fertilizer use, nitrogen fixation in agricultural systems, and the movement of nitrogen in and out of a region in food and animal feeds (Figure 9).

If we take these data and simply plot the sum of the total nitrogen inputs into a region from human activity, they are strongly correlated with the flux of nitrogen to the coast in rivers and from sewage treatment plants (Figure 10). The relationship is linear, and surprisingly strong. If humans burn more fossil fuel, or use more nitrogen in agriculture, or import more food into a region, more nitrogen will escape to the coast and pollute coastal ecosystems. Note that the slope of this line

	NOy deposition	Fertilizer	N fixation by crops	Net Import or export in foods	Total
North Canada rivers	70	160	30	-50	210
St. Lawrence basin	610	330	260	-30	1170
NE coast of US	1200	600	750	1000	3550
SE coast of US	1020	1170	370	450	3010
Eastern Gulf of Mexico	760	1260	250	580	2850
Mississippi River basin	620	1840	1060	-1300	2220
Baltic Sea drainages	480	1730	30	20	2220
North Sea drainages	1090	5960	5	-5	7050
NW European coast	1090	2870	50	-320	3700
SW European coast	460	3370	15	-65	3780

Figure 9. The sources of nitrogen to regions surrounding the North Atlantic Ocean. Reprinted from Howarth et al. (1996).

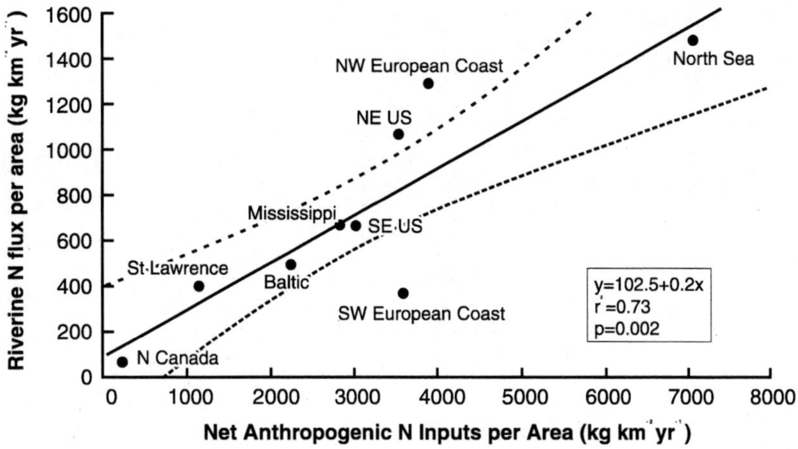

Figure 10. The relationship between nitrogen inputs to a region and the flux of nitrogen from a region to the coast for the temperate-zone regions surrounding the North Atlantic Ocean. Reprinted from Howarth et al. (1996).

is 0.20, that is, approximately 20% of the nitrogen mobilized by human activity reaches the coast, and the rest is retained in the landscape by various mechanisms.

Globally, fertilizer use is a bigger source of nitrogen use and pollution than is NO_y deposition from fossil fuel sources. However, as shown in Figure 2, much of the nitrogen from agricultural sources also flows through the atmosphere, both in oxidized forms, and as ammonium. I'll come back to the flux of ammonium in just a few minutes, but first I want to focus a little more on the issue of NO_y deposition in the northeastern United States. In the United States generally, most of the NO_y deposition comes from fossil fuel combustion, as opposed to some tropical areas where fluxes of NOx from agricultural fields are also significant sources for NO_y deposition. The regional importance of this NO_y deposition varies within the country, and is the largest single source of nitrogen pollution in the northeastern US, while being relatively unimportant in the Mississippi River basin (Figure 11).

Let's take a closer look at nitrogen sources in the northeastern United State, where deposition from fossil fuel sources is the major source of nitrogen under human control. A group led by my colleagues Beth Boyer and Nico van Breemen has made a careful analysis looking at the nitrogen flux out of a variety of watersheds in the northeastern United States, from Maine down to Virginia

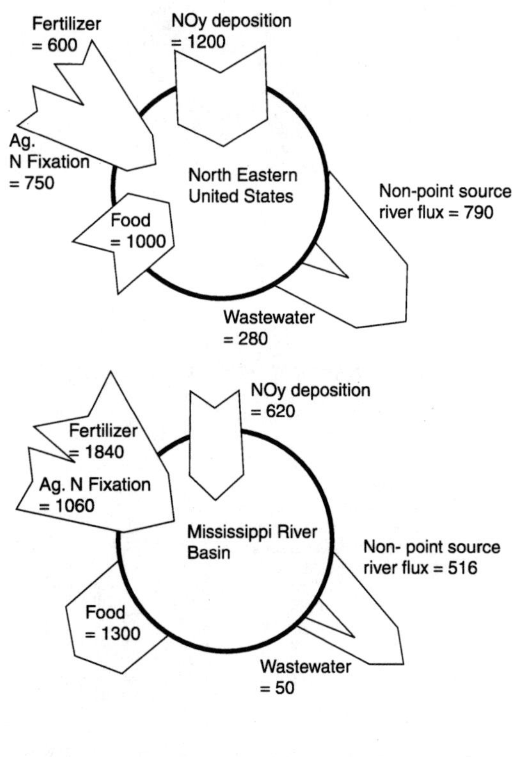

Figure 11. A comparison of nitrogen inputs and exports (in kg N km-2 year-1) for the northeastern United States and for the Mississippi River basin. Reprinted from the report of the National Academy of Sciences Committee on Causes and Management of Coastal Eutrophication.

(Figure 12). In some of these watersheds, NOy deposition is the major input, while other watersheds have significant use of nitrogen in agriculture and significant import of nitrogen in food (Figure 13). If you plot the sum of human-controlled nitrogen inputs to these watersheds, you again see that it is an excellent predictor of the flux of nitrogen out of the watershed in rivers (Figure 14), much as we saw earlier for the course regional-scale analysis.

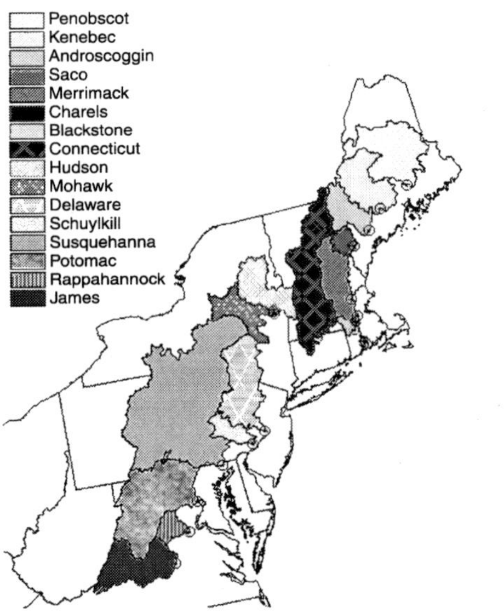

Figure 12. The major watersheds of the northeastern United States. Reprinted from Boyer et al. (2002).

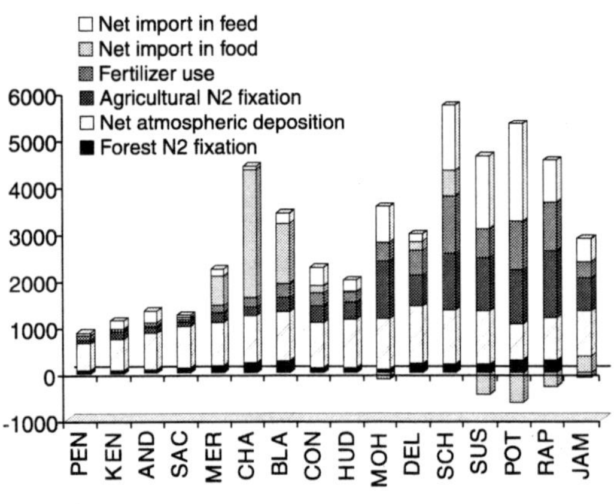

Figure 13. Inputs of nitrogen to the major watersheds of the northeastern United States (nitrogen inputs in kg/km²/yr). Reprinted from Boyer et al. (2002).

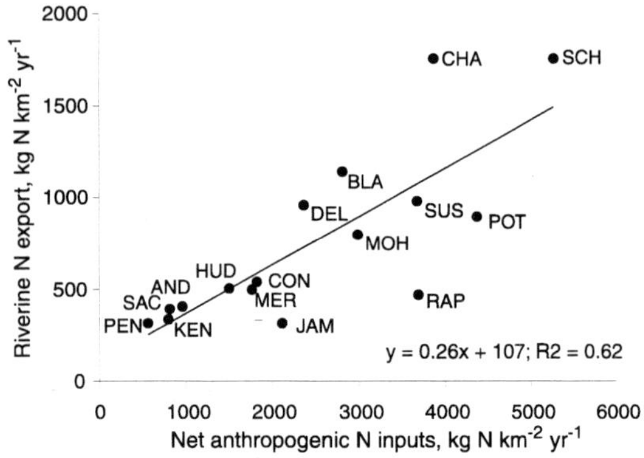

Figure 14. The relationship between nitrogen inputs per area and nitrogen flux in rivers for the major watersheds of the northeastern United States. Reprinted from Boyer et al. (2002).

One of the implications of this analysis is that a significant amount of the nitrogen deposition that falls onto the landscape, including onto forests, must be exported to rivers. This conclusion surprises a lot of scientists who have worked on the biogeochemistry of forests, because we have this paradigm, which is a reasonable paradigm, that forests do a good job of retaining nitrogen deposition until some point where they become saturated. If you add more and more nitrogen, the forest is able to retain it, since production of the forest is limited by nitrogen. This is the paradigm, until you add so much nitrogen, the forest is no longer limited by nitrogen and it starts to leach out as nitrate.

However, we now see that forests may be leakier of nitrogen even before they become nitrogen saturated. The analysis by Nico van Breemen and his group is that for the forested landscape of the northeastern United States as a whole, approximately one quarter of the nitrogen entering the forest is leached to streams and groundwater, with some of this leaving as dissolved organic nitrogen, but most probably leaving as nitrate (Figure 15).

I want to say a little bit about ammonia. While NOy deposition in the United States has been fairly steady over the past 15 years or so, the deposition of ammonia and ammonium is probably increasing. In our NAS committee report, we looked very carefully at what the fluxes of ammonium are through the atmosphere, what the deposition is and what the sources are. We

N balance for forests in northeastern U.S. (kg N $km^{-2}yr^{-1}$)

Inputs	
Deposition	1070
N fixation	240
	1310
Sinks and exports	
Tree/plant storage	460
Tree harvest	260
Soil storage	100
Denitrification	195
Leaching (75% NO_3)	295
	1310

Figure 15. The inputs and fate of nitrogen to the forests of the northeastern United States, on average. The area considered are the watersheds shown in slide 12. Data are from van Breemen et al. (2002).

concluded that, for the United States, 92 percent of the ammonium flux comes from animal waste.

We are having a large-scale increase in industrial farm animal production in the United States, and that is what is contributing to the increased ammonium flux we are seeing. As I mentioned above, the deposition of ammonium globally is similar to or perhaps even larger than NOy deposition. In the United States, this is not true, and NOy deposition from fossil fuel combustion is probably some 4-fold greater than is ammonia and ammonium deposition. But the ammonia source is rapidly growing and deserves attention.

Our NAS committee, relying on the work of the International SCOPE Nitrogen Project, tried to estimate what the nature of the overall human alteration of the nitrogen cycle might be. We looked at what a pristine flux of nitrogen to coastal ecosystems might be, if there were no human activity in the landscape, no NOy deposition, and no agriculture. We concluded that for northern Canada, there has been relatively little human influence, but for the northeastern United States, humans have probably increased this nitrogen flux by at least 10-fold and, in parts of Europe, by even more than that.

Much of this change has been recent. Some of this has been going on since the start of the agricultural and industrial revolutions, but it has greatly accelerated over the last several decades, and we can see that in a variety of data sets. Fluxes in many rivers in the northeastern United States and the flux in the Mississippi River (Figure 16) have all increased as more nitrogen has been mobilized in their watersheds by human activity over the past many decades.

Let me just conclude with some of the major conclusions from our National Academy report. If any of you are interested in learning more about atmospheric deposition and coastal nitrogen pollution, I would urge you to take a look at our report.

First, the nitrogen cycle is being altered more rapidly than the cycle of any other major element globally.

As a result of that, something like two-thirds of the coastal rivers and bays of the United States are now moderately to severely degraded from nutrient pollution.

Our committee concluded, on that basis, that this is the largest pollution problem in the coastal waters of the United States.

The extent of the nitrogen alteration varies from region to region, in the United States and globally. There has probably been a 10-fold increase on average for the nitrogen flux to coastal ecosystems in the northeastern United States, and much of that increase has occurred over the last four decades.

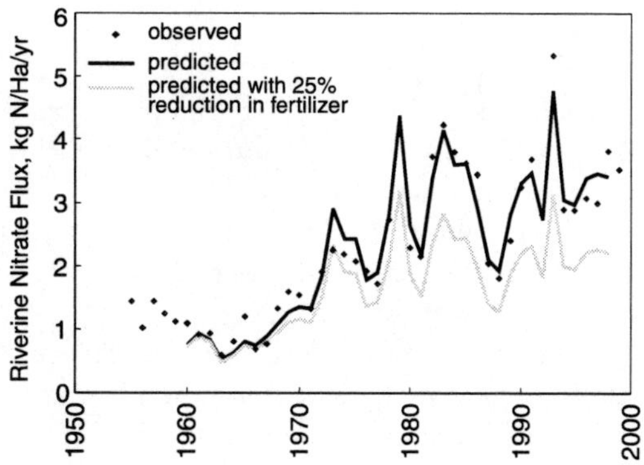

Figure 16. The flux of nitrate down the Mississippi River over time. The diamonds show actual data. The dark line shows model predictions, based on estimates of nitrogen input to the basin and variation in climate over time. The light line shows an estimate of what the nitrogen flux would have been had 25% less fertilizer been used in the basin. Modified from McIsaac et al. (2001).

Non-point sources are the dominant source of nitrogen in all regions of the world, at the scale of large regions. Non-point sources also dominate in many individual coastal ecosystems, including Chesapeake Bay and the dead zone in the northern Gulf of Mexico. However, sewage inputs are the dominant source of nitrogen to some coastal systems, such as Long Island Sound.

In some regions, the major source of nitrogen contributing to non-point fluxes is from fertilizer. Other agricultural sources are also major contributors. Animal waste from industrial-scale operations stand out in this regard, and are major contributors of nitrogen to the atmosphere as ammonia, as well as directly to coastal waters.

NO_y deposition from fossil fuel sources is the major nitrogen source in some regions, including the northeastern United States. Our committee report highlighted that many existing models and reports on the importance of NO_y deposition to coastal nutrient pollution, in the opinion of our academy study, have underestimated the extent, perhaps by a factor of up to two to three, for a variety of methodological issues.

It is a huge contributor, even more of a problem than one would have guessed, say, five or ten years ago. I will stop there and take questions, if there is time.

QUESTIONS

AUDIENCE: There is no reference to organic nitrogen. Could you identify some of the principal chemicals and what their sources might be?

MR. HOWARTH: The question is, what are the sources of the organic nitrogen and what are the chemical forms we are seeing.

The quick answer to that is I believe there is insufficient study. There are several people who have been looking at that over the last decade, but it has only been over the last decade, and it is a complex question.

In terms of the depositional sources, a lot of the material appears to be relatively low weight amine-like things, which may well have sources from these animal wastes and similar sorts of things as well.

Some of it is natural. For instance, pollen would be a contributor to the particulate phase of the organic flux.

In terms of breaking out how much of the organic deposition is natural versus from human sources, I don't think there is a strong enough consensus to say.

If we look at the landscape, what is moving, say, from forests, it would be a variety of higher weight molecular compounds, and those are very poorly characterized.

We do know, from mass balance, that once they hit larger rivers, some of this organic nitrogen is converted to nitrate and other inorganic forms, so some is labile on a time scale of days to weeks to months.

MR. BULGER: The American Fertilizer Institute thinks that most applications of nitrogen are five to 10 times what plants could actually use under those agricultural conditions. Do you think that is true? That is a huge amount of wasted nitrogen.

MR. HOWARTH: Our NAS committee took a fairly careful look at the fate of fertilizer, and concluded that over half of the nitrogen inputs to agricultural fields in the US is actually removed from the field in harvested crops. That indicates farmers are doing a pretty good job, and in fact are using fertilizer fairly efficiently.

There is room for improvement, though. Some evidence suggests that farmers in the mid-west fertilize on average at rates around 30% higher than necessary for the optimum economic return. They see little if any additional crop yield for this extra use of fertilizer, but it gives then some sense of security in knowing they are maximizing crop yield. They could probably reduce fertilizer use by 30% and see little loss of yield, while greatly reducing the loss of nitrogen to downstream ecosystems.

Clearly, we need some change in agricultural incentive policies to discourage over-fertilization.

MR. STOSS: Fertilizers in residential use, does the fertilizer data include recreational uses, such as on golf courses and residential like lawn care chemicals?

MR. HOWARTH: It does, and there we probably are seeing massive over-fertilization on golf courses and lawns. On a national scale, they are still relatively small users compared to agriculture, but locally they are going to do damage to coastal ponds. It is a big problem in some parts of the country.

MR. KNAUER: Doug Knauer with the Wisconsin Department of Natural Resources. [Question off microphone.]

MR. HOWARTH: There has been a 30 to 40-year tradition of looking at iron as a limiting nutrient as well as nitrogen and phosphorous and silicon and several other things.

Anything I say, I am sure, would be subject to some debate in some circles. Having qualified it in that way, let me say I think there is a pretty strong consensus that most of these coastal systems, certainly the ones in the temperate zones of the United States, are strongly nitrogen limited.

A couple of them are phosphorous limited. Some of them switch seasonally between nitrogen and phosphorous and there might be an additional slight effect of iron, but it isn't a major influence.

On the other hand, if you move to some of the oceanic areas of the world where iron is much lower, where there is a fair amount of iron coming from land sources, that iron becomes much more limiting, which is the basis for that argument. In these coastal systems, nitrogen really is the key.

MR. KRANOWITZ: Jeremy Kranowitz from the Izaak Walton League. At the CLEAR(?) network meeting at the end of the year, there was some discussion about nitrogen deposition and they were trying to identify power plants specifically as sources.

They said that in estuaries, by their best estimates, it was something like one to four or five percent. Do you have data that talks more about that the majority is from non-point sources, but do you have data specifically about power plants?

MR. HOWARTH: On average for the United States, the combustion of fossil fuels is contributing just over 25% of the input of nitrogen to coastal waters, from both direct deposition onto the waters and from deposition onto the terrestrial landscape with subsequent flux through rivers to the coast. As I mentioned, there is significant variation, so the NOy deposition term is more important than this in some regions such as the northeastern US and less important in others such as the Mississippi River. But on average, approximately one quarter of the nitrogen pollution is coming from fossil fuel combustion.

The US EPA estimates that power plants and other stationary sources of combustion are responsible for approximately 40% of the NO_x load to the atmosphere, and therefore for 40% of the NO_y deposition from fossil fuel sources.

If you combine these estimates, you conclude that approximately 10% of the nitrogen pollution in the coastal waters of the US, on average, come from power plants.

I think the estimates you heard of one to four percent are low, and are probably based on some of the older analyses.

It is a rapidly changing field. One of the problems is that the first estimates of what the contribution of atmospheric deposition would be to a nutrient polluted estuary are only 25 years or so old. They are good estimates, but they weren't based on a large amount of detailed monitoring data.

We still don't have good monitoring data in a lot of these coastal areas because the depositional networks were set up to look at the acid effects, not our coastal effects.

So, we don't have good handles on deposition in many areas, although people like Joe Scudlark and others are trying to improve that.

The other problem is that once the nitrogen hits the landscape, you need some sort of estimate of what its mobility gives off.

The older models – by older models I mean those that are more than four or five years old – are based on that model that nitrogen isn't going to leave the forest to any significant extent until we add so much that it saturates the needs of the forest.

The current evidence is that that is a wrong model, that nitrogen will be leaving at much, much lower levels of deposition. So, it is a much bigger problem than would have been estimated, even a few years ago.

MS. CHAN: Melissa Chan with the Department of Energy. I was wondering if you found that most of your nitrogen pollutants in the coastal areas were from U.S. sources, or whether nitrogen is a trans-boundary pollutant like mercury and can come over the oceans, or is very mobile in that way.

MR. HOWARTH: The United States produces plenty of nitrogen pollution. I think we are probably our major contributor.

Nitrogen is capable of moving through the atmosphere up to around 1,000 kilometers or so. The aerosol particle is an ammonium sulfate particle.

Most of the nitrogen in the atmosphere is being deposited fairly close to sources, and most of the things that are moving in streams and rivers are just moving at the scales of those rivers or coastal currents.

I think it is safe to say that the majority of nitrogen is coming down relatively close to home and we should clean up close to home.

Forest and Terrestrial Systems

Greg Lawrence – US Geological Survey

What I am going to be talking about this afternoon has been touched on in some of the other talks. I am going to be giving a little bit more detail, and I am also going to try to highlight some of the scientific progress, since 1990.

That was brought up earlier this morning, and scientists do feel that we have made quite a bit of progress since 1990, the last big NAPAP assessment, particularly in terms of understanding the extent of effects.

We have that better defined now, and also in terms of understanding the processes that have been affected by acid rain. That is extremely important in terms of evaluating the potential for recovery.

The topic is a little more narrow than shown in the handouts. I am going to specifically be talking about forests, soils and trees. This is where the majority of the terrestrial effects research has been done.

If we use 1990 as our baseline, we can say, since then, that we have developed a much better understanding of soil processes and how changes in soils are linked to changes in surface waters.

In 1990, there had not been good documentation of changes occurring in soils as a result of acid rain, something that had been suspected since the 1970s, but by 1990, still really hadn't been proven.

Also, in the 1990 assessment, the only tree effect that they talked about was high elevation red spruce, which was felt was clearly the result of acid rain, but the mechanisms weren't all that well understood at that time.

I will jump right in first with the soil side of things. You can look at the effects of acid rain on soil, really, in terms of three primary changes.

Again, these have already been touched on earlier today. The first one is depletion of calcium from the soil. This is very important in terms of evaluating recovery potential. I will get into that a little bit more. It is something we weren't aware of in 1990.

The second change – mobilization of inorganic aluminum is something we actually were aware of as early as 1980, but how that mobilization of aluminum interacts with calcium in the soils is something we have a little better handle on today.

I need to make sure that I am clear on the distinction of why I am referring to it as inorganic aluminum.

As was mentioned earlier, aluminum is very common in most soils, but if it is not mobilized, it is not harmful. In some cases, it can be naturally mobilized by organic acids, but what you form is organic aluminum, and that form generally is not toxic to aquatic life and doesn't cause a problem for tree growth.

Now the third change, accumulation of sulfur and nitrogen. In 1990 the widely held view was that deposition of sulfur – I am referring to northeastern soils that don't have the absorption capacity that the southeastern soils have, that Art Bulger mentioned earlier – the sulfur coming in as sulfuric acid basically passes through the soils and goes right into the streams.

It was also felt that the nitrogen being deposited, in terms of its effects on surface waters, was a similar direct effect, primarily the result of nitric acid building up in the snow pack in the spring, and then being released quickly as the snow melted, causing acidification episodes.

We know now that in both the cases of sulfur and nitrogen, that the system is quite a bit more complicated than that, and I will talk a little bit more about that as well.

I would like to go over why calcium is important in natural ecosystems, particularly terrestrial ecosystems.

Calcium is essential for wood formation in trees. Trees have a very high demand for calcium. If they don't have enough calcium, they don't grow at as fast a rate.

Calcium in the soil also is very important. It is the primary element that neutralizes acidity, whether the acidity was generated through natural organic acids in soil or by acid rain. If there is sufficient calcium in the soil, the acidity will be fully buffered.

Also, something that hasn't been touched on

much, calcium is an essential nutrient for aquatic plants and animals.

We talk about the effects of acid rain on aquatic ecosystems in terms of toxicity and so forth, but related to the issue of recovery, we have less and less calcium going into our aquatic systems, and that is undoubtedly having a more subtle effect on the productivity of these systems.

Now, I mentioned that we suspected a problem with acid rain effects on soils as early as the 1970s, and some of you are probably wondering after more than 20 years, why didn't you figure it all out sooner.

The biggest reason was a lack of historical data. Soil science in this country is a fairly new science, and there really was no pre-acid rain data that we could compare to.

This was changed by a study done out of the University of Pennsylvania. They were able to repeat the sampling of one of the earliest forest soil studies in the country, which was conducted in the Adirondack Mountains.

They determined, using the original laboratory methods, that there was about this much calcium, referred to as acid-extractable calcium, in the Adirondack soils between 1930 and 1932 (Figure 1).

That individual point right there represents a large number of sites, over 50 sites scattered throughout the Adirondacks.

The group at the University of Pennsylvania went back and re-sampled these sites, duplicated the original methods and came up with the 1984 value.

You can see a drop from about 80 to 50, indicating fairly strong evidence that there is less calcium in the Adirondack soils than there was in the 1930's.

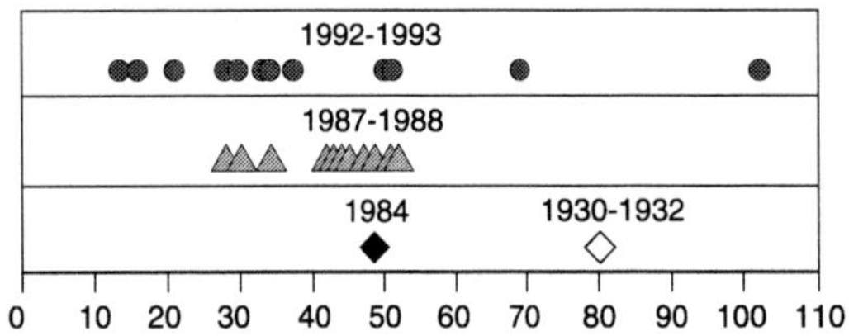

Figure 1. Acid-extractable calcium in millimoles per kilogram.

Also shown on this figure is a study up on the top bar, that was done in the 1990s to evaluate what the range in calcium concentrations was throughout the northeast.

The sites were specifically selected to see how broad a range we had. You can see that the value in the Adirondacks falls pretty much in the middle of the range. So, that Pennsylvania study wasn't presenting data that was atypical to the northeast. Results of an additional study are also presented there in the third bar.

That really is the closest thing we have to being able to evaluate what soils look like before acid rain in this country.

Even though we don't have much historical data in the United States, we have been getting a better handle on the processes involved, how these soils may be changing and how the calcium in different horizons relate to each other.

This particular graph shows data collected at those same spruce sites that were selected over the northeast to determine the range in soil variability (Figure 2).

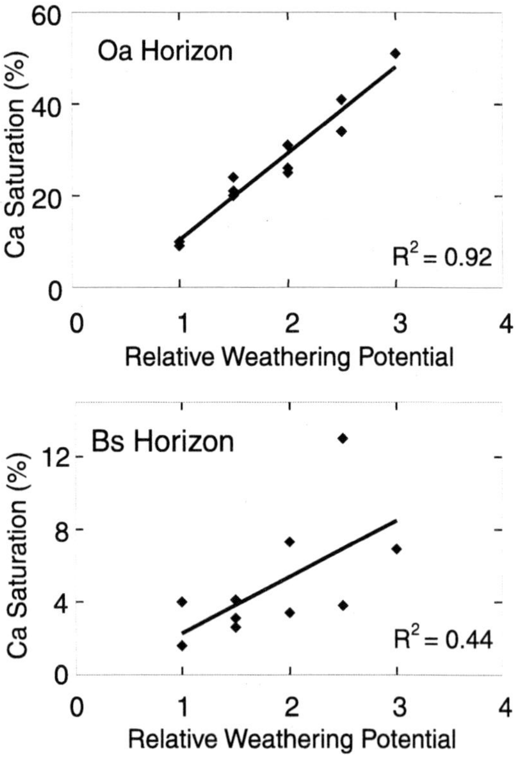

Figure 2. Oa and Bs horizon of Northeastern spruce tree sites.

This is the amount of calcium in the soil plotted versus relative weathering potential. I think most of you are familiar with the fact that differences in mineralogy affect the availability of calcium.

Certain types of rocks produce calcium that is available for plants and for leaching into surface waters, at higher rates than others.

The weathering potential, or the calcium availability, was plotted versus the amount of calcium. This is in the O horizon, the top surface horizon that is formed by organic materials, leaves and branches falling down and so forth.

The calcium that you see here was actually taken up by roots into the biomass and then deposited to form this OA horizon.

You can see that the amount of calcium is very, very strongly related to the weathering potential, how much calcium can be released from the rocks at this site. This is what you would expect, although this is an organic horizon.

In the mineral horizon directly below it, which is the ultimate source of this calcium, you would expect to see a very similar relationship, but actually you don't.

There is a correlation there, a positive correlation, but it is much weaker than it is for the upper organic layer.

This is getting into the details a little bit further, but basically what this suggests is that the calcium in this mineral horizon, in the B horizon, has been depleted to a level where the mineralogy is no longer as clearly reflected.

As you lower the calcium to a certain level you begin to mobilize aluminum.

The aluminum starts to control the chemistry and basically the aluminum buffers any further decreases in calcium. So, you get into that range where aluminum is mobilized and you lose the signal between the calcium and the type of parent material that is there.

The reason this happens in the mineral soil rather than the organic layer above, is that the organic layer is where there is very, very intense root activity, a recycling of the calcium.

These sites are all pretty low in calcium, so as soon as the plant material decomposes and releases some calcium, there are plenty of roots there to grab it, and that prevents it from leaching out of the soil.

In the deeper soil, this recycling by roots is not nearly as strong. Therefore, we end up with the differences between the two horizons.

I am going to get a little bit more technical here and talk about aluminum mobilization. So, as we are depleting the calcium, we get to a point where the pH of the soil decreases and aluminum enters soil solution.

Basically, this can be looked at in terms of the difference between the base cations and the acid anions. Among the base cations, calcium is typically the most abundant. So, we usually talk about base cations in terms of calcium. Magnesium is also important, particularly for tree growth, but it is not as abundant, potassium as well. In a situation that is well buffered, the sum of the base cations are much higher than the acid anions and aluminum just stays put and doesn't cause a problem. As you get closer and closer between these two amounts, that is when you begin to see aluminum appearing in the soil solution.

This is the sort of result you get in terms of aluminum to calcium ratios. Quite a bit of work has been done in the laboratory to evaluate the effects of aluminum to calcium ratios on growth of various tree species. It has been difficult to do *in situ* studies because of the nature of trees being large and growing so slowly, but the greenhouse experiments pretty clearly show that if you get an aluminum to calcium ratio of greater than one, you are going to be impairing the growth of the tree.

Here are average values for all those northeastern spruce sites. In the OA horizon we have a range from 0.1, which is very healthy, to 2.2, which is not (Table 1).

Then you look at the B horizon, and, as I said, the B horizon has been affected to a greater extent, and you see that even the lowest number is still over that threshold of 1.

I want to look at it in terms of what happens to the soil water when it moves into the streams. You heard about how aluminum is toxic to fish.

As this difference between the base cations and acid anions decreases from high numbers down toward zero, the inorganic aluminum concentra-

Table 1. Aluminum to calcium ratio (mol/mol).

	Oa Horizon	B Horizon
Spruce Stands		
Soil Solution	0.1 - 2.2	2.3 - 12
Exchange sites	0.1 - 3.8	3.0 - 35
Hubbard Brook		
Exchange sites	0.3	5.7
Winnisook		
Exchange sites	1.9	24

tion increases abruptly and you cross this toxic threshold.

This is the threshold for brook trout, which is one of the most acid tolerant species. So, once you get across that line, you are really going to have problems with your fish populations.

Now I would just like to shift to the third item I had listed in terms of soil changes, the effects of acid rain on sulfur and nitrogen in forest soils.

We knew that sulfate could accumulate in the southern forests through absorption, but it was really not well understood that sulfur could accumulate in soils of the northeastern region.

Most of the attempts at doing sulfur balances through the 1980s showed that the sulfur in, more or less balanced the sulfur out.

This graph shows that the total sulfur in soil versus the atmospheric deposition is actually very strongly correlated (Figure 3).

There does seem to be a mechanism for accumulation of sulfur within the soil, but the sulfur doesn't necessarily stay there indefinitely. It does seem to be gradually working its way out, particularly in situations where you have an extended dry period followed by some rain. That seems to release some of this sulfur and, every time it does, you get a very severe acidification episode.

We look at nitrogen. Nitrogen is a little bit different. Under pristine conditions, it is the growth limiting nutrient in forest ecosystems.

It also is accumulating in forest soils to the point where some of these forests do not seem to be very effective at retaining nitrogen, or at least not as effective as they used to be, in which case it behaves similarly to sulfur. You get a dry period followed by a wetting up and you can release a lot

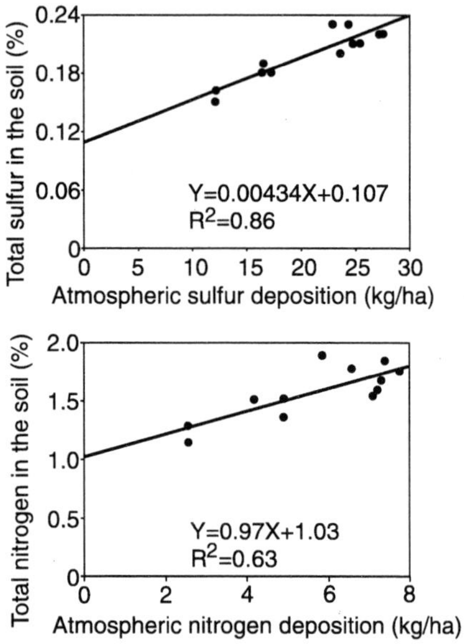

Figure 3. Soil chemistry in spruce forests (NY, VT, NH, ME).

of nitrate as nitric acid, and you get your acid episodes in stream water. These can occur any time of the year, and they still do occur during spring snowmelt as well.

Just to summarize the current status of the watershed budgets for sulfur and nitrogen, because they do look a little different than we used to think they did, the outputs of sulfur are currently exceeding the inputs.

That means if we shut off any inputs of sulfur and acid rain tomorrow, we would still be losing some sulfur into the surface waters as sulfuric acid. As we are doing that, we are continuing to put a drain on the base cations.

The same thing with the nitrogen situation. As the nitric acid is formed and leaches into the surface waters, it is contributing to the calcium depletion problem.

If we want to think about recovery in terms of soils, we need to put this information into context. It is a difficult concept to define because we really don't know what the situation was prior to acid rain, but we are getting a pretty good understanding of the changes in soils that have occurred.

In regard to the magnitude of the changes, there is still a fair amount of uncertainty. The timing of the changes, did they happen quickly or did they happen over a 50-year time period, is also quite uncertain.

We do know that the changes have occurred, and we do have some understanding of the mechanism. As a result, the models currently, such as the one that Dr. Driscoll presented this morning, present a much lower level of uncertainty than the models did back in 1990.

In 1990, the models did not assume changes in soils. Now they are incorporating these changes in soil, so we are making progress.

That is not to say that with Dr. Driscoll's model, all the uncertainty is gone. There is still a fair amount of uncertainty but there has clearly been progress.

This third item here, how much change has occurred through natural processes or land management, non acid rain issues. We know these are also factors.

Early on in acid rain research, it was felt that soils really didn't change much and that changes would occur typically over centuries, not decades.

We know that is not the case and that factors other than acid rain can change soils very significantly over decadal time scales. That also plays into any sort of assessment to determine recovery potential.

Now I will switch to tree responses. I am really going to limit it to just two trees. Work has continued on red spruce, as you heard, and sugar maple, but it really hasn't expanded beyond that point. So, we really can't talk about forest ecosystem effects, but we can talk about two very important tree species in the forest.

This plot shows, in the top box, the incidence of winter injury, reported winter injury in the northeast (Figure 4). Extremely cold temperatures end up killing the current year needles of the red spruce trees.

Some pretty conclusive evidence has demonstrated that this is the result of acid rain, through interactions of the needles with acidic mist and fog and so forth.

We also can say with a fairly high level of confidence that the changes in soils that I have been talking about have also reduced the stress tolerance of red spruce trees and contributed to some of this.

We have a good record of the fact that the red

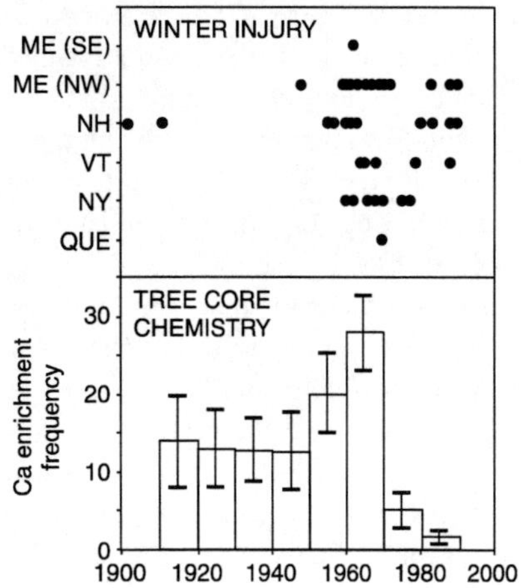

Figure 4. Relationship between treecore chemistry and winter injury in red spruce trees (NY, VT, NH, ME).

spruce trees have responded physiologically to changes in soils. That is what is shown in this lower panel.

Without getting into the complexities of this vertical axis, it is basically an indication of the amount of calcium retained within the wood as the tree grows.

This is a summary of 300 to 400 trees collected over the northeast, looking at the calcium incorporation over these different decades.

You can see that things run along pretty consistently right up until 1940, and then we see a big jump for a couple of decades, and then a big drop.

At the early stages of acid rain, calcium actually was more available. The hydrogen ion was actually freeing up the calcium in the soil, making it easier for the tree roots to take it up. So, the higher calcium availability was reflected in the calcium in the tree tissue.

As the acids continued to be deposited on the soils, the calcium continued to be leached out at a higher rate than it was being added through the breakdown of the rocks and minerals. You end up getting lower levels of available calcium. This is a pretty strong indication that these trees have been growing in soils that have been changing.

The situation in Pennsylvania is that the trees need a certain level of calcium or magnesium. If they don't get enough, they are much more susceptible to stresses. In this case, defoliation from native insects is a big stress. Studies conducted by the U.S. Forest Service have shown a strong link among concentrations of calcium and magnesium in the soil and leaves, and unusually high mortality of sugar maple trees. This relationship appears to be confined to ridge tops, where soil base concentrations tend to be lowest.

If there is sufficient magnesium, or at least higher levels of magnesium in the foliage, trees can recover from the defoliation. If they have extremely low levels, that is where you get the dead sugar maples. So, acid rain doesn't kill the trees directly, but it reduces their ability to deal with stresses.

I just want to show one graph (Figure 5). The scientists from the U.S. Forest Service have developed a way to measure stress in trees before they die, which is actually a very important tool.

These are results from northeastern red spruce stands. The concentration of putrescine is related to the ratio of Al to Ca in the soil, the higher the ratio, the higher the putrescine concentration. A value of this ratio of greater than 1.0 has been shown in laboratory studies to stress red spruce seedlings. This figure presents evidence from the field that this ratio is also linked to stress in mature trees.

Going from a low slope position to a ridge top, this is typically where we see the worst effects. The ridge tops have the least amount of calcium. They probably had the least amount of calcium before acid rain. Therefore, they are the most sensitive.

It is clear that the trees are under the greatest amount of stress in these soils. Also, I would like to point out that there is some data there for YB, which stands for yellow birch. So, we are beginning to take a look at a new tree species.

Just to summarize, then, before I stop – and you have heard some of this already – 20 to 50 percent of the canopy spruce have died in high elevation sites in the Adirondack, Green and White Mountains, but the spruce die back is not limited to high elevations. Pretty much throughout its range we can find it.

It is partially the result of acid fog, partially the result of calcium depletion in soils and, as I just got done saying, the sugar maple die back is also linked to insufficient calcium in the soil. Thank you.

Questions

AUDIENCE: I wonder if you have any data as to acid death mortality rates versus acid deposition rates?

MR. LAWRENCE: Mortality rates for trees?

AUDIENCE: Correct.

MR. LAWRENCE: Yes, that is part of any study to evaluate unusual causes of mortality. I can throw this up (Figure 6). This is just a very crude way to get at that.

This is a healthy sugar maple stand in Pennsylvania. There will be a certain percentage of trees that are dead through natural processes, with not a lot of light coming through the canopy and a situation where things look pretty green.

In these situations where you have the sugar maple die back – this was shown earlier – you end

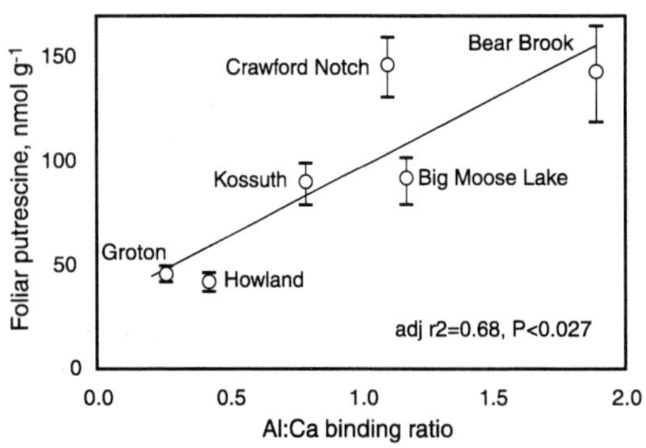

Figure 5.

Figure 6.

up losing 50 to 100 percent of the sugar maple trees and there is no regeneration going on.

AUDIENCE: Could you elaborate on that? I have information that says they are dying at the rate of 160 percent, and the absolute rate is 50 percent.

MR. LAWRENCE: I guess I would have to look at that. The numbers, I don't really understand the numbers. Forests don't die at a 50 percent rate.

AUDIENCE: Over 100 years they don't. [Remainder of question off microphone.]

MR. LAWRENCE: The trees in the forest will not reach the same age they used to; is that what you are saying?

I am not familiar with any studies to directly look at that. It is a very interesting question. One thing I could comment a little bit about, and I am more of a soil chemist, but my forest colleagues feel that, for example, with the spruce situation, the spruce that have died tend to be of a certain size rather than of a certain age.

This was part of the controversy early on. They said, well, maybe the spruce are all reaching their natural life span, but that really wasn't the case. It was more a matter of size and the fact that these trees grew to a large size under certain soil chemistry and the conditions changed too quickly for them to adapt.

That explains why, if you go to these sites that have had extensive spruce die back, you will see spruce regeneration.

Those trees probably won't ever achieve the growth rates of their parents and how long they end up growing is a little hard to say right now. I think it is probably safe to say that that site is not as productive. There is still a lot of work that needs to be done with trees.

AUDIENCE: Along those lines, I am under the impression that the calcium requirement in a small spruce is the same per square inch as a larger tree. [Remainder off microphone.]

In view of the calcium requirement, do you think these trees will grow very – you say slowly, but will they have a chance of really replacing the trees on those sites?

MR. LAWRENCE: That is a very good point. If my Forest Service colleague was here, Walter Shortle he could give you more details.

That is exactly what he says, and that is what they use to explain why these large spruce trees died, because all of a sudden they couldn't get enough calcium. On the basis of that, we probably wouldn't expect to see a spruce forest replace the ones that died of the same vigor, size, stand density, et cetera.

MS. MC KENNA: Mary McKenna from Howard University. I know you said you were not looking at the forest ecosystem as a whole, but how much evidence is there that the understory layer is playing a role here, and perhaps not allowing more calcium to be held in, for example. [Portion off microphone.]

MR. LAWRENCE: I don't have any direct knowledge of understory studies to specifically look at that. The understory is certainly important and there are other factors, particularly in the northeast, that are coming into play, things like deer browsing, that is also stressing the understory and the regeneration of these trees.

Studies have been done to exclude the deer and they are still having trouble regenerating. In terms of the herbs themselves, I am not sure – I am not aware of any studies but there is certainly potential for interaction there.

Human Health: Effects of Ambient Air Particulate Matter

Morton Lippmann – New York University School of Medicine

ABSTRACT

The clearly identifiable human health effects of ambient air pollution follow their inhalation into the respiratory tract. In eastern North America, most of the acidic air pollution is associated with sulfuric acid aerosol, which is formed in the atmosphere as ultrafine particles (d < 0.1 μm) through the oxidation of sulfur dioxide emitted during fossil fuel combustion, and which is gradually agglomerated into fine particles (0.2 < d < 1.0 μm) and also neutralized to ammonium sulfate by reaction with ammonia gas, a product of anaerobic decay at ground level. Short-term peak exposures to ambient fine particles can cause asthma exacerbations and affect the lungs ability to clear itself of inhaled particles. Long-term chronic exposures are closely associated with premature mortality, increased rates of emergency department and hospital admissions, respiratory symptoms, and lost-time from work and school.

BACKGROUND

The quantitative associations between health and air pollution can be traced back to England in the 1950s and, until the mid-1980s, we had most of our health effects defined by the experience in London and other parts of the United Kingdom.

Beauty is in the eye of the beholder. Claude Monet spent three years in London – 1901 to 1904 – and produced over 100 oil paintings of the London scene because he was fascinated by the interplay between the diffused light and the coal smoke particles. There were periodic pollution episodes in the late nineteenth and early twentieth centuries where many people died during the heavy London fogs. However, prior to the December 1952 episode, the public health data were quite limited.

The December 1952 episode couldn't be ignored because, in retrospect, it was determined that 4,000 people had died. During this fog episode people couldn't see more than a few feet in front of their faces. Furthermore, the effects were not limited to humans; prize cattle died at the Smithfield cattle show, and their lung edema was attributed to the acidic nature of the fog aerosol. There was a royal commission report (Ministry of Health, 1954) that led to the U.K. Clean Air Act, which banned soft coal burning, leading to an initial replacement with coke, and later with natural gas.

Figure 1 illustrates data from the last London

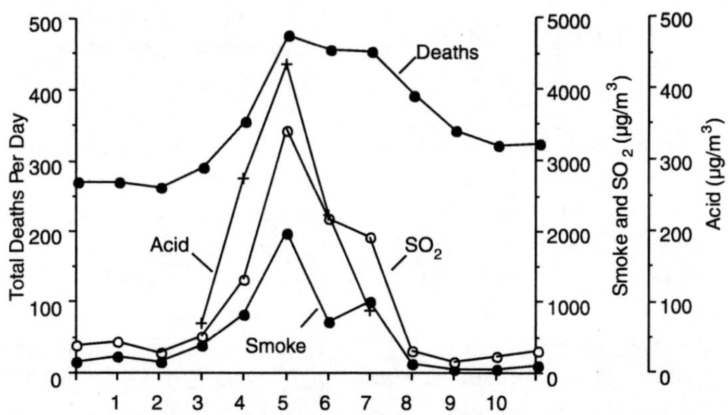

Figure 1. December 1962 London pollution episode.

fog episode (December 1962) where one didn't need complex statistical analyses to see that there was an association between pollution, as it was being measured, and mortality. In the 1962 episode, the rise of the pollution was also clearly associated with a rise in the daily death rate, but the number of excess deaths was only about 700. The typical London fog pollution episodes lasted a few days. As in the previous mortality episode 10 years earlier, the death rate didn't come down as promptly as the pollutant concentrations. So, there was acute mortality with a temporal lag of a few days to a week or more, i.e., some people who were fatally affected didn't succumb immediately.

The 1962 fog episode provided the only data from that era where they had the daily measurements of sulfuric acid, as well as their classic indices of electrical conductivity of a scrubber solution (expressed as SO_2), and of particles, expressed as Black Smoke (blackness on a filter paper). The analyses suggested, as did the analyses of the earlier episode, that what was causing mortality was the strong acidity associated with the reaction, in the fog droplets, of SO_2 being oxidized and converted to sulfuric acid.

Beginning in the mid-1980s, in the United States, we began to see published reports of some epidemiological studies that have provided the basis for our newer thinking about the influence of ambient air particles on health. They provided a basis for the revised ambient air standards for the particulate matter (PM) in ambient air that I will review later in this presentation.

Current Understanding Of Ambient Air Particulate Matter

Let us consider the physical aspects of the situation that we are dealing with. The sulfur dioxide (SO_2) gas that is emitted from a power plant stack moves downwind, and its concentration diminishes gradually with time and distance. One factor accounting for reduced concentration is the ground uptake of SO_2 on vegetation and other surfaces. Another factor is aerosol formation. What does that mean? It means that there is oxidation of the SO_2 to form SO_3, a highly hygroscopic vapor. Since there is always plenty of water vapor in the air, very fine (ultrafine) droplets of sulfuric acid form in huge numbers, over a million per milliliter. They rapidly coagulate, producing fine particles having diameters between about 0.1 and 1 µm that have a long atmospheric lifetime.

Thus, there is a persistent airborne source of SO_2 that creates fresh acidic aerosol over a period of days and hundreds or even thousands of kilometers downwind, with concentrations that are relatively uniform on a regional basis. This is illustrated in Figure 2, which depicts the results of some ambient air sampling conducted by NYU for another purpose. Hydrogen and sulfate ions were measured on a daily basis in the aerosol in both Buffalo and Rochester, NY over a six-week period. These cities are downwind of numerous power plants in the midwestern United States.

The figure shows that, over this 60-mile separation, there was a close correspondence in both sulfate ion and hydrogen ion concentrations, and that the measurements made in these two cities were as close as you would get with co-located samplers, i.e., running two samplers side-by-side. Thus, we can conclude that these fine particles are pervasive and widespread, at least over relatively flat land, and we don't need a lot of monitors in a metropolitan region to determine the sulfate ion concentration.

In fact, the fine particles, as represented by sulfate ion, are about 90 percent as high in most indoor environments as outdoors, because fine particles penetrate indoors. Thus, we are not just exposed to outdoor generated fine particles outdoors, but indoors as well; this helps us understand why epidemiology using outdoor measurements are relevant to exposures that may occur in indoor environments.

Now, let's consider the nature of airborne particles. Figure 3 is a depiction of the particle mass distribution. When we measure particles distribution according to their diameters in micrometers (aka microns), we find that most of the individual particles are very small (< 0.1 µm). However, these very numerous ultrafine particles contribute very little to the mass of the airborne particulate matter (U.S. EPA, 1996).

Most fine particles (diameters between 0.1 and 2.5 µm) were formed in the atmosphere from

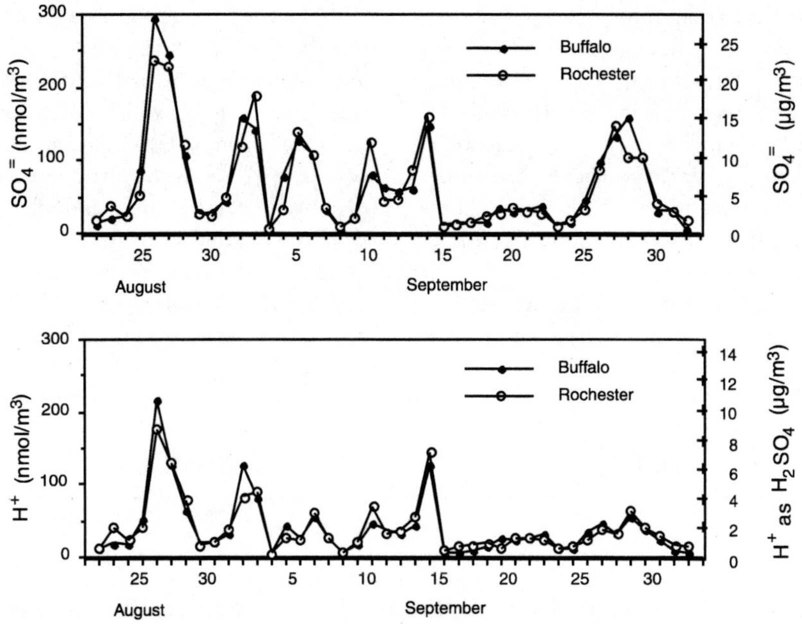

Figure 2. Intercomparison of Rochester, NY and Buffalo, NY sulfate and daily acid aerosol concentrations (August 22 – October 2, 1990).

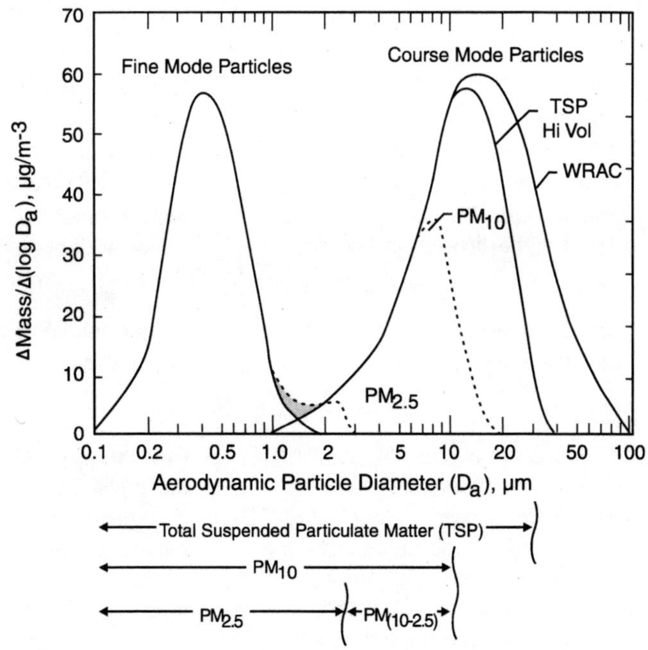

Figure 3. Representative bimodal mass distribution as a function of aerodynamic particle diameter for Phoenix, Arizona, showing effect of size-selective sampling inlet on mass collected for (a) wide-ranging aerosol classifier (WRAC), (b) standard total suspended particulate (TSP) high-volume sampler, (c) sampler following EPA's (PM_{10}) criteria for thoracic dust, and (d) sampler following EPA's criteria for fine particulate matter ($PM_{2.5}$). (Source: U.S. EPA, 1996).

gaseous precursors as ultrafine particles. Following the coagulation of the ultrafines, which takes place within about a half hour of their formation, the resulting aerosol has a mass median diameter of about 0.3 to 0.5 µm, with only a small fraction of fine particles having diameters above one micron. However, when we consider the total surface area of the particles in the ambient air, most of it is in the fine particle mode.

When we look at Figure 3, which shows aerosol distribution in terms of mass, which varies with the cube of diameter, we see a bimodal distribution. The coarse mode particles are soil-like and are produced by mechanical processes. The fine mode particles, which dominate in terms of surface area, are known as the accumulation mode because particles in this size range are too small to deposit by sedimentation, and too large to continue to grow by coagulation.

The chemistry of the two mass modes is radically different. As indicated in Table 1, the coarse particles are essentially soil particles or

Table 1. Comparison of ambient fine and coarse mode particles. Source: EPA Paper (1996).

	FINE MODE	COARSE MODE
Formed from:	Gases	Large solids/droplets
Formed by:	Chemical reaction; nucleation; condensation; coagulation; evaporation of fog and cloud droplets in which gases have dissolved and reacted.	Mechanical disruption (e.g., crushing, grinding, abrasion of surfaces); evaporation of sprays; suspension of dusts.
Composed of:	Sulfate, $SO_4^{=}$; nitrate, NO_3^-; ammonium, NH_4^+; hydrogen ion, H^+; elemental carbon; organic compounds (e.g., PAHs, PNAs); metals (e.g., Pb, Cd, V, Ni, Cu, Zn, Mn, Fe); particle-bound water.	Resuspended dusts (e.g., soil dust street dust); coal and oil fly ash; metal oxides of crustal elements (Si, Al, Ti, Fe); $CaCO_3$, NaCl, sea salt; pollen, mold spores; plant/animal fragments; tire wear debris.
Solubility:	Largely soluble, hygroscopic and deliquescent.	Largely insoluble and non-hygroscopic.
Sources:	Combustion of coal, oil, gasoline, diesel, wood; atmospheric transformation products of NO_x, SO_2, and organic compounds including biogenic species (e.g., terpenes); high temperature processes, smelters, steel mills, etc.	Resuspension of industrial dust and soil tracked onto roads; suspension from disturbed soil (e.g., farming, mining, unpaved roads); biological sources; construction and demolition; coal and oil combustion; ocean spray.
Lifetimes:	Days to weeks	Minutes to hours
Travel Distance:	100s to 1000s of kilometers	< 1 to 10s of kilometers

mineral ash particles, and are basic, and light in color. The accumulation mode is acidic and also oily because some of them result from photochemical reaction processes. That is why we now make a distinction between fine and coarse particles, in terms of their sources and their potential for producing adverse health effects.

Review Of Ambient Air Quality Standards For Particulate Matter

With that background information, we can now consider how air quality measurements have been made. We have had an evolution in terms of the methods used to measure mass distributions. Initially, in the United States, we tried to measure total suspended particulate matter (TSP). We used what was essentially a big vacuum cleaner, and sucked up whatever we could onto an eight-by-ten inch filter, and used the resulting gravimetric concentration as an index of health risk.

The decision to use TSP as an index was a serious mistake, but it became the basis for the original (1971) particulate matter ambient air standard. It was a poor choice because the largest particles that were collected were too big to get into the respiratory system, and yet they could dominate the mass concentration on a windy day.

Thus, in 1987, EPA switched their measurement index to PM_{10}, which could properly be called thoracic particles. At about 10 microns and smaller in aerodynamic diameter, you can inhale these PM_{10} particles into the thoracic part of the respiratory tract, where they can produce some health effects.

PM_{10} samplers use an inertial separating system to prevent the entry of the rocks as they are being sampled. In 1997, the PM ambient air standard was once again revised to include separate limits for fine particles ($PM_{2.5}$), since the fine particles were more closely associated with mortality and other health effects than was PM_{10}. While most of the fine particles are smaller than 1 μm, the sulfate particles, which are a major mass component of $PM_{2.5}$ in the eastern United States are hygroscopic and grow into dilute solution droplets. If you want to capture them under humid conditions, you really have to include particles up to 2.5 μm to make sure that you captured all the hygroscopic fine particles.

There are both daily and annual average concentration limits. For the daily maxima, allowances need to be made for exceptional weather patterns. Thus, the limits are expressed as the fourth highest concentration over a three-year period, as described in Table 2. The revised PM_{10} standards, and new $PM_{2.5}$ standards promulgated by EPA in 1997, survived legal challenge, and were upheld by the Supreme Court.

EPA, by changing the basis for a PM_{10} exceedence in a community, in effect reduced the number of people impacted because, in the west,

Table 2. 1997 revisions: U.S. National Ambient Air Quality Standards (NAAQS).

	1987 NAAQS	1997 NAAQS	
Index pollutant	PM_{10}	PM_{10}	$PM_{2.5}$
Annual av. concentratio n limit (μg/m^3)	50	50	15
Daily concentration limit (μg/m^3)	150	150	65
Basis for excessive daily concentration	4th highest over 3 year period	>99th percentile av. over 3 years	>98th percentile av. over 3 years
Number of U.S. counties expected to exceed NAAQS	41	14	~150
Number of people in counties exceeding NAAQS	29x10^6	~9x10^6	~68x10^6

on windy days with high levels of soil-like particles, there were legal exceedances that were not really considered much of a health hazard.

The stringency of the PM_{10} standard was thus reduced, and the most stringent limits are now in the annual average limits of the fine particle ($PM_{2.5}$) standard. The scientific basis lies in that mortality is most closely associated with fine particles, and we now have 15 micrograms per cubic meter as the limit for the annual average in a community.

The 65 $\mu g/m^3$ daily limit for $PM_{2.5}$ is a relatively relaxed standard. When the annual $PM_{2.5}$ standard is implemented, it can be expected to more than double the number of people living in areas of PM exceedence unless, of course, ambient air levels change in the meantime.

Health Effects Of Ambient Air Particulate Matter

What is the evidence for health effects? It is largely epidemiological associations between measured air quality parameters and mortality and morbidity indices. Figure 4, from EPA's PM Criteria Document (U.S. EPA, 1996), shows that there is a significant daily mortality risk associated with PM_{10} that is not greatly influenced by the presence of criteria pollutant gases.

Hopefully, at some time in the future we will get past reliance on these rather simple gravimetric indices of exposure to ambient air particles. Having a non-specific particle standard is hardly different in concept from having a non-specific gas standard, lumping ozone, CO, SO_2, and NO_2 together. Unfortunately, we don't yet have the basis for deciding which components of the particle mixture are most closely associated with the health effects.

The candidate chemical components of ambient PM currently being most seriously considered as likely causal factors include: hydrogen ion content, the number concentration of ultrafines, the soluble transition elements – iron and vanadium in particular, and perhaps the

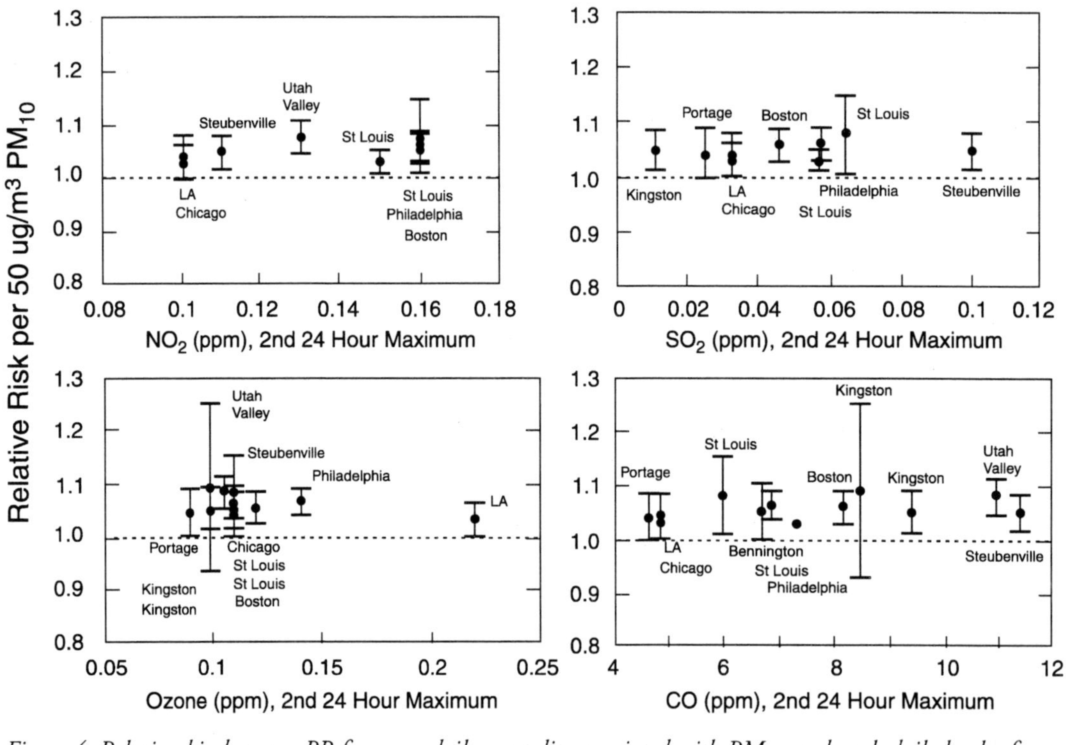

Figure 4. Relationship between RR for excess daily mortality associated with PM_{10} and peak daily levels of other criteria pollutants. (Source: U.S. EPA, 1996).

peroxides. Epidemiological evidence is, however, lacking because we don't have monitoring data for these components.

The new PM monitoring networks, i.e., the Super Sites and the speciation sites that EPA has set up in connection with the need to monitor $PM_{2.5}$ concentrations will be providing some of the evidence that will be needed in the near future, so that the next time that an ambient air particulate matter standard comes up for review, beyond this current round, we may be able to establish a better index than fine particle mass.

In the recent years, there has been a lot of relevant health research to consider.

One of the recent landmark studies, clearly, is the Health Effects Institute (HEI) sponsored study by Samet et al. (2000), which described associations between daily excess mortality in 90 U.S. cities in relation to daily levels of PM_{10}. In this study, using standardized data on both measured PM_{10} and mortality, there was a highly statistically significant indication of an association. This study provided further evidence than we had early in the 1997 standard setting exercise that daily mortality is more closely associated with thoracic particles than with other air criteria pollutants. This judgement was based on multiple regression analyses with the other criteria pollutants. When ozone, or SO_2, or CO, or NO_2, were included into a multiple pollutant regression, there was very little shift in the PM coefficient. This strengthens the basis for the need to regulate particles.

The Samet et al. (2000) study also examined whether there was a uniform risk coefficient around the country. They divided the U.S. data for the 90 cities into geographic regions. The increase in mortality per 10 micrograms per cubic meter increase in PM_{10} was greater in the northeast than in the other regions of the country.

This suggests that composition does make a difference, and provided some support for the hypothesis that aerosol acidity, which is higher in concentration in the northeast than elsewhere in the U.S., may be an important risk factor. In any case, smaller, but statistically significant effects were seen in all regions of the country. It is just the magnitude that may be different in different regions, when using gravimetric mass as an index of risk.

We also have an enhanced data base on the associations between daily average PM concentrations and daily rates of hospital admissions for various respiratory diseases. The highest excess risk per unit of particles is for asthma. Pneumonia is next highest, and chronic obstructive pulmonary disease generally also shows significant excesses.

While the excess risks are small, they are coherent with the mortality effects (Bates et al., 1992), i.e., some people may be especially susceptible in some way or another, by advanced age, by disease, and are unable to cope with these inhalation challenges.

In very recent years, there has been a huge shift in research emphasis toward studies of the association of PM and cardiovascular disease. While the relative risks are highest for respiratory diseases, as might be expected for an inhaled pollutant, the total numbers of people affected appears to be greater for cardiovascular disease because there is a larger underlying prevalence of such disease in the population as a whole.

Many recent studies show associations between hospitalizations for all cardiovascular disease, and congestive heart failure in one study, heart failure and ischemic heart disease in another study.

Now, while the studies on acute effects are most abundant and highly consistent, and do tell us that there are health effects, the most stringent of the new standards is focused on the annual average concentration, which is the hardest to achieve (see Table 2).

Why is that? The cohort studies of annual mortality rates of the populations in the Harvard Six Cities (Dockery et al., 1993) and the American Cancer Society cohort (Pope et al., 1995) indicated that the mortality rates in different communities varied in relation to both fine particles (50 communities) and their sulfate content (151 communities). Their findings for sulfate and TSP are illustrated in Figure 5. It is notable that the cumulative impact of excess mortality was about three to four times greater than that indicated by summing the daily excesses from the daily mortality studies. In the economic benefit/cost analyses performed for the Clean Air Act between 1970 and 1990, 90 percent of the benefits that could be both quantitated and monetized were associated with the mortality end point, using the mortality coefficient for $PM_{2.5}$ established for the ACS cohort (EPA, 1997). One aspect of this study was its analyses of longevity reduction. For the population average, this was in the order of one to two years. One can also look at these data in another way. For example, one can estimate that, for those more sensitive people who were affected, they may, on average , have lost about 15 years of life span (Brunekreef, 1997).

Because of the importance of their cohort mortality studies, the HEI was asked to sponsor a re-analysis of the Harvard six cities and ACS cohort studies. They did it by a completely independent team and came up with essentially the same results (HEI, 2000).

So, while EPA stuck its neck out in basing the 15 microgram PM 2.5 annual average PM limit on these two annual mortality cohort studies in establishing the $PM_{2.5}$ limits in 1997, their judgement has been confirmed by the HEI reanalysis.

Further Research On Health Effects Of Ambient Air Particulate Matter

Further study of these cohorts is currently underway. Since they were last studied as a population, enough years have elapsed that perhaps the number of people who have died in the interim is up by a factor of about three, and we can get much more specific in terms of components and causes for that excess annual death rate.

The Harvard School of Public Health is undertaking further analysis of their Six-Cities cohort, and we, at NYU, in collaboration with Dr. C. Arden Pope III at Brigham Young University, are doing the follow up analysis of the American Cancer Society cohort.

Thus, considerable progress is being made in clarifying the relationships between exposures to airborne particulate matter, but we will need to know much more before we can optimize source controls to reduce the health and environmental impacts of both acidic sulfates and other fine particles.

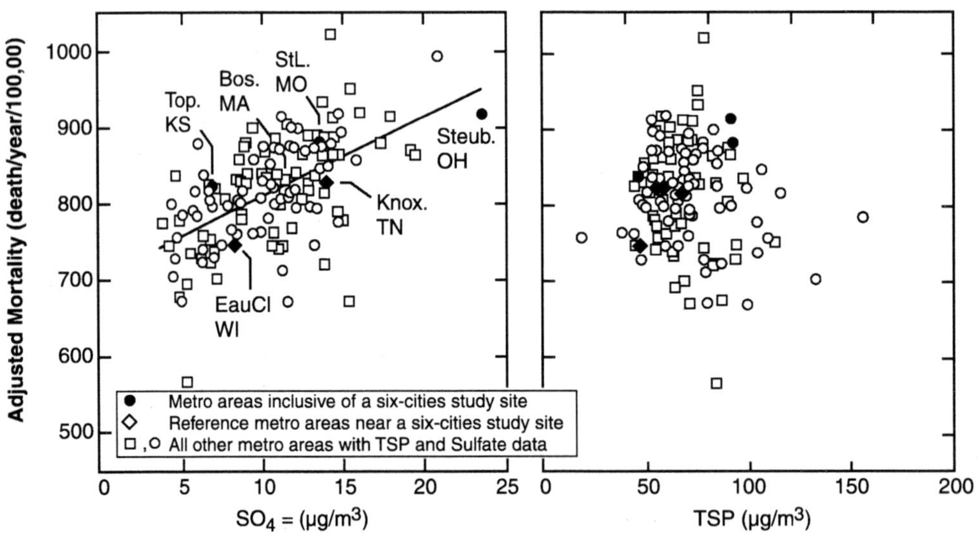

Figure 5. Age-, sex-, and race-adjusted population-based mortality rates for 1980 plotted against mean sulfate air pollution levels for 1980. Data from metropolitan areas that correspond approximately to areas used in prospective cohort analysis. (Adapted from Pope et al., 1995).

References

Bates, D.V. 1992. Health indices of the adverse effects of air pollution: The question of coherence. Environ. Res. 59:336-349.

Brunekreef, B. 1997. Air pollution and life expectancy: Is there a relation? Occup. Environ. Med. 54:781-784.

Dockery, D.W., C.A. Pope, X.P. Xu, J.D. Spengler, J.H. Ware, M.E. Fay, B.G. Ferris, and F.E. Speizer. 1993. An association between air pollution and mortality in six U.S. cities. N. Engl. J. Med. 329 24:1753-1759.

Health Effects Institute. 2000. Reanalysis of the Harvard Six Cities Study and the American Cancer Society Study of Particulate Air Pollution and Mortality. A Special Report of the Institute's Particle Epidemiology Reanalysis Project.

Ministry of Health. 1954. Mortality and Morbidity During the London Fog of December 1952. Her Majesty's Stationary Office, London.

Pope, C.A. III, M.J. Thun, M.M. Namboodiri, D.W. Dockery, J.S. Evans, F.E. Speizer, and C.V. Health, Jr. 1995. Particulate air pollution as a predictor of mortality in a prospective study of U.S. adults. Am. J. Respir. Crit. Care Med. 151:669-674.

Samet, J., F. Dominici, S. Zeger, J. Schwartz, and D. Dockery. 2000. National Morbidity, Mortality, and Air Pollution Study. Part 1: Methods and Methodologic Issues. Health Effects Institute Research Report No. 94.

U.S. EPA. 1996. Review of the National Ambient Air Quality Standards for Particulate Matter: Policy Assessment of Scientific and Technical Information. OAQPS Staff Paper. Office of Air Quality Planning and Standards. EPA-452\R-96-013.

U.S. EPA. 1997. The Benefits and Costs of the Clean Air Act 1970 to 1990. EPA Report to Congress. Office of Air and Radiation. Office of Policy. (Oct. 1997).

Questions

MR. OREN: Craig Oren, Rutgers. Do you see any indication that there is any kind of so-called threshold level of particulates that is, in some sense, safe, or does the relationship between the concentration and effects exist, even at very low concentrations of particulate matter?

MORTON LIPPMANN: The data, which are abundant, and which are quite consistent, do not indicate any threshold. Thus, EPA cannot set a standard to fully protect the public health with a margin of safety, because we, as a society, can't get down below the non-anthropogenic background of fine particles from various natural sources.

On the other hand, the available evidence doesn't prove that there isn't a threshold.

MR. SALKOVITZ: I am Dan Salkovitz, Virginia Department of Environmental Quality. I was looking at some of the data that you put up and it just occurred to me, have you done, or do you have information about the synergistic effects, for example, high fine particulate in combination with, say, ozone? What threshold levels – maybe they would be lower. I don't know. Do you have any information about that?

MORTON LIPPMANN: There are lots of data on ozone. Ozone appears to be independently associated with mortality, but to a much lesser extent than PM.

In terms of ozone and physiological measurements, the responses look linear down to background levels. So, there is no evidence of a threshold for the physiological responses to short-term ozone exposures.

When you look at some specific ozone health effects, such as hospital admissions for respiratory disease, ozone has a significant effect. In fact, if you can separate the effect of the PM and the ozone in the statistical models, ozone has a bigger effect on hospital admissions than PM does, although they both influence it. That is unlike mortality, where PM drives it, and ozone has a much smaller effect.

Synergism can't be ruled out. On the other hand, the kinds of data that show that the gaseous pollutants do not materially change the coefficient for particles reduces the likelihood that it is a major factor.

MR. POIROT: Rich Poirot, Vermont air program. You had indicated that the controlling standard for the fine particle standard was the long-term annual average one, as a response to evidence for chronic effects.

The Canadian standard, I believe, is one that is based on a short-term metric. Could you comment on the difference between those two and what are some of the implications?

MR. LIPPMANN: Well, first of all, I could say that we are right and they are wrong. Seriously, the data are very hard to interpret. Different people look at it and come up with different reasons.

The Canadian reliance on the daily mortality data can be justified based on the greater amount and consistency of the database. There are more than 30 valid studies of daily mortality which are roughly consistent, and there were only two readily interpretable cohort studies for the annual mortality effect.

A major advantage of the daily mortality data is that you don't have to worry about region-to-region differences in factors like smoking, the presence of varying levels of other pollutants, the differences in ethnic composition, and dietary preferences that go with them. You are looking at the same population on a day-to-day basis, only a small fraction of whom succumb, too small to affect the next day's result.

When you are looking at multiple city results on an annual basis, all of these other risk factor considerations could be confounders that could explain, theoretically, the mortality differences from region to region.

What made the two cohort studies so important was that they had individual data on influential factors, including diet, smoking, and occupation in the cohort data base, and could show that they didn't influence the mortality coefficients.

Also, there had been about eight cross-sectional studies of annual mortality using PM_{10} or sulfate as the index that didn't have data on individual risk factors. These previous studies were discounted because they lacked the ability to correct for these factors. In fact, they had shown the same kind of impact as the cohort studies, but people said that all these other factors could explain it. Well, they didn't.

In fact, when you consider the data from the two cohort studies that had data on the individual risk factors other than pollution, those other factors didn't explain the excess mortality. In my view, these previous cross-sectional study results gave greater credence to the results of the two cohort studies, because these two cohort studies showed that the diet, the ethnicity and so forth, even including smoking, didn't make much difference.

Therefore, the consistency of the estimated effect from the cohort studies with those produced by the prior cross sectional annual mortality studies was notable. Furthermore, the HEI re-analysis gives them a much greater weight. The re-analysis looked at the data base in every possible way and came to the same conclusion.

I can understand the reluctance of other authorities to rely more on the annual mortality data, but I think they are wrong.

Materials and Cultural Resources

Mary F. Striegel – National Center for Preservation Training and Technology

I am from the National Center for Preservation, Technology and Training, and many of you may have never heard of our center. So, I am going to tell you a little bit about it.

The national center was established by Congress in 1992 and is part of the National Park Service. It is a research and development facility located in Natchitoches, Louisiana. We have a mission to promote the application of science and technology to the preservation of our nation's cultural resources and cultural patrimony.

Within the center we have four main components. That includes a research component, a training component, an information management component and we have an environmental and materials research program. We also have a newly formed federal preservation institute that is under the national center.

We look at problems that are within the fields of historic architecture, historic landscapes, archeology, materials conservation and ethnographic studies and historic interpretation.

I am going to talk about a few topics today. I am going to talk about our philosophy of research. I am going to give you an overview of pollutant-caused stone decay. I will be talking about a few of our recent experiments, and then I am going to tell you what we have learned.

At NCPTT, or the national center, we talk about a knowledge continuum. What we are trying to do is to understand the universal laws of nature. That is what we are all trying to do as scientists. In order to do that, some of our highest understanding comes from what we observe, identify, describe and test.

We use that information to develop technology, the application of science to broader issues, including issues that are industrial and commercial in nature. Ultimately, we want to provide knowledge to a user for consumption.

Now, at NCPTT, we try to span this whole spectrum or continuum of knowledge, from understanding basic principles from science, developing new technologies and applications, moving to disseminating that information and ultimately using it. You, in a sense, are some of my end users because you need to know about stone deterioration and how air pollution affects them.

When we think about the relationship between air pollution and stone decay, we know it is a complex situation. We have many of the individual pieces described, but we don't always end up with a clear picture. It is a lot like working a jigsaw puzzle.

Sometimes we have better pictures than others, depending on the amount of effort and resources that we put in place.

With a lot of energy, we can sometimes get a better picture, but we may not always have the ability to get the big picture. That is the way it is with our air pollution interactions with stone.

We know that outdoor man-made air pollution is affecting our cultural resources and it is causing decay (Figure 1). We know that sulfur dioxide is one of the most important pollutants affecting stone, and we know that combustion processes lead to the production of large quantities of sulfur dioxide in the air.

When we have a simple model, we can talk about sulfur dioxide reacting with material such as stone by two main processes. One process is through dry deposition, where the gas is deposited directly on the stone, and the second is through wet deposition (Figure 2).

Once the pollutant has deposited onto the stone, it can interact to create an alteration crust

Figure 1. Air pollution.

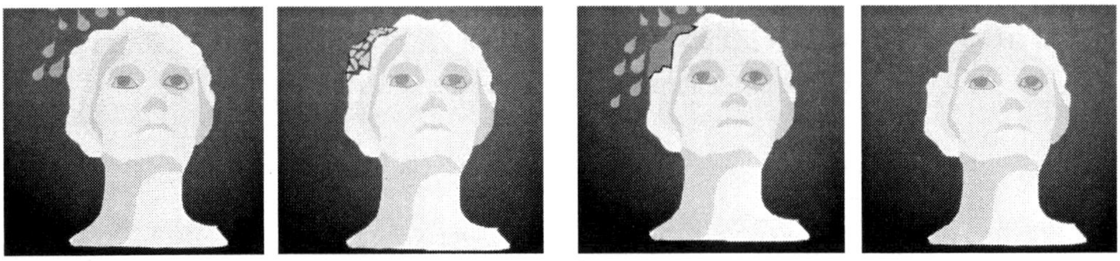

Figure 2. Deterioration of limestone caused by wet deposition of sulfur dioxide.

on the stone, and then that alteration crust is usually calcium sulfate, which is more water soluble than the stone itself and, with subsequent rains, the alteration crusts are moved.

What we end up with is a permanently altered sculpture. We have heard a lot of talk today about recovery. When a sculpture or a building or a stone is altered, it is permanently altered. We can't talk about recovery. It is gone.

Now, we have also heard a lot today about dose response functions. I am going to talk to you just a little bit about dose response functions. This is my personal opinion. You may agree or disagree.

We are going to talk about stone as a receptor, a complex receptor. It is heterogeneous. It has many different minerals in it. It is porous. When we talk about a dose response function, we want to say, if the stone sees air pollution, such as sulfur dioxide, nitrogen oxides, ozone, or particulate matter, it is doing to have a damage response, something that we can measure.

Those changes may be chemical changes or they may be physical changes. What we want to know is, given a dose, we have a response. But we can't get as true dose response relationship because of the complexity of the system.

First of all, stone has a very long response time. It sees air pollution today. It may react 25 years from now. It may react 50 years from now. It may react 100 years from now.

I am showing you two examples of stone decay (Figure 3). The results of gypsum formation and the incorporation of pollution onto the stone can result in spalling, delamination and total loss of the stone.

On the right-hand side you see a Funerary sculpture that is located in Cahill Cemetery in Louisville, Kentucky. This sculpture was featured in an article in National Geographic in 1978. At that time, it had a head and shoulders on it. This photograph was taken by me 20 years later and the head and shoulders are completely gone.

Now, gypsum is the resulting product of the interaction of sulfur dioxide with the calcite or calcarious materials of stone. It forms a needle-like porous structure that can trap particulate matter (Figure 4). It also acts like a sponge to absorb water and increase the surface time of wetness.

The studies that we have done in the past and are currently doing include field studies, case

Figure 3. Stone deterioration.

Figure 4. Gypsum crust formation.

studies and lab studies. The field studies were part of the NAPAP studies. We have had current research going on since 1990 on the Cathedral of Learning building, which is in Pittsburgh. This is work done by Cliff Davidson at Carnegie Mellon University. In 1936, the air pollution concentrations in Pittsburgh were much higher than they are today. By the time of the completion of the building, it was completely soiled from top to bottom.

Today the air pollution concentrations have lowered significantly, but we still have a deposition process competing with an erosion and a washing away. So, we see changes in soiling patterns.

We also have laboratory studies. For example, I will be talking to you a little bit today about our studies on the surface morphology and porosity of stone and how that affects deposition and what that means to ultimately trying to talk about dose response functions.

Some of our work focuses on how we can use this information to develop better treatments for buildings, sculptures and monuments, so that we can get a better handle on how to prevent the damage rather than lower the pollution (Figure 5). This is our baby, as we call it. It is an environmental exposure chamber constructed by the National Oceanic and Atmospheric Administration. It allows us to test the uptake of pollution onto materials. We can control temperature, humidity, wind speed and turbulence within the sample and can measure directly the deposition of air pollution and particularly sulfur dioxide onto materials.

One of the studies that we have done was to look at the effects of texture on pollution uptake to different types of limestone. We looked at four different types of limestone because we wanted to find out if there were differences between types. We looked at several different types of textures. We looked at very smooth surfaces and we looked at very rough surfaces.

We imposed the texture on the samples by etching them and then we characterized surface parameters such as the traditional roughness values, R^a and 17 other parameters. We also used mercury porsimetry to determine porosity within the stone, and we used nitrogen absorption to determine surface area on the very surface of the stone and also on the pore network.

This is probably the most important slide that I am going to show you. That is, we took our rough and our smooth samples. We exposed them eight different times for 24-hour exposures (Figure 6).

Now, the important thing to note is that the very first exposure and the second exposure have a much higher deposition of air pollution onto the stone than subsequent runs. What that tells us is that we can get sulfur dioxide onto the stone, but we have a lot of trouble getting it off. It gets into the stone. It gets into the spongy pores of the stone and it stays there.

The reason that is important is because it ultimately leads to a memory effect. I heard someone speaking about soils earlier. He said if you took all the sulfur dioxide in the air away tomorrow, you would still have problems. With stone, if you take all the sulfur dioxide away tomorrow, you are still going to have problems.

The results of our work show that we can identify trends within each stone type. For example, with the Salem limestone, the rougher

Figure 5. NCPTT exposure chamber

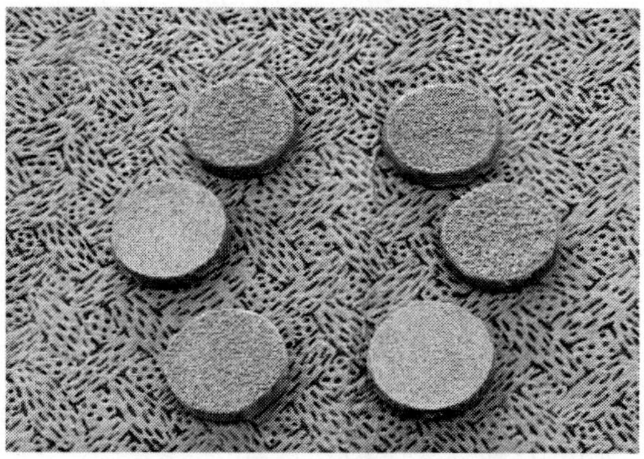

Figure 6. Texture on stone.

stone, we will have more deposition that the smoother stone (Figure 7 and 8).

Surface roughness isn't the same across stone types. For example, a smooth Cordoba cream limestone may have a much lower deposition than a rough Salem limestone. Surface roughness itself isn't the controlling or dominant factor. It may affect the deposition but only within stone types, not across stone types.

We did learn something that was very important to us. That is, sample porosity and the number of valleys that we see on a stone surface will enhance deposition while the protruding piece will limit deposition.

Now, why is that important to us? Well, when we go out and clean our buildings, let's say we water wash them or we sand blast them or we use a chemical cleaner. Every one of those treatments changes that stone surface and ultimately sets up a different type of deposition.

What are the implications for us? Well, really, what we are seeing is that the deposition of sulfur dioxide is a complex phenomenon that involves many environmental and material factors.

Porosity of the stone is a dominant factor within the deposition process. Once sulfur dioxide gets into the pores, it is difficult to get out. This leads to that memory effect that I talked about.

The take home message that I want you to know is that we are continuing to look at the stone systems and trying to describe the uptake of air pollution. I don't think we are going to get to dose response functions, but it just may require new approaches to increase our understanding.

Our understanding of stone decay is limited by the long response time of the stone and, to be frank with you, one of our biggest problems is that our resources are limited, there are very few scientists worldwide looking at this problem, and this limits our ability to make further progress.

I thank you for your attention and am willing to answer any of your questions.

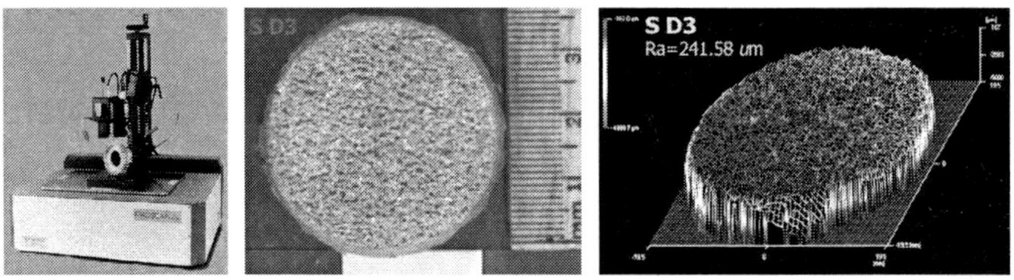

Figure 7. 3-D laser profilometry.

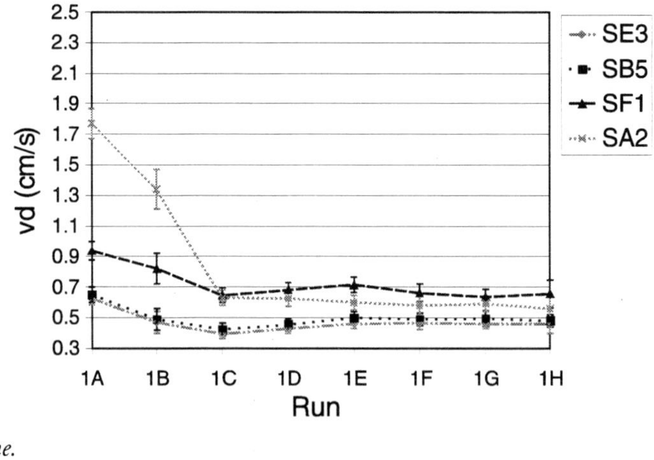

Figure 8. Salem limestone.

Questions

MR. WHITE: I am Jim White from the Center for Environmental Information. I understand that in Europe in old buildings such as cathedrals there is a lot of deterioration of stained glass windows. Have you done anything at all with glass?

MS. STRIEGEL: I haven't been working with glass, but there are groups in Germany and in Austria that are looking at the effects of air pollution on glass.

There is a group sponsored by the UNECE on long-range transboundary air pollution and they have been looking at material effects. The last meeting that I know of was in Berlin and they reported some work on stained glass there.

MS. BAUM: I am Ellen Baum from the Clean Air Task Force. I have seen the work that has been done in D.C., where buildings have been affected, and the National Park Service believes this due to acid deposition. I understand you're saying that there are very limited resources to study this.

Is this just bits and pieces or is there a bigger piece of information so that we can understand where these impacts are being felt outside of Washington?

MS. STRIEGEL: From the early 1980s until about 1995, the work was done in Washington, D.C. and was supervised by Susan Sherwood, who some of you may well know. In 1995, the program was transferred to Natchitoches, Louisiana. We have approximately 45 projects that we have operated, but not all of them are focused on understanding deposition.

We have turned our resources to focusing on treatments and mitigation and how to prevent the damage to the stone. We know the damage is going to happen. We just don't know the time frame and we also don't know at what level.

We think there is not a threshold, that any sulfur dioxide is ultimately going to lead to some problems.

MR. OOTEN: My question is this. Do you have data, or has anyone aggregated data on the economic aspects of this materials degradation, the cost of replacement or however you would go about measuring it. .

MS. STRIEGEL: Some of the early efforts were to look at a national inventory of stone material and then look at the cost of replacing component parts within a building. So, for example, what would it cost to replace a lintel, what would it cost to replace a door frame. Those efforts didn't lead to good evaluation.

There was a study about 1996 or 1997 out of a group from Colorado, I think from Boulder, Colorado. They decided they would survey the general public. They took a recognizable sculpture and they digitally manipulated it to age it. They had it in about four or five stages of deterioration.

They asked the general public how much they would pay to keep it pristine, to keep it at each of the different stages. That has probably been the most successful method of valuing the cultural resource.

MR. LIPPMANN: We had a prime example here in Washington. The Washington Monument was recently restored. That is clearly an example of stone deterioration that needed fixing. Do you have an idea of how much was spent on that?

MS. STRIEGEL: Unfortunately, I don't know what the cost of that project was. I can try to find that information and get back to you.

MR. LIPPMANN: I am just raising it as an example. We probably spent a lot of money.

MS. STRIEGEL: There was also a lot of money spent on the Statue of Liberty in 1986. If any of you recall that, that was a national campaign to raise the money for that restoration effort.

MR. LIPPMANN: The Lincoln Memorial, too.

MR. PENNINGTON: When I build my cabin on the top of the Blue Ridge Mountains, what should I build it out of? There is sandstone and limestone up there and some granite down the road.

MS. STRIEGEL: The granite might be nice. If you are going to choose a limestone, I recommend Salem limestone from Indiana, and polish it smooth.

Special Program - Acid Rain: Are the Data and Information Problems Solved?

Frederick W. Stoss – Sciences Librarian, Science and Engineering Library, University at Buffalo

ABSTRACT

The topic, generally known as acid rain, presents a series of data and information challenges to researchers, policy makers, and librarians and information specialists. Acid rain became one of the first and at the time one of the largest and most vexing environmental problems of a global scale in terms of both its origins and in its solutions. The transport of air pollutants over long distances had been postulated as early as the 17th Century. Scientific study of the chemical composition of the Earth's atmosphere began in earnest in the later half of the 19th Century and continues today. This paper examines the challenges for the identification and acquisition of acid rain data and information and several trends related to research and policy aspects of acid rain.

INTRODUCTION

Like many of today's environmental problems the causes and sources of acid rain have no respect for political boundaries. The polluting sources of acid rain may be found hundreds or thousands of miles up-wind from the ecosystems suffering adverse effects. The long range transport of pollution in the Earth's atmosphere requires inter-regional and international cooperation and exchanges of research data and technologies to identify the causes, effects, remediation, and reduction or, hopefully, the elimination of the causes of acid rain.

In addition to the exchange of scientific and technical research data is the need to broadly disseminate the resulting data and information bases among all parties involved in the study of the problem of acid rain: the interdisciplinary communities of researchers, policy makers, managers, administrators, educators, and students. This dissemination of multi-disciplinary data and information is necessary to stimulate and sustain the exchanges across disciplines for the ideas and proposals to reduce the causes and effects of acid rain.

There are four major data and information challenges revealed as a result of multi-disciplinary studies related to acid rain: the growth of the data and information base reporting the results of research on the topic; the rapid growth and dispersion of the information base among the multi-disciplinary communities generating and consuming this data and information base; the rapid changes in information technologies required to identify, acquire, archive, disseminate and utilize the data and information resulting from research and policy analyses; and the need to assure equity in the access to this data and information across disciplines, lines of work, corporate or individual lines of thought or need.

ACID RAIN INFORMATION: STUDY OF RESEARCH IN A HISTORICAL PERSPECTIVE BUILDS AN INFORMATION BASE

Scientific and technical information is derived from data and recording the results of research from laboratory and fields studies, monitoring campaigns, modeling strategies and other activities. The topic of acid rain has a rather long information trail and a rich history revealed in its literature.

Ellis Cowling (1) provided the first major historical review of the acid rain problem, tracing its origins to studies by Evelyn in 1661 and Graunt in 1662 noting the influences of industrial air emissions to plants and humans and the transboundary transport of air pollutants between England and France. Cowling attributes the first use of the term "acid rain" to R.A. Smith in an 1872 book *Air and Rain: The Beginnings of a Chemical Climatology* (London, Longmans, Green), who noted chemical variations in the chemistry of rainwater and the influence of the combustion of coal, direction of the wind, and proximity to the sea as affecting the damage to plant life and materials exposed to rainwater.

The early decades of the 20th Century saw a small number of scientific articles continuing to monitor the acid content of rain in Australia (2), the United States (3), Uithuizermeeden, Groningen (4), and Russia (5). These papers reflect an effort to record basic rain water chemistry with implications for variations due to natural (meteorological) or human (combustion of carbon-based fuels, e.g., wood and coal). Cowling (1) provides a chronology of increased monitoring of rainwater in Europe and increased studies relating the effects of acidification to biotic and abiotic components of the environment.

The decades of the 1950s and 1960s saw a continuation of research dealing with the chemical composition of rain and the atmosphere. Much of this research focused increased interest of the implications of acidification of rainfall as a result of industrialization and rain water predominantly in Europe (6-7). These European research initiatives lead to one of the first major general reviews of the phenomenon linking increased acidification of rain and other forms of precipitation attributed to the generation of sulfur dioxide in the industrialized regions of West Germany and England (8). John Reuss of the U.S. National Ecology Research Laboratory (Corvallis, Oregon) authored one of the earliest U.S. Technical Reports linking carbon dioxide—carbonate equilibrium and its relationship to the ecological effects of increased acidity of acid rainfall (9). In the same year F.B. Hill proposed the need for increased information to more accurately model the impacts of sulfur dioxide and ecosystem responses to environmental dosing of sulfur compounds in the environment (10). At the close of the 1960s Svante Oden described the effects of changing acidity of precipitation in Sweden as a result of industrial pollutants in England and their transport over space and time and their impacts to the environment (11). Cowling (1) notes that a series of lectures by Oden at the 1971 International Limnological Congress in Winnipeg, Manitoba stimulated both scientific and public scrutiny of acid rain and it effects on the North American continent.

The 1970s was marked by a rapid increase in the gathering of preliminary background information on the phenomenon of acid rain. As a result of this increase in research and investigation, it was during the decade of the 1970s that the first trends of a rapid growth in the literature related to acid(ic) rain and acid(ic) precipitation (12-17) was seen in 1976. Publication of the *1st International Symposium on Acid Precipitation and the Forest Ecosystem* (18) held in 1975 and the publication of a triple, special issue of the journal, *Water Air and Soil Pollution* (19) marked the beginnings of a period of rapid growth and accumulation of literature related to acid(ic) rain and acid(ic) precipitation. It was during this period that researchers such as Gene Likens, F. Herbert Borman, and Charles Cogbill (14-17) noted patterns similar to those cited by Oden in rainwater chemistry and the environments of North America, especially in the Northeast U.S. and Southeast Canada.

Coordinated Research in the 1980s

In 1977 the President's Council on Environmental Quality began calling for the consolidation of the existing monitoring and research programs such as the National Atmospheric Deposition Program and the Multistate Atmospheric Power Production Pollution Study. President Jimmy Carter announced in August 1979 the creation of a 10-year initiative to study the causes and effects of acid precipitation in Title VII of the Energy Security Act (P.L. 96-264), the Acid Precipitation Act of 1980. A major provision of the Acid Precipitation Act was the creation of the National Acid Precipitation Assessment Program (NAPAP) to define an ongoing, 10-year research initiative to provide a solid scientific understanding for the development of sound environmental policies that were needed to address the issues related to acid rain.

It was during this period that an era of coordinated interdisciplinary research was started and took on its own transboundary aspects with conferences focusing on the international exchange of ideas, resources, and research results. The media increased its news coverage of the phenomenon of acid rain, which until the late 1970s had remained more of an academic or scientific curiosity. As media coverage increased so did public concerns and demands for answers to the issue of acid rain.

This period of coordinated scientific research had a unique feature in providing the scientific and technical underpinnings for the development of policies that would ameliorate, reduce, or otherwise eliminate the threats to the Earth's ecosystems resulting from the effects of acid precipitation. The coordination of research clearly focused on the need for interdisciplinary initiatives that involved not only the physical and life sciences, but also included the social and policy sciences as well. The relationship between scientific research and policy developments took on new meanings as a debate over science-driven policies or policy-driven science emerged amid the political climate of the 1980s in the U.S. and elsewhere.

It is also noted that during the early 1980s attempts were made to develop specific information services and programs for acid rain. The Center for Environmental Information (CEI, Rochester, New York) secured funding for the development of the Acid Rain Information Clearinghouse (ARIC) in 1982 and developed a series of reference and referral services, conducting regional and international multi-disciplinary conferences, the development of a dedicated research library collection for acid rain literature, and began publication of the highly acclaimed current-awareness publication, the *Acid Precipitation Digest*. ARIC also served as the New York component of the New York-Quebec Acid Rain Documentation Center, which facilitated the development of the *ACIDOC* bibliographic database, which is still a viable information database operated by the Quebec Ministry of Environment.

Continuation of a Research Agenda and Implementing Resulting Policies

Authorization of the Clean Air Act Amendments of 1990 coincided with the end of the 10-year mandate of NAPAP. Implementation of the provisions of the Clean Air Act, a continuation of the long-term monitoring and research programs established by NAPAP, and further refinements of research agendas would become the driving forces of acid rain study at the close of the 20th Century.

During this period the growth of acid rain information continued as projects initiated during the NAPAP era publish the results of their research and monitoring programs. While a majority of this current information base remains in the physical and life sciences (biology, chemistry, ecology, geology, physics), it is during the 1980s and 1990s that we see the emergence of a solid literature base developing in the social sciences (economics, education, geography, information and data management, law, management, policy science, psychology). There appear two rather distinct clusters within the social science research. The first cluster of social science literature published in the 1980s begins to formulate the economic aspects of the effects of acid rain to ecosystems and materials. To a larger extent is the emergence of policy and legal strategies required to address the precursor pollutants causing acid rain, most notably sulfur dioxide and oxides of nitrogen. A second clustering of social science literature in the 1990s and continuing into the first decade of the 21st Century are articles discussing the effectiveness of various abatement strategies outlined in the Clean Air Act Amendments of 1990 <http://www.epa.gov/airmarkt/arp/regs/caaa.html>, and the effectiveness of emissions trading scenarios to reduce the overall generation of sulfur dioxide.

A Survey of Acid Rain Literature, Part 1. Growth of Information

To examine the scientific literature base of acid rain, the *Chemical Abstracts* portion of the CAS database, *SciFinder Scholar*, which was used to collectively gather bibliographic data on the concepts of acid(ic) rain (rainfall, rain water, etc.) and acid(ic) deposition. *SciFinder Scholar* was selected as one of several bibliographic data bases from which to examine the simple search strategy of "acidity of air pollution or acid rain" provided the most direct and comprehensive retrieval of related information. Selection of the *Chemical*

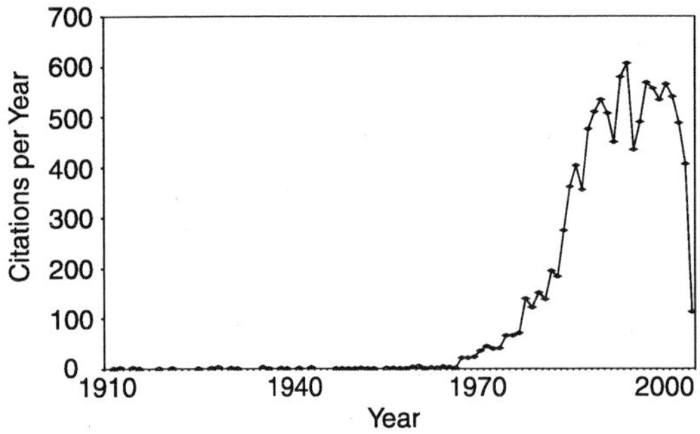

Figure 1. Growth of acid rain literature, 1907-2002 (June), Chemical Abstract and SciFinder Scholar.

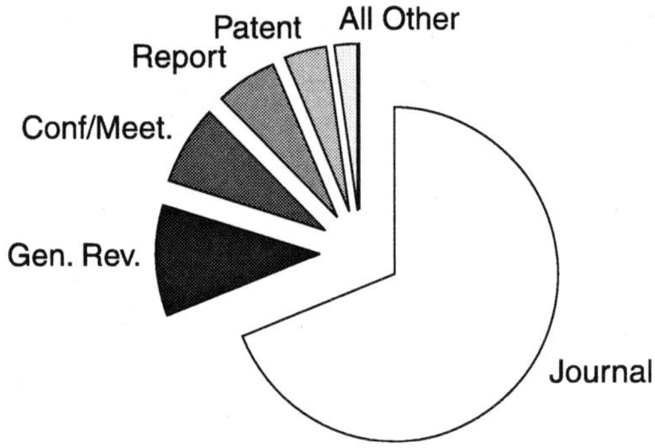

Figure 2. Types of acid rain literature published 1907-2002 (June), Chemical Abstracts and SciFinder Scholar.

Abstracts portion of *SciFinder Scholar* was for its abilities to sort and analyze components of the bibliographic data retrieved and the comprehensive treatment of the scientific aspects of the subject matter covered by *Chemical Abstracts*. Figure 1 shows the growth of the literature base of 11,206 citations retrieved on these topics from 1907 to June 2002.

Of this total, 8,841 citations were published as journal articles (79%). Figure 2 shows the distribution of published acid rain literature from the SciFinder Scholar database, and clearly shows the dominance of the journal article as the primary source of information and the primary vehicle for

Table 1. List of the 20 journals most frequently publishing articles about acid(ic) rain or acid(ic) precipitation. Data retrieved from the Chemical Abstracts portion of the SciFinder Scholar database from 1907 to June 2002.

Number of Articles Retrieved	Journal Name
611	Atmospheric Environment and Atmospheric, Part A
461	Water, Air, and Soil Pollution
276	Proceedings of the Annual Meting of the Air and Waste Management Association (and the Proceedings of the APCA Annual Meeting)
203	Environmental Science and Technology
149	Environmental PollutioN
122	Journal of Geophysical Research (Atmosphere)
106	Science of the Total Environment
83	Studies in Environmental Science
77	Journal of Atmospheric Chemistry
73	Geophysical Research Letters
64	Journal of the Air and Waste Management Association and the Journal of the Air Pollution Control Association
64	Kinky Jujutsu
58	Keening Exude Xebio
56	Journal of Aerosol Science
54	Chemosphere
54	Journal of Environmental Quality
52	Haunting Kexue
51	Zhongguo Huanjing Kexue

disseminating information on the topic. The journals publishing the most number of articles are listed in Table 1.

The research Agenda set forth by NAPAP provided a fertile environment for the publication of resulting research results and helped to sustain the proliferation of literature for more than 20 years. A closer examination of the impacts of NAPAP on the generation and subsequent publication of scientific and technical information is found in the production of U.S. Federal technical reports and output of graduate research measured by the production of doctoral dissertations and theses on the topics related to acid rain. Again, the data was extracted from the pool of bibliographic data obtained from the SciFinder Scholar database and are presented in Figure 3. It appears that NAPAP funding and support stimulated graduate research and as the overall quantity of NAPAP sponsored research declined, so did the number of technical reports and scholarly dissertations.

Dispersion of Acid Rain Across Disciplines

An analysis of the output of literature searches done in the Institute for Scientific Information Web of Science database allows users to review the contents individually or as the combined three resources comprising the Web of Sciences: the Science Citation Index, the Social Science Citation Index, and the Arts and Humanities Citation

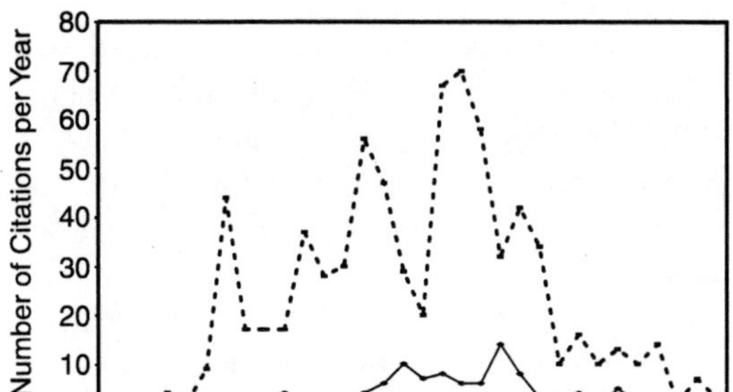

Figure 3. Growth of technical reports (top) and dissertations (bottom) in Chemical Abstracts, SciFinder scholar, 1970-2002.

Index. A total of 5,224 items were retrieved from a combined search of terms acid(ic) rain, acid(ic) precipitation, or acid(ic) air pollution(tants) from 1977-2002 (the period to which access was limited). Of this total, 4,968 citations are found in the *Science Citation Index* (covering the physical, applied, and life sciences), 418 citations in the *Social Sciences Citation Index* (covering legal, policy, business/economic, and education aspects), and 27 citations in the *Arts and Humanities Citation Index* (including articles on ethics, effects to art works and statuary, contribution of poetry and prose, history and philosophy of science). A more exhaustive survey of the acid rain literature can be constructed by use of the bibliographic databases described in the next section.

Bibliographic Control of Acid Rain Literature

The environmental literature from various multi-disciplinary research agendas around the world appears in a very large number of bibliographic databases (20-22). The major bibliographic utilities from which acid rain literature can be identified and accessed are described here.

ABI/Inform covers the business, management, policy and regulatory compliance aspects of scientific issues and topics, including those related to acid rain. This database also provides good coverage of the various trade publications in the energy, environmental, and regulatory segments of business and industry.

ACIDOC covers all aspects related to acid precipitation from its source to its control. General works; environmental effects: effects on aquatic ecosystems, effects on terrestrial ecosystems, human health effects, ecosystem sensitivity, effects on materials, monitoring; air and atmospheric processes: emissions, meteorology, transportation, deposition, monitoring; socio-economic aspects; political aspects: law, regulations, government policy, international relations, economic effects, education, public participation; mitigative and corrective measures: emission reduction, chemical treatments, liming, biological methods.

AGRICOLA is the U.S. National Agricultural Library's AGRICulture OnLine Access database covers the literature of agriculture, including agricultural and forest ecology, environmental quality and impacts associated with agriculture (including aquaculture) and forestry practices, environmental and human health aspects of environmental change (including the causes and effects of acid rain to agriculture and forestry).

Aquatic Biology, Aquaculture & Fisheries Resources is a combination of the world's premier databases covering fisheries and aquatic sciences. It combines all relevant portions of *Aquatic Sciences & Fisheries Abstracts* with the FISHLIT, CABI and AGRIS Fisheries & Aquaculture databases' subsets, and resources from *Fisheries Review*, Fish & Wildlife Reference Service.

ASCE's Civil Engineering Database is a Web-based database produced by the American Society of Civil Engineers that provides citations to all ASCE publications, including those devoted to environmental engineering, including acid rain and air pollution control.

Biological Abstracts/BIOSIS Previews provides the full range of topical coverage related to the biological and life sciences, including biological and biochemical aspects of ecology, effects of biological and chemical agents on biological systems, biological impacts of climatic changes, environmental biology, pollution biology (water, air, and soil). The effects of acid rain to biota and ecosystems are covered in a comprehensive manner.

Chemical Abstracts (see *SciFinder Scholar*)

Compendex*Plus covers over 2,600 journals and conferences as well as non-governmental reports in all areas of engineering including environmental engineering. Environmental topics covered include the physical and chemical aspects of pollution, environmental restoration and remediation, risk assessment and management, health and safety, hazardous waste management and impact analysis, and energy exploration, delivery, use, and management.

Current Contents Connect (ISI) provides multi-disciplinary bibliographic coverage of the contents of over 8,000 major journals and scholarly reviews including the literature of ecology, environmental science (biology, chemistry, engineering, geology, physics), and environmental management and policy (sociology, business, economics). (see also *Science Citation Index*, *Social Science Citation Index*, and the *Web of Science*)

Digital Dissertations (*Dissertation Abstracts/ Index*) indexes doctoral and masters theses from all disciplines including those related to scientific, technical, policy, social, and legal aspects of air pollution, acid rain, and related studies.

DOE Reports Bibliographic Database contains citations for Department of Energy (DOE) sponsored scientific and technical reports. Environmental coverage is directed primarily to ongoing DOE-sponsored research in ecology and ecological restoration, global environmental change, acid rain and the ecological and environmental aspects of energy science and policy.

DOE Information Bridge is a publicly accessible website that provides access to the full-text and bibliographic records of DOE research and development reports in the sciences with multi-disciplinary coverage of environmental and energy sciences. The database includes over 2 million searchable pages in over 27,000 reports produced by the DOE and the extended DOE contractor research community released to the public since January 1996. An expanded and enhanced version of this database is available for DOE contractors.

Dow Jones Interactive provides access to the full text of current and retrospective news stories, periodical articles, and financial data from several thousand sources covering virtually every subject. Although often associated with business information, Dow Jones Interactive is much more than a business database, with coverage of environmental issues such as hazardous waste remediation, regulatory compliance, health and safety issues, and environmental policy.

Ecology Abstracts focuses on the interactions of organisms with their environment and provides detailed coverage of the impacts of natural and human interactions with ecosystems and the environment, including environmental change and destruction.

Energy Science and Technology *Database* abstracts and indexes basic and applied scientific and technical research literature. *It* was designed for use by government managers and researchers at the national laboratories, and other technical/ research efforts sponsored by the U.S. Department of Energy and includes coverage of ecology, including the assessment of energy technology on the environment.

Environment Abstracts/Environmental Universe covers a broad range of issues and topics related to the environment and the management and use of natural resources. Coverage includes: air pollution, acid rain, environmental education, environmental law and policy.

EPA National Online Library System (OLS) consists of several databases with information on books, reports, and audiovisual materials on a variety of topics. OLS contains the holdings for most of the 28 EPA libraries at the EPA Headquarters (Washington, D.C.), the ten EPA Region offices, and EPA laboratories. It also contains records for EPA documents distributed through the National Technical Information Service (NTIS).

ERIC, the database from the Educational Resources Information Center, is the largest education database in the world with coverage of over 800 periodicals and non-journal publications (research papers, guides, government reports, speeches and more). Topics range across all aspects of educational research and practice and cover conservation, energy, environmental, nature, and outdoor education with ample coverage related to acid rain.

GeoArchive is a major geoscience database covering more than 5,000 serials, books from more than 1,000 publishers, conferences, doctoral dissertations, and technical reports as well as over 15,000 geological maps. Provides coverage of geoscience, hydroscience, and environmental science topics.

GeoRef, produced by the American Geological Institute database, is the most comprehensive database in the geosciences, covering all aspects of geology and geography including the earth sciences and the environment.

GPO Monthly Catalog is a comprehensive listing of federal government publications. It attempts to index all the unclassified publications of federal government agencies. These include reports (policy, assessments, technical), booklets, popular publications, and Congressional documents related to ecology and the environment (Senate and House hearings, committee reports). The legislative chronologies related to acid rain are found in this database.

InfoTrac OneFile covers articles from more than 1,500 scholarly, trade, and general-interest periodicals, and provides access to articles for the most recent six months of the *New York Times*. Subject coverage includes humanities, social sciences, general sciences, and current events including broadly defined coverage of ambient, home, and workplace environments. *OneFile* was formerly known as *Expanded Academic Index*.

Marine, Oceanographic & Freshwater Resources is a bibliographic database on marine, oceanographic, and related freshwater resources. Providing coverage on international marine and oceanic information, as well as substantial coverage of estuarine, brackishwater, and freshwater environments,

MEDLINE database produced by the National Library of Medicine provides comprehensive coverage of the biomedical literature including all aspects of environmental health and medicine related to acid rain and its precursor air pollutants, sulfur dioxide and oxides of nitrogen.

NOAA Library Catalog is a searchable interface to the holdings of the National Oceanic and Atmospheric Administration's library network. Collections consist of reports, special datafiles and datasets, documents, articles, and audiovisual resources related to ocean science and atmospheric sciences.

NTIS is the National Technical Information Service, the U.S. government's central source for reports written on scientific, technical, engineering, and related business information produced by or for the U.S. government.

Pollution Abstracts is the leading resource for environmental engineers and scientists, hydrogeologists, geologists, chemists, industrial engineers and hygienists, corporate regulatory affairs managers, waste managers, and government officials concerned with pollution.

SciFinder Scholar is a bibliographic and reference research tool based on an enhanced compilation of Chemical Abstracts and MEDLINE. Users explore this collection by research topic, substance identifier, chemical structure, chemical reaction, or author's name. It includes materials on the chemical aspects of earth sciences and geology, ecology, environmental impact, pollution, and toxicology.

USGS Online Catalog is the online catalog of the USGS library and includes over 200,000 records for books, periodicals, maps, microforms, videocassettes, CD-ROMs, and pamphlets. Online citations describe items that have been added to the library since 1975 plus all of the publications of the USGS. To find information about maps and related data for geology, natural hazards, earth resources, geophysics, geochemistry, geochronology, paleontology, and marine geology, look at this resource site.

Web of Science is comprised of the Science Citation Index, Social Sciences Citation Index, and Arts & Humanities Citation Index. The database covers journal articles in all subject areas including environmental science and engineering.

Acid Rain and the Internet

The topic of acid rain emerged before the popularization of the Internet and the creation of the World Wide Web (WWW). This does not mean the topic of acid rain is not found in the Internet or the WWW. While it is completely beyond the scope of this present paper to provide a comprehensive inventory of the acid rain-related online sites, it is possible to highlight several of the more useful sites.

An analysis of the raw retrieval of information retrieved by several major Web search engines provided the results listed in Table 2. Table 3 shows the breakdown by domains for the 257,900 matches made by the HotBot search engine

Clearly, there is an abundance of resources on the topic of acid rain on the Internet. A primary feature of the Internet is the ease with which scientific and technical data can be brought to the individual, classroom, library, or organization. There are, however, many words of caution about the Internet and the WWW, all of which can be summarized as, "Buyer Beware!" and "Sorting the Wheat from the Chaff." It is not nearly as easy to delineate what is "good" data or information or what is "bad" data or information on the Internet.

Table. 2 Search results for acid(ic) rain, acid(ic) precipitation, or acid(ic) air pollution(tants) on the Internet.

Search Engine	Number of matches
AltaVista	1,261
Google	8,000
HotBot	257,900
Northern Light	13,254

Table. 3 Domain breakdown for Hot Bot search engine results.

Domain	Number of Matches
.gov	17,400
.org	45,200
.edu	59,700
.mil	<500
.com	5
.us	8,100
.ca	12,900
.uk	9,200
.se	>900

The University of South Carolina's Beaufort Library has a very good tutorial for searching the World Wide Web for relevant resources, "Bare Bones 101: A Basic Tutorial on Searching the Web http://www.sc.edu/beaufort/library/bones.html. Librarians at the University at Buffalo created "How to Find What You Want on the World Wide Web" http://ublib.buffalo.edu/libraries/search/searchint.html as an interactive tool for locating and evaluating resources obtained on the WWW.

Provided below is a short and *representative sample* of a few dozen of the hundreds of high-quality acid rain web sites available for researchers, policy makers, educators, students, librarians, managers, and administrators to use when looking for data and information resources on the topics related to acid rain.

National Acid Precipitation Assessment Program (NAPAP)
http://www.oar.noaa.gov/organization/napap.html

NAPAP remains an interagency program to coordinate scientific research, environmental monitoring, and integrated assessments for examining the effects of acid precursors, sulfur dioxide and oxides of nitrogen. NAPAP extends its network among individual states, universities, private sector interests in the energy industries, and among six Federal Agencies: Department of Agriculture, Department of Energy, Department of the Interior, the Environmental Protection Agency, the National Aeronautics and Space Administration, and the National Oceanic and Atmospheric Administration. NAPAP was reauthorized under Section 901, j of Title IX of the Clean Air Act Amendments of 1990 (P.L. 101-549), which also included Title IV, the Acidic Deposition Control Program. NAPAP's biennial reports provide the scientific underpinnings linking ongoing research and monitoring activities to policy initiatives In addition to NAPAP's assessment documents and reports is the Tracking and Analysis Framework, a modeling framework developed to access, inform, and guide U.S. regulatory policies on emissions of acid rain precursors. The TAF Model can be navigated by going to its Web site at http://www.lumnia.com/taf/start.htm. Table 4 lists acid rain-specific Web sites maintained by federal departments of agencies participating in NAPAP.

The National Atmospheric Deposition Program/ National Trends Network (NADP/NTN)
http://nadp.sws.uiuc.edu/nadpoverview.asp

NADP/NTN was established in 1978 as a network of 22 monitoring stations in the U.S. Today, more than 200 stations are found in the continental U.S., Alaska, Puerto Rico, and the Virgin Islands. In addition to monitoring precipitation for hydrogen (acidity as pH), sulfate, nitrate, ammonium, chloride and base cations (e.g., calcium, magnesium, potassium, and sodium), more than 35 sites are now participating in the Mercury Deposition Network (formed in 1995 to collect weekly samples for mercury content). From this site it is possible to link to a variety of NADP data products:

- Weekly and daily precipitation chemistry data
- Monthly, seasonal, and annual precipitation-weighted mean concentrations
- Annual and seasonal deposition totals
- Mercury deposition data
- Daily precipitation totals
- Color isopleth maps of precipitation concentrations and wet deposition
- Site photos, maps, and descriptive information
- Quality assurance data and information

Data are available for individual sites or as a series of isopleth maps (2000 data) as gif images and Adobe PDF (portable document format) maps (there are a total of 22 maps, one for each parameter).

The Atmospheric Integrated Research Monitoring Network (AIRMoN) is part of the NADP/NTN and provides a higher resolution of temporal acid rain data. A variety of AIRMoN data products can be retrieved at http://www.arl.noaa.gov/research/programs/airmon.html. Data provided includes daily wet and dry deposition and various inter-laboratory programs in the U.S. and Canada.

Man and the Biosphere Program
http://www.mabnetamericas.org/

As stated in the introduction from the Man and the Biosphere Home Page, "The MAB Program is an interdisciplinary research effort directed toward providing information for the solution of natural resources and environmental issues. As an intergovernmental program, MAB presents an opportunity for international cooperation and a focus for the coordination of related programs aimed at improving the management of natural resources and the environment."

Table. 4. Acid Rain Web sites for Federal NAPAP or Acid Rain Departments and Agencies.

EPA	http://www.epa.gov/airmarkets/acidrain/
NOAA	http://www.oar.noaa.gov/organization/napap.html
USGS/DOI	http://bqs.usgs.gov/acidrain/

MAB is an international cooperative of environmental research organizations operating in the following Regional Networks:

- AfriNet
- ArabNet
- East Asia
- EuroMAB
- MABNetAmericas
- IberoMAB
- USMAB

The Biodiversity Resources for Inventorying and Monitoring (BRIM) component of MAB is a major component of MAB and is a collaboration of U.S. MAB and the Information Center for the Environment to facilitate collection and dissemination of biological inventory and monitoring information derived from protected areas worldwide. Components of BRIM include the MABFlora and MABFauna software programs to standardize methods in the collection, storage, and retrieval of bioinventory information.

"The mission of the United States MAB Program is to explore, demonstrate, promote, and encourage harmonious relationships between people and their environments building on the MAB network of Biosphere Reserves and interdisciplinary research. The long-term goal of the U.S. MAB Program is to contribute to achieving a sustainable society early in the 21st Century. The MAB mission and long term goal will be implemented, in the United States and internationally, through public-private partnerships and linkages that sponsor and promote cooperative interdisciplinary research, experimentation, education and information exchange on options by which societies can achieve sustainability." (Adopted by the U.S. National Committee MAB, July 26, 1995)

The U.S. MAB is operated by the Department of State, Bureau of Oceans and International Environmental and Scientific Affairs. A July 1, 1998 directive describes the main functions of biosphere reserves as:

1. conservation of important biological resources;
2. development of environmentally sound economic growth;
3. support of environmental research, monitoring, education, and training;
4. as a framework to bring people together to accomplish the above three.

The United States has 47 biosphere reserves with 99 administrative units. The *protected areas* and *managed use areas* are owned/administered by the National Park Service; U.S.D.A.-Forest Service; state, county, or city governments; U.S. Fish and Wildlife Service; The Nature Conservancy; universities; private owners; U.S.D.A-Agricultural Research Service; National Oceanic and Atmospheric Administration; Bureau of Land Management; Tennessee Valley Authority; and the Department of Energy. A list of U.S. Biosphere Reserves includes the following locations (and sponsoring agencies) is appended to this paper as Appendix A.

Each Biosphere Reserve maintains its own research and monitoring networks, programs, and services, and is responsible for its site's data and information management and sharing strategies.

Among the MAB Biosphere Preserves, the Adirondack Park Agency (APA), is one of the major acid rain data repositories in the entire U.S. The Web site for access to the APA acid rain data is http://www.northnet.org/adirondackparkagency/. The Adirondack Park Agency oversees the research and monitoring activities within the Adirondack Park, including a growing body of GIS Coverages, that now includes maps, land use data, legal jurisdictions and land designations, and wetlands designations. Much of this GIS data is now available from the APA in CD-ROM formats.

Influences on Wetlands and Lakes in the Adirondack Park of New York State – A Catalog of Existing and New GIS Data Layers for the 400,000 Hectare Oswegatchie/Black River Watershed is one of the specific Adirondack Park reports which includes access to data related to ongoing acid rain monitoring. The following data layers are divided into thematic groups as they appear on the final report poster. Links for each data layer lead to a graphic image of the data, to the topic in the final report, metadata when available, and to other related documents.

- Vegetative Cover

    - Upland Cover from TM Data
    - Wetlands
- Geology/Soils
    - Bedrock Geology
    - Surficial Geology
    - Bedrock Acid Neutralizing Capacity
    - General Soils—Parent Material
    - Elevation
- Land Use
    - Adirondack Park Land Use & Development Plan
    - 1989 Residential Areas
- **Hydrology/Atmospheric**
    - Watersheds and Adirondack Long Term Monitoring Waters
    - Mean Annual Precipitation (1951 to 1980)
    - Total Nitrogen Deposition
- **Landscape Disturbance**
    - State Forest Acquisition
    - 1916 Fire Protection
    - November 1950 Blowdown
    - Forest Damage Assessment of July 1995 Blowdown
    - Landscape Disturbance Composite (to 1950)

One of the major acid rain projects taking place in the Adirondack Park is the ongoing monitoring network established and continued by the Adirondack Lakes Survey Corporation (ALSC). The ALSC is a not-for-profit corporation established in 1984 as a cooperative agreement between the Empire State Electric Energy Research Corporation (ESEERCO) and the New York State Department of Environmental Conservation (NYSDEC) to determine the extent and magnitude of acidification of lakes and ponds in the Adirondack region. Presently, the ALSC is moving into the NYSDEC's Division of Air Resources and receives joint funding from the New York State Energy Research and Development Authority (NYSERDA) and the USEPA. Data available include physical attributes of wetland and watershed areas, data about the 1950 blowdown (resulting in excessive forest damage), the 1916 wild fires, the 1995 microburst wind storm that ravaged large portions of the western Adirondack region, various land use data, bedrock and surficial geologic data, soil parameters, and atmospheric data (precipitation, nitrate and sulfate data). These data are provided in polygon and raster format for GIS coverages.

Their monitoring work began with an extensive baseline survey conducted from 1984 – 1987 that culminated with the 1990 Adirondack Lakes Survey Report <http://www.adirondacklakessurvey.org/execsum.htm>. Since that time acid rain monitoring surveys, fisheries and watershed projects have continued. The most intensive of these is the Long-Term Monitoring Project <http://www.adirondacklakessurvey.org/ltmpage.htm>, which began in June 1992 and continues presently. A special highlight of this website is the month by month chemistry update for two key monitoring lakes, Big Moose Lake and Willys Lake <http://www.adirondacklakessurvey.org/monthly.html>. The *Adirondack Lakes Survey: 1987 Report* is a 3-volume set of data collected in the Lower Hudson River region.

The original *Adirondack Lakes Survey* is a 19-volume set of the original 1985-1989 monitoring activities where extensive data were collected in hundreds of Adirondack lakes and ponds and select waters in the Hudson River watershed. The following data was collected for each lake tested: location and status, bathymetric map, morphometrics, macroinvertebrate macrophytes, fisheries, chemical/physical parameters, and management history arranged by numerical listing of lakes within each major Adirondack watershed. An additional 3 volumes were added in 1989 for watersheds of the Lower Hudson River. In all, these Adirondack data represent about 20 linear shelf-feet of loose-leaf datasheets. As mentioned above only small sets of these data have been converted into CD-ROM or digital formats.

Long Term Ecological Research (LTER) Network
http://lternet.edu

LTER is another collaborative research effort that links more than 1100 scientists and students in ongoing investigations of ecological processes covering long-term temporal scales and broad geographic scales. The LTER was established in 1980 by the National Science Foundation to support dedicated, long-term ecological research in the U.S. There are 24 LTER Sites in the U.S. <http://www.lternet.edu/sites/>:

- Andrews LTER
- Arctic Tundra LTER
- Baltimore Ecosystem Study
- Bonanza Creek LTER
- Central Arizona – Phoenix
- Cedar Creek LTER
- Coweeta LTER
- Harvard Forest
- Hubbard Brook LTER
- Jornada Basin
- Kellogg Biological Station
- Konza LTER
- Luquillo LTER
- McMurdo Dry Valleys
- Niwot Ridge LTER
- North Temperate Lakes
- Palmer Station
- Plum Island Ecosystem
- Sevilleta LTER
- Shortgrass Steppe
- Virginia Coast Reserve
- Florida Coastal Everglades
- Georgia Coastal Ecosystems
- Santa Barbara Coastal

As with the U.S.MAB Biosphere Reserves, each LTER site should be examined for its own research, monitoring, and data and information management activities. You will note that there is an overlap in the reporting jurisdictions of the LTER Network Sites and the U.S. MAB Biosphere Preserves, such as is the case with the Hubbard Brook Ecological Study site.

Among the LTER sites, The Hubbard Brook Ecosystem Study <http://www.hubbardbrook.org/ > facility is, perhaps, the best known for its compressive and pioneering acid rain research. HBES is a long-term ecological research project located at the Hubbard Brook Experimental Forest (HBEF), a 3,160 hectare reserve located in the White Mountain National Forest, near North Woodstock, New Hampshire. The HBEF was established by the USDA Forest Service in 1955 as a major center for hydrologic research in New England. A large repository of "Research and Data" is available from their Web site and presents a description of long-term data sets and the on-site research activities of the HBES:

- Stream Flow and Chemistry
- Atmospheric Inputs (precipitation and deposition)
- Vegetation
- Weather and Climate Data
- Watershed Budgets
- Soil Conditions and Chemistry
- Animals
- Lake Chemistry
- Paleoecology

Data are provided in both print/hard copy reports and publications as well as online data files and data sets. Images (photographs, aerial photos, and graphics), GIS coverages, and publications are also provided by the HBES.

Additional Internet Web Sites and Resources

US CURRENT pH MAP: You can view a map of the most current data (latest is for Nov. 26 through Dec 23, 2001) at: http://water.usgs.gov/nwc/NWC/pH/html/ph.html

US pHMAPS—1994-2000: You can view a map of the US pH averages for 1995 (change the date in the html to 1994 and that year is also available) at: http://nadp.sws.uiuc.edu/

CANADA: Canadian data can be viewed at http://www.ns.ec.gc.ca/aeb/ssd/acid/index.html

EUROPEAN FORESTS: Maps showing damage to European forests can be found at http://www.soton.ac.uk/~engenvir/environment/air/acid.how.big.problem.html

NORWAY: Maps showing increasing damage to Norway's rivers are at: http://www.grida.no/soeno95/acidrain/effects.htm

Conclusion

The information base related to acid rain will continue to grow in future decades. This growth most likely will decrease in magnitude over the years as research initiatives begun more than a quarter of a century ago come to a close and as monitoring programs are incorporated into more broadly defined global environmental change programs. Access to real-time, on-demand data related to all aspects of acid rain science and policy will continue to be rigorously pursued and the growth of such data repositories will increase and become more organized. Digitization of historical data will continue at its present painfully slow place unless funding and technical support is provided for such print-to-electronic conversions to take place.

A glimmer of hope resides in the summary statement on Information Systems presented in *Recommendations for Improving the Scientific Basis for Environmental Decisionmaking*, a report from the first National Conference on Science, Policy, and the Environment, held by the National Council for Science and the Environment (December 7-8, 2000, Washington, D.C.). Among the recommendations was the call for the creation of a National Environmental Information Infrastructure to support "intensified public information programs and environmental education and training initiatives (K-Adult). The NCSE recommendation for Information Systems is as follows:

"The United States should develop a National Environmental Information Infrastructure (NEII) that crosses scientific domains (i.e., applied, physical, information, natural and social sciences, engineering), sectors of the economy (i.e., private, academic, government), and lines of work (i.e., research, education, advocacy, communication, information)." http://www.cnie.org/2000conference/13.cfm

This recommendation is based on six components:

1. National Environmental Information Infrastructure
   The NEII should be an open architecture for network development, with appropriate computer power [data, information, and knowledge management], and user interface.
2. Ensure Data
   The federal government and its partners should work to ensure data (information) availability, quality, and preservation through this architecture.
3. Tool Development
   The federal government should promote the development of tools to make information available to a multiplicity of users at varying geospatial scales and time frames.
4. Data Evaluation
   Creation of the NEII should begin with a comprehensive analysis to evaluate the many data and information repositories and data support systems at global/international, national/federal, state/provincial, and local levels. They should be evaluated in regard to:
   - acquisition of data (and information)
   - management of data
   - integration and analyses of data
   - dissemination of data
   - examination of cross-cutting issues to examine organizational roles of data producers, providers, and users.
5. Multi-Stakeholder Advisory Board
   There should be a multi-stakeholder advisory board (e.g., data producers/providers, data/information managers, data/database vendors/providers, librarians and other information providers, and various user communities) to examine the creation of a central, comprehensive environmental information infrastructure that would:
   - coordinate efforts across scientific domains, industries, and institutions
   - provide access to and communication of data and information for multiple categories of end users
   - promote the use of environmental indicators and methods of advanced environmental accounting
   - emphasize the need for a U.S. commitment to support environmental data and information systems and management
   - identify environmental decision areas that currently lack robust supporting data and information resources
   - examine educational opportunities and training in scientific, policy, and information technology (including librarians)
   - examine policies for discussion of publication, dissemination, and "digestion" of data and information
   - identify and network among repositories of human and organizational expertise and resources.

6. Outcome

Completion of these tasks would serve to:

- define an environmental information infrastructure that is timely, adequate, and comprehensive
- address the need to develop services, products, and programs that are efficient, economic, and equitable.

This recommendation would certainly address nearly all of the data and information challenges that currently envelope the issues related to acid rain (and a host of other global environmental problems and concerns). Further refinement of this recommendation is needed and might include the expansion of the geosciences-topical framework of the existing World Data Center complex to include the creation of individual WDCs for Ecological Data. Programs such as the National Biological Information Infrastructure in the Department of the Interior's U.S. Geological Survey (already an active partner in the WDC-A complex in the U.S.) would seem to be well-suited to carryout this much needed ecological data coordination and management function. The next step in this process may have been undertaken at the Second National Conference on Science, Policy and the Environment, whose recommendations for Information Systems Interoperability in Support of Effective Environmental Decisionmaking was featured in *Recommendations for Achieving Sustainable Communities: Science and Solutions, A Report from the second National Conference on Science, Policy, and the Environment* http://www.cnie.org/NCSEconference/2001Conference/report/page.cfm?FID=1692, and appended to this paper.

References Cited

(1) Cowling, Ellis B. 1982. Acid precipitation in historical perspective. *Environmental Science and Technology.* 16(2):110A-123A.

(2) Anderson, V.G. 1915. Influence of Weather conditions on amounts of nitric acid and of nitrous acid in the rainfall near Melbourne, Australia. *Quart. J. Roy. Met. Soc.* 41:99-122.

(3) Gerlach, R. 1914. Proof of the origin if smoke acids in rain-water flowing down tree trunks by means of an automatic separator and the influence of these acids on the soil. Expt. Sta. Record 32: 422.

(4) Hudig,J. The amount of nitrogen as ammonia and nitric acid (and nitrous) acid in the rain water collected at Uithuizermeeden, Groningen. J. Agr. Sci. 4:260-69.

(5) Vtuynii, I. 1912. Amounts of chlorine and sulfuric acid in rain water. J. Chem. Soc. 100(II):432.

(6) Witt, Ernst. 1969. Lethal effect of exhaust gases on the water fauna. Zentralbl. Arbeitsmed. Arbeitsschutz. 19(9):264-65.

(7) Nyberg, Alf. 1970. Weather changes important for the acidity of rainfall. Swed. Forsk. Framsteg. 1:18-21.

(8) Reiquam, H. 1970. European Interest in acidic precipitation. Review. AEC Symposium Series 22: 289-92.

(9) Reuss, John O. 1975. Chemical-biological relations relevant to ecological effects of acid rainfall. NTIS Report PB-244409, 53. pp.

(10) Hill. F.B. 1975. Some information needs for air quality modeling. Brookhaven National Laboratory Report BNL-25041 8 pp.

(11) Oden, S. 1968. *The Acidification of Air and Precipitation and Its Consequences in the Natural Environment,* Ecology Committee Bulletin, No. 1. Stockholm, Sweden:Swedish National Science Research Council

(12) Beamish, Richard J. 1972. Acidification of the La Cloche Mountain Lakes, Ontario, and resulting fish mortalities. J. Fish. Res. Board Can. 29(8):1131-43.

(13) Askne, C., Brosset, C. 1972. Determination of strong acid in precipitation, lake water, and airbourne matter. Atmos. Environ. 6(9):695-96.

(14) Likens, Gene E., Bormann, F. Herbert, and

Johnson, Noye M. 1972. Acid rain. Environment. 14(2):33-40.

(15) Likens, Gene E., and Bormann, F. Herbert. Acid rain: Serious regional environmental problem. Science. 184(4142):1176-79.

(16) Cogbill, Charles V., and Likens, Gene E. 1974. Acid precipitation in the northeastern United States. Water Resour. Res. 10(6):1133-37.

(17) Fairfax, J.A.W., and Lepp, N.W. 1975. Nature (London). 255(5506):324-25.

(18) Dochinger, L.S., Seliga T.A., Eds. 1976. Proceedings of the 1st International Symposium on Acid Precipitation and the Forest Ecosystem, Columbus, Ohio, 1974. Published as U.S. Department of Agriculture, Forest Service, General Technical report NE-23. 1074 pp.

(19) Water, Air, and Soil Pollution. 6(2-3-4), 1976.

(20) Stoss, Frederick W. 1991. Environment Online: The Greening of the Databases, Part 1, General Interest Databases. Database. 14(4):13-27

(21) Alston, P. Gayle. 1991. Environment Online: The Greening of the Databases, Part 2, Scientific and Technical Databases. Database. 14(5):34-52.

(22) Alston, P. Gayle and Stoss, Frederick W. 1992. Environment Online: The Greening of the Databases, Part 3, Business and Regulatory Information. Database. 15(4):17-35.

(23) Ercegovac, Z. 1992. Environmental Research Communication Studies and Information Sources. Annual Review of Information Science and Technology. 27:173-225.

Appendix A. U.S. Man and the Biosphere Preserves and Their Sponsoring Agencies

Key:

NPS – National Park Service, Department of Interior
NOAA – National Oceanic and Atmospheric Administration, Department of Commerce
FWS – U.S. Fish & Wildlife Service, Department of Interior
FS – Forest Service, Department of Agriculture
ARS – Agricultural Research Service, Department of Agriculture
DOE – Department of Energy
BLM – Bureau of Land Management, Department of Interior
Private – Denotes Some Form of Private Ownership
Complex –Denotes Multiple Ownership

U.S. MAB Sites:

- Aleutian Islands National Wildlife Refuge (FWS)
- Beaver Creek Experimental Watershed (FS)
- Big Bend National Park (NPS)
- Big Thicket National Preserve (NPS)
- California Coast Ranges Biosphere Reserve (8 Units)
    - Elder Creek Area of Critical Environmental Concern (BLM)
    - Heath and Marjorie Angelo Coast Range Preserve (University of California)
    - Jackson Demonstration State Forest (California Department of Forests)
    - Landels-Hill Big Creek Reserve (University of California)
    - Redwood Experimental Forest (FS)
    - Redwood National Park (NPS)
    - North Coast Redwoods District State Parks (3 Units)
    - Del Norte Coast Redwoods State Park and Alternate Site (California Department of Parks and Recreation)

  - Jebediah Smith Redwoods State Park and Alternate Site (California Department of Parks and Recreation)
  - Prairie Creek Redwoods State Park and Alternate Site (California Department of Parks and Recreation)
  - Western Slopes of Cone Peak, Los Padres National Forest (FS)
- Carolinian-South Atlantic Biosphere Reserve (11 Units)
  - Blackbeard Island and Wolf Island National Wildlife Refuges (FWS)
  - Cape Lookout National Seashore (NPS)
  - Cape Romain National Wildlife Refuge (FWS)
  - Capers Island Heritage Preserve (South Carolina Department of Natural Resources, Marine Resources Division)
  - Cumberland Island National Seashore (NPS)
  - Hobcaw Barony (North Inlet) (Complex Private)
  - Little St. Simons Island (Private)
  - Santee Coastal Reserve (South Carolina Department of Natural Resources)
  - Tom Yawkey Wildlife Center (South Carolina Department of Natural Resources)
  - Washo Reserve (The Nature Conservancy)
- Cascade Head Experimental Forest & Scenic Research Area (FS)
- Central California Coast Biosphere Reserve (See Golden Gate Biosphere Reserve)
- Central Gulf Coastal Plain Biosphere Reserve
  - Apalachicola National Estuarine Research Reserve and Alternate Site (NOAA)
- Central Plains Experimental Range (ARS)
- Champlain-Adirondack Biosphere Reserve (3 Units)
  - Adirondack Park Agency (Private)
  - Green Mountain National Forest (FS)
  - Mount Mansfield State Natural Area (Vermont Agency of Natural Resources)
- Channel Islands Biosphere Reserve (2 Units)
  - Channel Islands National Marine Sanctuary (NOAA)
  - Channel Islands National Park (NPS)
- Congaree Swamp National Monument (NPS)
- Coram Experimental Forest (FS)
- Denali National Park and Biosphere Reserve (NPS)
- Desert Experimental Range (ARS)
- Everglades National Park (with Dry Tortugas National Park) (NPS)
- Fraser Experimental Forest (FS)
- Glacier Bay-Admiralty Island Biosphere Reserve (2 Units)
  - Admiralty Island National Monument (FS)
  - Glacier Bay National Park and Preserve (NPS)
- Glacier National Park (NPS)
- Golden Gate Biosphere Reserve (13 Units)
  - Bolinas Lagoon Preserve and Cypress Grove Preserve (Private)
  - Bodega Marine Reserve (University of California)
  - Cordell Bank National Marine Sanctuary (NOAA)
  - Farallon National Wildlife Refuge (FWS)
  - Golden Gate National Recreation Area (NPS)
  - Gulf of the Farallones National Marine Sanctuary (NOAA)
  - Jasper Ridge Biological Preserve (Coordinator) (Stanford University)
  - Marin Municipal Water District (Marin County, California)
  - Mount Tamalpais State Park (California Department of Parks and Recreation)
  - Point Reyes National Seashore (NPS)
  - San Francisco Peninsular Watershed (City of San Francisco, California)
  - Samuel P. Taylor State Park (California Department of Parks and Recreation)
  - Tomales Bay State Park (California Department of Parks and Recreation)
- Gray's Reef National Marine Sanctuary

(NOAA)
- Guanica Commonwealth Forest Reserve (Puerto Rico Department of Natural Resources)
- H.J. Andrews Experimental Forest (FS)
- Hawaiian Islands Biosphere Reserve (2 Units)
    - o Haleakala National Park (NPS)
    - o Hawaiian Volcanoes National Park (NPS)
- Hubbard Brook Experimental Forest (FS)
- Isle Royale National Park (NPS)
- Jornada Experimental Range (ARS)
- Konza Prairie Research Natural Area (The Nature Conservancy)
- Land Between the Lakes (Tennessee Valley Authority)
- Luquillo Experimental Forest (FS)
- Mammoth Cave Area and Barren River Area Development District, BRADD (NPS and Complex)
- Mojave and Colorado Deserts Biosphere Reserve (5 Units)
    - o Anza-Borrego Desert State Park and Alternate Site (California Department of Parks and Recreation)
    - o Death Valley National Monument (NPS)
    - o Joshua Tree National Monument (NPS)
    - o Philip L. Boyd Deep Canyon Desert Center (University of California)
    - o Santa Rosa Wildlife Management Area, San Bernadino National Forest (FS)
- New Jersey Pinelands Biosphere Reserve (Pinelands Commission, Complex)
- Niwot Ridge Biosphere Reserve (University of Colorado)
- Noatak National Preserve (2 Units)
    - o Gates of the Arctic National Park (NPS)
    - o Noatak National Preserve (NPS)
- Olympic National Park (NPS)
- Organ Pipe Cactus National Monument (NPS)
- Rocky Mountain National Park (NPS)
- San Dimas Experimental Forest (FS)
- San Joaquin Experimental Range (ARS)
- Santee Coastal Reserve (South Carolina Department of Natural Resources)
- Sequoia-Kings Canyon National Parks (NPS)
- South Atlantic Coastal Plain Biosphere Reserve
- Southern Appalachian Biosphere Reserve (5 Units)
    - o Coweeta Hydrologic Laboratory (FS)
    - o Grandfather Mountain (Private)
    - o Great Smoky Mountains National Park (NPS)
    - o Mt. Mitchell State Park (North Carolina Department of Environmental Health and Natural Resources)
    - o Oak Ridge National Environmental Research Park (DOE)
- Stanislaus-Tuolumne Experimental Forest (FS)
- Three Sisters Wilderness, Deschutes National Forest (FS)
- University of Michigan Biological Station (University of Michigan)
- Virgin Islands National Park and Biosphere Reserve (NPS)
- Virginia Coast Reserve (The Nature Conservancy)
- Yellowstone National Park (NPS)

Appendix B. Information Systems: Interoperability in Support of Effective Environmental Decisionmaking.

A recommendation made at the National Conference on Science, Policy, and the Environment, December 6-7, 2001, Smithsonian National Museum of Natural History, Washington, D.C.

Introductory Remarks

Building on the recommendations of the first National Conference on Science, Policy and the Environment, the Information Systems Breakout Session confirmed that data and information are the transportation between science and decisionmaking for a sustainable environment.

The US is moving forward aggressively in

building a national environmental information infrastructure, although many challenges remain. National borders do not limit many of these challenges.

Recommendations

The World Summit on Sustainable Development should advance the following approaches:

1. Global Environmental Information Commons – Nations need to work together to create a Global Environmental Information Commons as a cooperative venture among nations, including all sectors of society (government, non-governmental organizations, industry, academia and the public). The Commons would include data, tools and infrastructure.
2. Open Information Culture – Decisionmakers and scientists need to work toward creating an open information culture where the needs of the scientists and generators of data are valued. The culture needs to embrace the accessibility of scientific data to the many potential user communities including policymakers, resource managers, researchers, students, teachers and citizens. Stronger interfaces must be built among data generators, managers and users.
3. Common Data Definitions – Scientists and users need to work on globally common data definitions so compatible information can be

SESSION IV: Acid Rain Linkages - Ozone, Hazardous Air Pollutants, Particulate Matter, Eutrophication, and Climate Change

Tamara Blett – National Park Service

My name is Tamara Blett with the Air Resources Division of the National Park Service, and I am pleased to be here with you this morning to begin the next session, Session IV, Acid Rain Linkages.

This morning we are going to explore sources of impacts of other air pollutants and air pollution-related phenomena in conjunction with acid rain. USA Today has been kind enough to help introduce this topic, by printing an article this morning on linkages between ozone, PM and sulfate. I think it is a very timely topic.

We have some excellent speakers ready to assist with this discussion. Dr. Paulette Middleton of the Rand Corporation will be talking about sources of pollutants, and Dr. James Galloway from the University of Virginia will be discussing impacts from those pollutants. Ozone, hazardous air pollutants, particulate matter, eutrophication and climate change are all extremely complex topic areas in their own right, and it would certainly take much more time than we have here to cover them thoroughly.

It is important to view acid rain precursors, such as SO_2 and NO_x, in context with other pollution sources and to address acid rain impacts in context with other air pollution impacts society must face, like eutrophication of waterways and climate change. Scientists, land managers, policy makers, environmental advocacy groups, and the public are going to be increasingly involved in dialogue about the way in which source emissions reductions can be made to mitigate individual and cumulative impacts from all types of air pollutants.

These impacts have serious implications for the life support system on which this planet relies, as they impact the web of life in the soil, air and water that sustains ecosystems in humans.

That is the context in which I would like to begin this session.

I would like to start this session by turning it over to Paulette Middleton from the RAND Corporation, who will be discussing an overview of the sources.

Linkages Among Sources and Impacts

Paulette Middleton – RAND Corporation

I appreciate the opportunity to participate in this timely and worthwhile meeting. And I am particularly thrilled that we will be hearing from members of Congress later about their views on these important environmental issues.

Presentation Outline

- Scope of Issues Related to Acid Deposition
- Chemical Connections and Challenges
- **The People Factor—Pollution Solutions** *(The "People Factor" refers to how we are part of the problem and the solution.)*

Let's begin by reviewing what we know about sources-impact relationships. As we will conclude, these relationships are complicated but they provide opportunities for addressing acid deposition and related environmental problems at the same time.

Scope of Issue Related to Acid Deposition

Multiple Spatial Scales

- Multiple Temporal Scales
- Multiple Chemicals Emitted From Multiple Sources
- Complex Chemical Relationships
- Impact Synergies
- Human Choices and Conditions Drive Emissions

These inter-relationships among sources and impacts including the scales and synergies and the key chemical players are illustrated in the air quality summary below.

Scales

- Local: 0-20km, impacts dominated by chemicals from near-by sources and occur on short episodic as well as longer time frames
- Regional: <2000km, impacts dominated by chemicals emitted from tall stack sources and occur mainly on longer time frames
- Global: impacts dominated by unreactive chemicals transported to the stratosphere where they react to deplete ozone and/or interact with solar radiation, causing changes in surface temperature and precipitation

Impacts

Human and ecological health, visibility, materials damage, land and water use/quality, economic and social well-being, weather/climate, and others (Figure 1).

Chemicals

- Emitted: Sulfur Dioxide (SO_2), nitrogen oxides (NO_X), ammonia (NH_3), volatile organic compounds (VOC—over 600 individual ones, some are highly toxic), carbon monoxide (CO), carbon dioxide (CO_2), nitrous oxide (N_2O), methane (CH_4), Freons (CFCs), Fine and Coarse Particles (aerosols and particles both refer to atmospheric chemicals in liquid and/or solid phases—some aerosols contain toxic metals like mercury and lead. Fine are less than 2.5 microns in diameter)
- Formed: Ozone, Acids, Fine Aerosols, VOC (Precursors: SO_2, NO_X, VOC, NH_3)

Many emitted chemicals form other chemicals that also are harmful. In fact, some of the formed chemicals are even more harmful than their precursors.

Sources

- Major point sources (utilities, large industrial)
- Area sources (small industrial, agricultural,...)
- Mobile sources (e.g. autos, rail, aircraft, etc.)

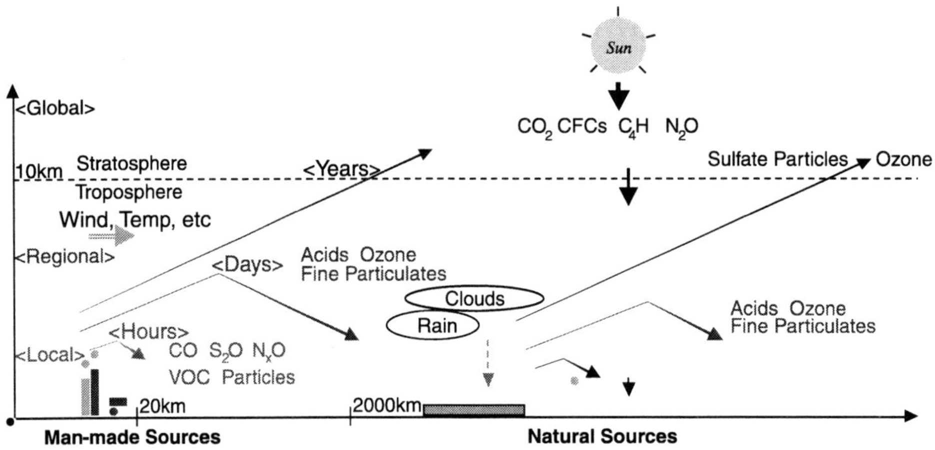

Figure 1. Air quality summary. ***Impacts:*** *human health, visibility, climate change, ecological, materials, land/ water use/quality, economic, and social.* ***Emission Driving Factors:*** *changes in human behavior, social, and political conditions, land and water use, natural processes, economic. Arrows indicate direction of interactions and spatial and temporal scales of processes noted by < >.*

- Natural sources (wildfires, trees, lightning, soil, volcanoes, sea spray,...)

Most categories contribute to most major pollutants, **but** relative importance of each source varies for each pollutant.

Relative Importance of Natural Sources

- Important contributors to: VOC, NH3, Particulates, climate change related gases (CO_2, Methane)
- Less Important to: SO_2, NO_X
- Not at all important for: CFCs

Natural sources make emission management challenging for most major concerns (Figure 2).

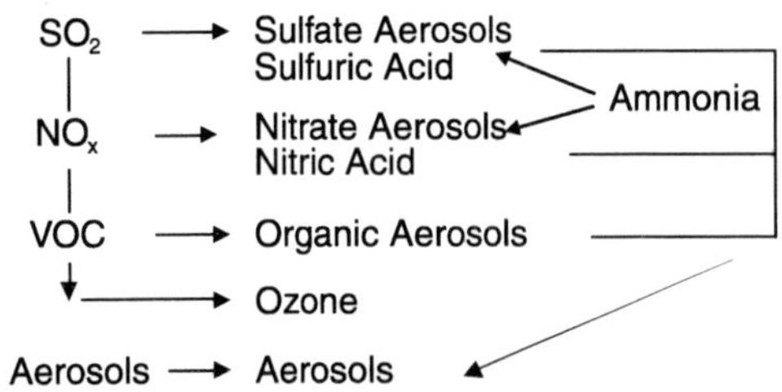

Figure 2. Atmospheric chemical complications - source-impact relationships are not simple. Emitted chemicals form chemicals. Aerosols are particles. Some VOC and some metal aerosols are on the "toxics" list.

Chemical Connections and Challenges

- SO_X/NO_X/VOC/Aerosols
- SO_X-NO_X
- VOC-NO_X

The basic set of pollutants, the "Gang of Four", are sulfur gases (mainly SO_2), nitrogen gases (mainly NO_x and NH_3), carbon gases (over 600 VOCs) and Particles (also called aerosols, mixtures of condensed sulfur, nitrogen and carbon gases and dust, minerals and other directly emitted aerosols)(Figure 3). The chemistry leading to the production of harmful products—acids, oxidants and fine particles (aerosols) is complex and inter-related making it difficult to make direct links between specific sources and specific impacts.

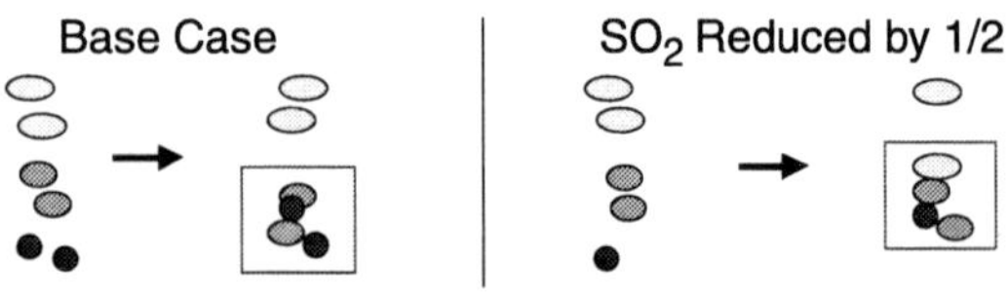

Figure 3. SO_X - NO_X - NH_3 air chemistry. Same amount of aerosol is formed even if SO_2 is reduced assuming O_X and NH_3 are constant.

This illustration shows how SO_2, NO_X and NH_3 are closely inter-related. Because of the chemical properties, NH_3 will stabilize sulfate first and then nitrate. When there is less sulfate available, NH_3 will stabilize more nitrate. The end result is that there can be the same amount of aerosol (acid or full neutralized aerosol) produced even if SO_2 is reduced, assuming that NO_X and NH_3 remain constant.

Implication of Sulfate-Nitrate Chemistry: Acid Rain Lesson and Challenge

- Lesson: Assessments show that acid deposition reduction is not as high as expected and that this is because NO_X emissions have not declined at the same rate as SO_2 emissions
- Challenge: Dealing with NO_X emissions means addressing mobile sources as well as major point sources

The Acid Rain Lesson and Challenge is illustrated in the following comparisons of changes in acidity (pH) and sulfate deposition from 1994-1997 (Figure 4).

- High reactivity of some naturally emitted VOC complicate emission management strategies aimed at reducing ozone (Figure 5)
- Role of selected VOC in fine particle production is important in many areas
- Relationship between VOC and NO_X emission management in reaching ozone reduction goals remains a challenge.
- Quantifying levels and locations of toxic VOC emissions is an important task.

Implications of S-N-C Chemistry, Emissions and Toxicity for Dealing with Aerosols

- SO_2-NO_X-NH_3 Chemistry must be considered when assessing pathways to formation of fine particles.
- NH_3 emission inventories need to be better understood.

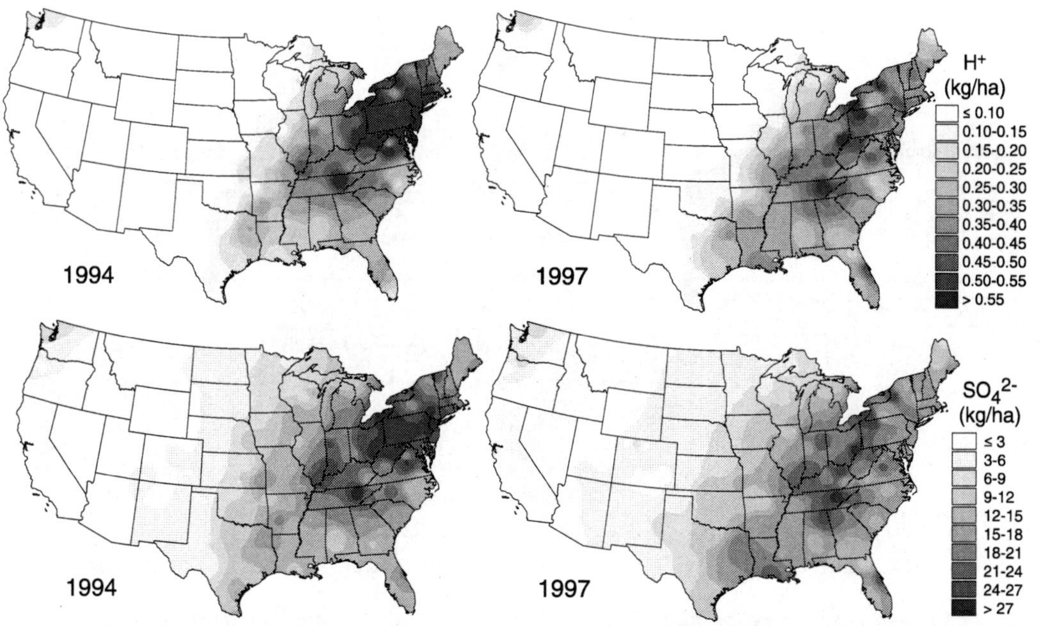

Figure 4. Examples of SO_2 - NO_3 complexity impacts on changes in pH and changes in sulfate deposition (1994-1997). Implications of VOC-NO_X chemistry for dealing with ozone and other issues. Source NADP/NTN.

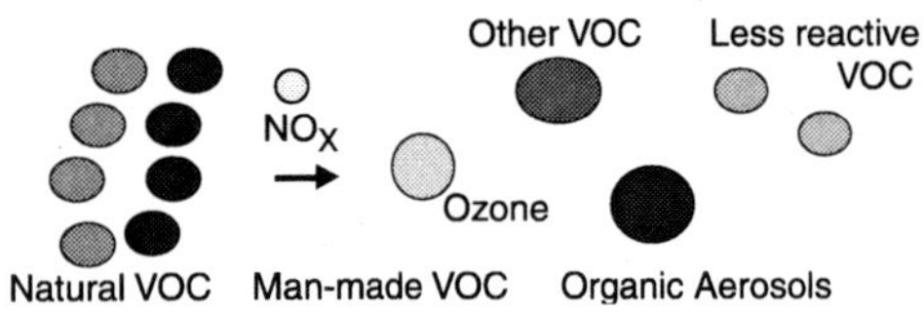

Figure 5. VOC - NO_X chemistry. The over 600 individual VOC differ in the relative importance of natural versus man – made sources, their reactivity (e.g., some VOC interact with NO_X to form ozone, some VOC lead to formation of toxic VOC), aerosol-forming capability and toxicity.

- VOC sources and reactivity differences mean that VOC classes and some key individual VOCs must be taken into account separately when assessing the role of VOC emissions in particle formation.
- Natural emissions of particles (e.g., wildfires, dust, etc...) are important and need to be properly quantified.
- Characterizing relative toxicity of all of the aerosol components remains an important challenge.
- Emissions, transport and fate of mercury and other highly toxic metal aerosols need to be better characterized.

People Factor – Pollution Solutions (Figure 6).

- Personal choices and emissions
- Energy and the Environment Example
- Toward a better life around the world

The relative benefits and costs of different choices made by individuals, communities, states, tribes, regions, nationals and international decision making groups can be illuminated by assessments that adequately characterize all of the important relationships that need to be considered (Figure 7). The National Acid Precipitation Assessment Program was one of the first major integrated assessment efforts. Such assessments are continuing to improve and capture more completely the complex interplay of human activities, natural processes and environmental and climate change.

Other issues, such as the use of environmentally harmful fertilizers in food production, and interconnections among these complex and interrelated issues need to be addressed together and throughout the world.

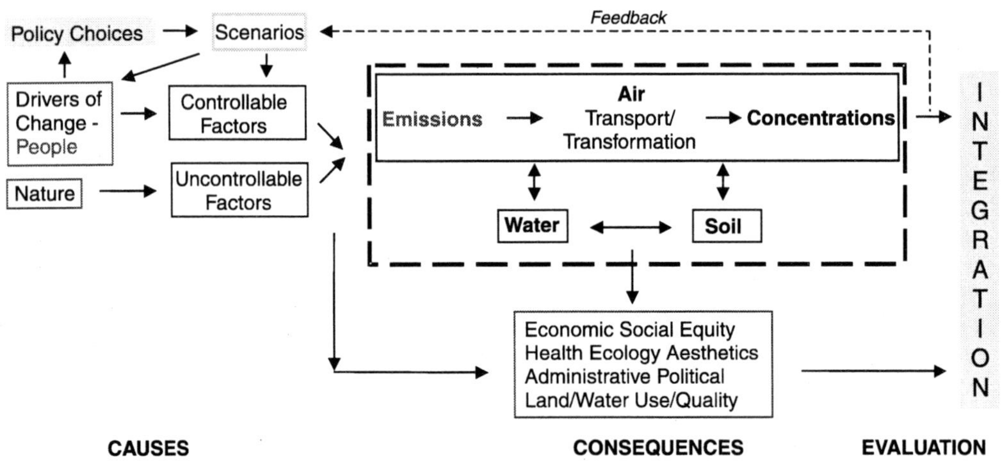

Figure 6. The people factor and emissions assessment system. People drive change through choices regarding energy use, food production, transportation industrial processing, etc.

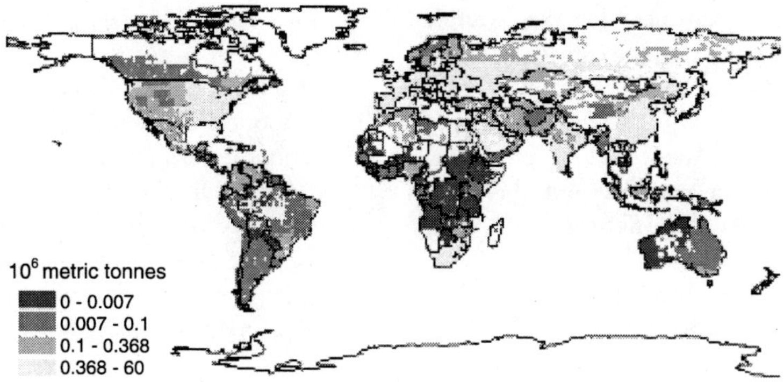

Figure 7. The global perspective - energy and the environment (1990 annual CO_2 emissions). Dealing effectively with the environmental impacts of fossil fuels as an energy source is a global concern. CO_2 emissions are a key indicator of fossil fuel. Ref: Global Emissions Inventory Activity (GEIA) at geiacenter.org.

Building Capacity to be Sustainable Worldwide

- Create sustainable strategies worldwide
- Communicate and enjoy positive impacts

Summary

- Understanding the relationships among sources, impacts, and driving forces in the US and worldwide is an ongoing process.
- Chemical complexity provides an opportunity for addressing multiple issues with the same emissions management strategy.
- All of us, as the fundamental drivers of change, can help form the truly sustainable strategies for the future.

For more discussion on these topics, please contact me at Paulette@rand.org.

Questions

[Unrecorded questions]

MR. LEWIS: I am Preston Lewis. I am with the New York State Department of Environmental Conservation. Would you tell us a little bit more about your thoughts on CO_2 and its influence on this whole mess?

MS. MIDDLETON: We can reflect on use of fossil fuel and the fact that it gives rise to CO_2 and the sulfur and nitrogen compounds that lead to acid rain. And fossil fuel also contributes to toxic chemicals of concerns such as mercury.

I think it is very important, from the standpoint of the climate change issue, for us to explicitly look at these issues and impacts associated with these multiple pollutants together. This is the intent of the 4-pollutant legislation.

Climate change and air quality are intimately connected. Even if there are uncertainties about the exact changes that might occur due to climate change, we definitely can note that addressing fossil fuels will have positive environmental effect in the near term as a result of reducing the other harmful pollutants associated with fossil fuel combustion So, we will be hedging our bets against undesirable climate change.

MR. POIROT: Rich Poirot, Vermont Air Program. Paulette, could you comment on the role that sulfate and nitrate and organic aerosols might play as a counterforce to global climate change?

MS. MIDDLETON: This is a good point to raise. Aerosols might be considered a counter reactant to global warming since they can lead to cooling. Having said that, I think the last thing any of us would want to do is advocate filling up the air with lots of aerosols to ameliorate climate change. Getting back to the big inter-connection picture, we must keep in mind that aerosols have several very important adverse effects that override possible "positive" impact on the climate side. There are tremendous health impacts. There are the important effects on visual air quality at national parks and wilderness. And, of course, we must remember that the gases that form the secondary aerosols, that is the SO_X and NO_X and VOC, are very harmful themselves.

MR. FRANKEL: Richard Frankel, GAO. Last year I heard that the Norwegians were developing a technique for controlling NO_X by doing an air separation and doing their combustion with pure oxygen.

What do you think about that? They are talking about doing it with gas, but what do you think about its adaptability to other fuels?

MS. MIDDLETON: I am not familiar with it. I would like to talk to you about it. I would say anything that is an innovation that gets us one step closer to our overall environmental goal by not creating other problems, is definitely worth looking at.

Overview of the Impacts

James Galloway – University of Virginia

In the beginning, there was acid rain, millions of years ago. Then it got better. Then it got worse as people came along. Then it got better, especially in 1990, with the Clean Air Act amendments. Now we see it is getting somewhat worse, as we increased our understanding of the effects of acid rain and, most of all, the linkages of acid rain with other issues. So, linkages among sources, agents of change and receptors, effects in parallel versus in series – I will come back to that. I am going to be talking about the future of space – i.e., there is more to the world than the United States – and what nitrogen mobilization as a pollutant of choice is going to look like in 2050.

In the following illustration Sources are on the left (Figure 1). Agents of change in the middle and Receptors are on the right.

We know about the effects of acid rain on fish and forest. We also know about the effects of ozone on two receptors, people and trees. In fact, if this were a time series, I would put the ozone one up first, because the photochemical smog in urban areas was the initial human health focus of this country, and the effects of ozone on vegetation was secondary

Somewhat surprisingly to me, a few years ago I realized, from other people's work, that there were significant effects of ozone on crops in the United States, and in China. There were estimates made of the dollar loss on agricultural productivity because of ozone damaging crop receptors.

SO_2, NO_X, and ozone, these are the effects and the linkages.

- As Air Pollutant:
  - o Adversely affects cardiovascular & respiratory health of humans, causing illness and death
  - o Increases oxidizing capacity of the atmosphere
- As Wet and Dry Deposition:
  - o Contributes to the acidification of lakes and streams
  - o Leaches nutrients from forest soils
  - o Damages buildings, monuments, bridges and auto paint
  - o Contributes to adverse health effects of nitrates in drinking water (blue-baby syndrome and potentially non-Hodgkin's lymphoma)
  - o Damages forests and crops
  - o Contributes to coastal and estuarine eutrophication

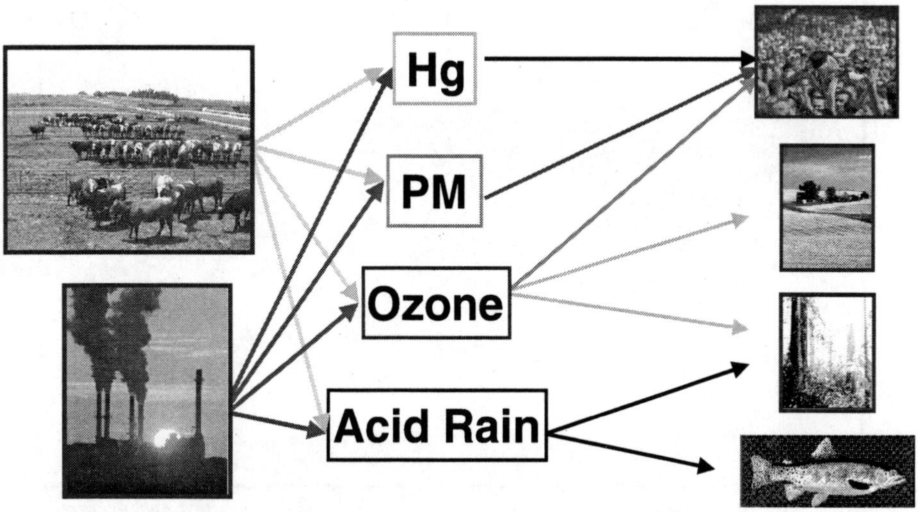

Figure 1. Sources, agents and receptors. Food and energy production are the drivers.

You can see that, as air pollutants affect human health and the oxidizing capacity of the atmosphere; the more ozone there is, the more chemical reactions will be sped up. Then when SO_2 interacts, you have this whole list of effects. I want to just draw your attention there to the crops and of course to the human health effects of the atmosphere.

In particulate matter we have, big chunks and small chunks, greater than 2.5 and less than 2.5 median diameter. Clearly the significant receptor for particulate matter is the human lung, the human respiratory system in combination with ozone, causing respiratory distress, as I understand was discussed expertly yesterday.

There are also the other particulate matter effects. This is a comparison figure from Cliff Davidson (Figure 2). There are mountains on the right and there are mountains on the left, but you just can't see them because of the particulate matter. We have all seen slides like this. We all know that, on a local basis, increased aerosol loading in the atmosphere has impacts on the visibility.

What we may not realize is that this is regional in scope. It is not just the urban areas. Visibility impairment occurs over large regions and, depending on where you are, is driven by organics, nitrates and sulfates (Figure 3).

With SO_2, NO_X and PM2.5 effects and linkages, we have air pollutants with health issues and the visibility issues. So, we can't talk about the impacts of SO_2 or NO_X on just health. We have to take into account the visibility aspects.

Figure 2. Visual impacts of particulate matter in the atmosphere as viewed from Cal Tech. Photo by Cliff Davidson.

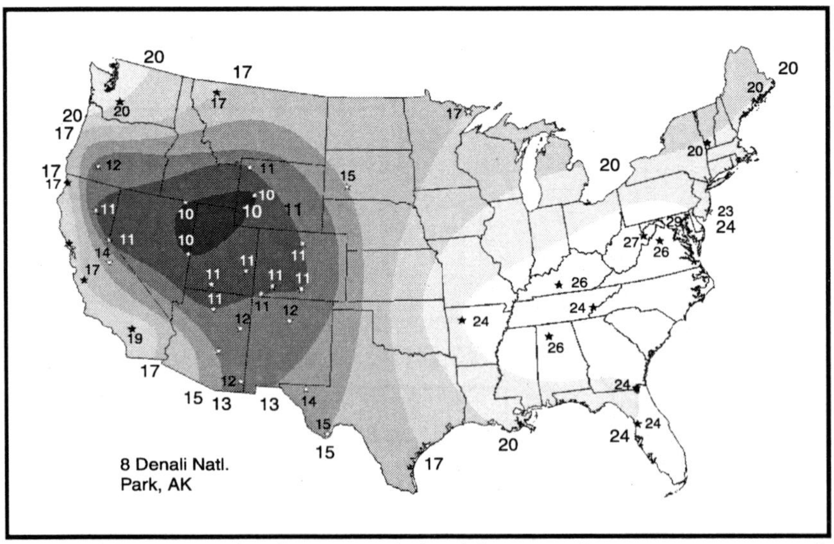

Figure 3. Average visibility, expressed as deciviews, calculated from aerosol concentrations measured in the IMPROVE monitoring program.

- As Air Pollutant:
  - o Adversely affects cardiovascular and respiratory health of humans, causing illness and death
  - o Impairs visibility and contributes to regional haze
- As Wet and Dry Deposition
  - o Contributes to the acidification of lakes and streams
  - o Leaches nutrients from forest soils
  - o Damages buildings, monuments, bridges and auto paint
  - o Contributes to adverse health effects of nitrates in drinking water (blue-baby syndrome and potentially non-Hodgkin's lymphoma)
  - o Contributes to coastal and estuarine eutrophication

Then, once these things leave the atmosphere, you have this sequence, cascade – I will be coming back to the word cascade, the concept of cascade later on – what happens when they are deposited to managed and unmanaged ecosystems.

Mercury, I have learned a lot about mercury in the last few days. Increased mercury emissions to the atmosphere from incineration and fossil fuel combustion impacts human health indirectly through the consumption of food and perhaps directly through inhalation.

- Similar emissions sources: in 1995 coal-fired utility boilers = 33% of Hg emissions; all coal-fired boilers = 46%
- Correlation between decreasing pH and increasing fish tissue methylmercury levels
- Primary human exposure route – fish consumption
- Fish advisories – 41 states; over 60% of US waterbodies
- Primary exposure impact – neurological impairment, particularly to developing fetus and young children
- NAS Study – up to 60,000 children born each year in US with Hg related neurological problems

You see that these four agents of change are impacting a series of receptors, from the natural to the managed to the actual human species. The linkages among SO_2 and NO_X and mercury are many. The point is that we have to be concerned about not just the sulfur and nitrogen that goes in the atmosphere but also mercury and, I venture to say, other trace metals that are acting together to change the status of the receptor, whether it is a lung, a forest, a freshwater or an agricultural crop.

Let's talk about the sources of these materials. I got my start in acid rain in 1974, really, and I would say it was power production, SO_2 and NO_X, that was contributing to acid rain. As our knowledge matured about this, we saw the linkages with ozone, particulate matter and, of course, mercury hitting a variety of these receptors.

We have to take into account the Haber process. Fifty percent of the world is here today because of this process that first went into industrial production in 1914. It is interesting to note that this process was created with food production as a secondary driver. It was Germany's need for munitions in World War I that really drove the Germans to develop, on an industrial scale, this process, a very interesting statement about society's need for resources, food versus weapons.

The basic reaction, N_2, hydrogen to ammonia ($N_2 + 3H_2 \longrightarrow 2NH_3$), is a very simple reaction. It took a long time to develop. In fact, I think it was 1780 nitrogen was discovered as an element, and it wasn't until 1880 that the significance of nitrogen in plant metabolism was understood through the nitrogen fixation process.

In 1898, Cook, who was the head of the British National Academy of Sciences said, the civilized world was running out of nitrogen. They were running out of the guano deposits and the nitrate deposits, primarily from the Pacific. So, they couldn't grow enough food. The point was, essentially two centuries ago, before all of us were born the world realized that nitrogen was limiting to the sustenance of people. So, it is very fortuitous that this reaction was developed. That is what allows us to mobilize nitrogen and put it in the environment.

Now, it is not just growing grain. This is the traditional method of making protein from the plant's resources. A cow with a rumen is able to take cellulose that we can't digest, and then convert that into milk and cheese that we can digest. So, a cow feeding on grass is really a protein generation factory. However, there are very few of the western world's cows, pigs, chickens, turkeys, et cetera, that use grass – essentially a resource we can't use. They mostly use grain and they are mostly fed in feedlot operations.

So, we add agriculture to this (Figure 1). This is key. Note those green arrows. The production of food increases the acidity of the atmosphere. The production of food increases the rate of acidification of our ecosystems. The production of food increases ozone concentrations in the atmosphere. The production of food increases $PM_{2.5}$ in the atmosphere. The production of food, primarily through mercury coating on seeds, increases mercury in the environment.

So, when we talk about linkages, when we talk about acid rain and what it is related to, we can't just talk about energy production and high temperature combustion. We have to talk about agriculture. To separate the two is a tremendous scientific mistake and, I would venture to say, a tremendous policy mistake. It has to be a holistic approach to nitrogen, sulfur, ozone, particulate matter and mercury.

I will stop preaching right now. Okay, the parallel linkages between sulfur and nitrogen, ozone, et cetera, et cetera. There are also effects in series.

- SO_2, NO_X, O_3, PM, and Hg have effects in Parallel in the atmosphere and in ecosystems.
- There are also effects in Series as N and S move through environmental reservoirs.
    - o The Sulfur Trickle
    - o The Nitrogen Cascade

I am going to, I hope, prove to you that sulfur is really quite a boring element and so is carbon. Nitrogen is something that, as a scientific body and as a policy body, requires significantly more attention.

The sulfur trickle. You convert sulfur to SO_2 in the atmosphere and there are significant health effects for sure. Once sulfur is deposited to an ecosystem, the major effects are in soil acidity increases, forest productivity decreases and increases.

Combustion converts reduced S to SO_2. The same sulfur atom in SO_2 can, in sequence:

- decrease atmospheric visibility/cool atmosphere
- increase soil acidity
- decrease forest productivity
- increase forest productivity
- decrease stream/lake biodiversity in headwaters

In 1976, Ellis Cowling told me," Jim, sulfur deposition to southeastern forests is good, because those forests, especially some in North Carolina, are sulfur deficient." So, sulfur deposition to forests can increase forest productivity.

Then, of course, as you acidify a soil, as you acidify the fresh water, you lose the biodiversity of those systems. After the headwater stream, sulfate essentially goes off our radar as far as changes in our environment, no river effects, no coastal effects, no more atmospheric effects, because it is essentially locked up in the oceans as sulfate, perhaps a very small recycling to DMS, but very small compared to the sulfate load.

We put all of them into the atmosphere and, in sequence, the nitrogen, as NO, can increase atmospheric ozone and decrease visibility before converting back to gaseous nitrogen.

Humans convert N_2 to NO_X. The same nitrogen atom in NO_X can, in sequence:

- increase atmospheric ozone
- decrease atmospheric visibility
- increase soil acidity
- increase forest productivity
- decrease forest productivity
- decrease stream/lake biodiversity
- decrease quality of drinking water
- increase coastal eutrophication
- decrease coastal biodiversity
- increase greenhouse warming potential

Before converting back to N_2

Now if, for every molecule of N_2 that we converted to NO_X or ammonia was somewhere along this process converted back to N_2 through the process of de-nitrification, that would be okay, but it is not. Reactive nitrogen from food and energy production is accumulating in the environment.

I want to expand on this cascade effect (Figure 4). So, we have energy production. You are fixing nitrogen. You are making reactive nitrogen. Humans, on a daily basis, create more reactive nitrogen than do natural terrestrial processes. We have overtaken nature in making nitrogen mobile in the environment and available for this cascade of effects.

So, you have ozone and then you have particulate matter with the health effects and visibility effects and you have a whole cascade of NO effects on forests, then on the soil, then groundwater and then stream. As nitrogen travels through these reservoirs, it has effects along the way.

There are residence time issues here. Nitrogen may stay in forests for a while. Once that forest decomposes, nitrate becomes mobilized and goes into the ground water and the stream and the river, and there are associated effects of that nitrogen.

Next, we have river effects, coastal effects, and coastal eutrophication. Some of this nitrogen in coastal ecosystems provides a tremendous service for us. About 90 percent of the nitrate, of the nitrogen that enters coastal ecosystems is denitrified to N_2, and very little reaches the open ocean. There is very little impact on the open ocean because our coastal ecosystems are acting as a filter, a sponge, and a barrier for that nitrogen to get in the open ocean.

While it is there, it can cause coastal eutrophication and, in the process of denitrification, a small amount of N_2O is formed which will have greenhouse effects and then stratospheric effects, of course, with ozone depletion in the stratosphere.

So, that is the nitrogen cascade from energy production. Now, you are probably guessing there is going to be an agricultural side to this.

Fertilizer production, the Haber process, creates ammonia. In fact, when you talk about creation of reactive nitrogen on a global basis by human action, about 80 percent of it is due to

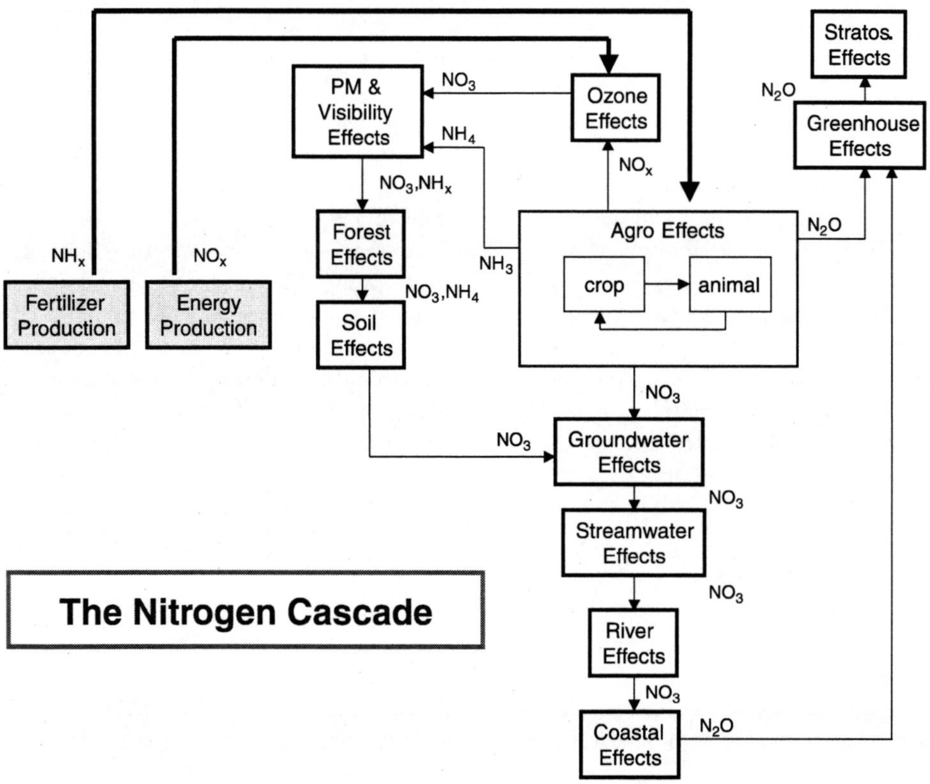

Figure 4. The nitrogen cascade.

food production and only 20 percent due to energy production.

This is a very important distinction to realize when we are concerned globally about nitrogen and its cascade and the linkages with other processes.

Of course, in the agricultural ecosystems, there is a recycling inside. The nitrogen goes to raise grain fed to animals, some of the residues go back to the crop, but there are four areas of leakages from agro ecosystems.

Nitrate goes into groundwater and then, once it gets through groundwater, you have this cascade of effects.

Then, ammonium enters the atmosphere. Then, once it is in the atmosphere, there are the effects on particulate matter concentrations and visibility, and then it goes into the terrestrial ecosystem where it can be quickly nitrified to nitric acid and then, again, cascade through these systems.

Once you hit the terrestrial ecosystem or the water body, it doesn't make any difference whether that nitrogen was from food production or energy production. The cascade continues.

The other NO impacts from fertilized fields are NO_X, and the impact on atmospheric ozone. Some people think that, on a global basis, the amount of NO emitted into the atmosphere from the agro ecosystems and other systems by the processes of nitrification and denitrification is greater than NO put into the atmosphere from combustion.

A few think the numbers are large enough for the two processes, energy production and food production to be about the same source strength as NO being emitted into the atmosphere on a global basis. Then we have N_2O emitted into the atmosphere from agro ecosystems and the same type of effects.

By the way, just let me say that every molecule of N_2 fixed by fertilizer production, every molecule of N_2 fixed by energy combustion or energy production reaches the environment eventually as a reactive form.

Yes, we eat nitrogen, but you know what happens to the nitrogen we eat. We don't care that we get all that atmosphere. It goes somewhere, and it is in the environment as a reactive nitrogen form and it is accumulating.

The future. This is the future. This is the past in the United States. These are SO_2 and NO_X and ammonia emissions, the limited data we have (Figure 5).

The point I want to make here is that this is the time line for SO_2 emissions. This axis is a little screwed up, 1860, the red line is 1980. This is the natural emissions, that little green bar.

It is great that there is less SO_2 going into the atmosphere in the United States because of the regulatory activity, but the change is small, recently, relative to the differences. Our ecosystems are changing now. My projections are that this is going to go up slowly, and I will show you later on why, on a global basis, that is relatively trivial.

Note the differential between total emissions and natural emissions. We are controlling the emissions of NO_X into the atmosphere in the United States.

Ammonia is the ignored nitrogen molecule, at least from the U.S. Government point of view (Figure 6). Our information on ammonia emissions is very poor, but we know enough to know that, again, this differential, ammonia emissions are large relative to what has occurred naturally. My projections and others show that this number

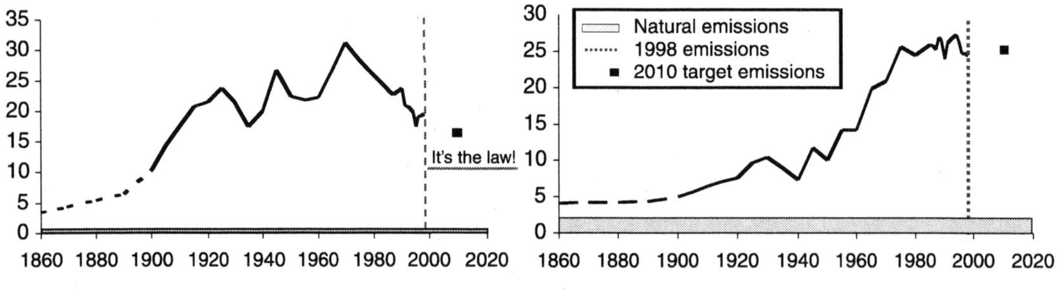

Figure 5. SO_2 and NO_X emissions for the United States (million short tons/year).

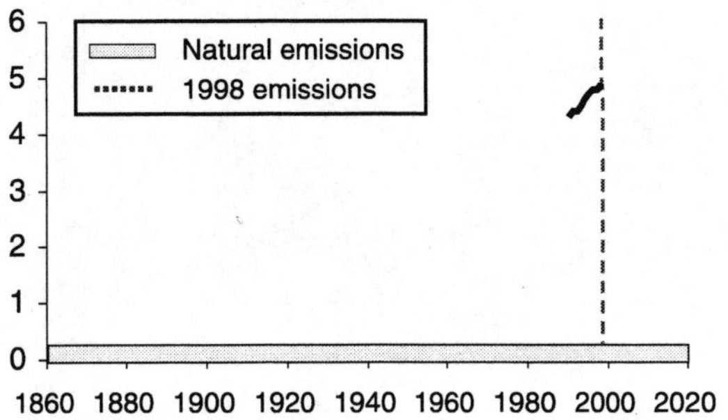

Figure 6. Ammonia emissions for the United States (million short tons/year).

is probably going to keep on going up in the United States.

I think, speaking of the western world, here are some projections for nitrogen mobilization rates (Figure 7). Now, this is not atmospheric emissions. This is the creation of nitrogen through production of fertilizer and production of energy. The point is, that from 1990, to 2020, there was a slight increase in the United States, a slight increase in Europe. However, keep in mind that as far as the world population goes, we are a pretty small piece of the pie.

These are the projected populations of North America in the year 2020, but look what is happening in the world (Figure 8). Most of this will be occurring in Asia. It has 60 percent of the world's population now. That percentage will be about the same in the year 2020 but there are going to be a lot more people. So, let's put Asia into the picture. The production of nitrogen as a reactive element, in the year 2020, is going to be driven by Asia (Figure 9)

I am delighted that we have this conference

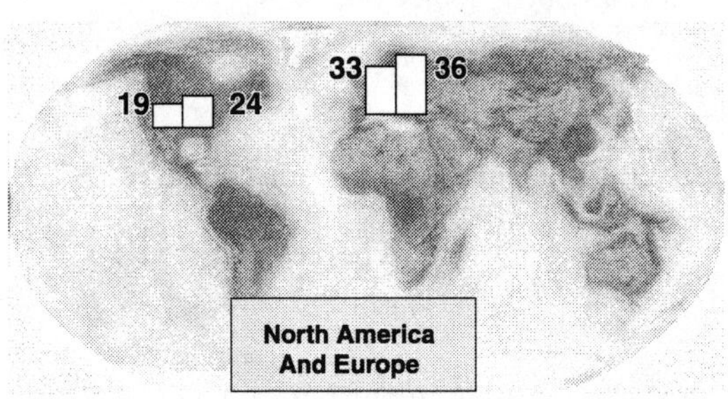

Figure 7. N mobilization rates (Tg N/yr) for North America and Europe. Values for 1990 (left) and estimated for 2020 (right).

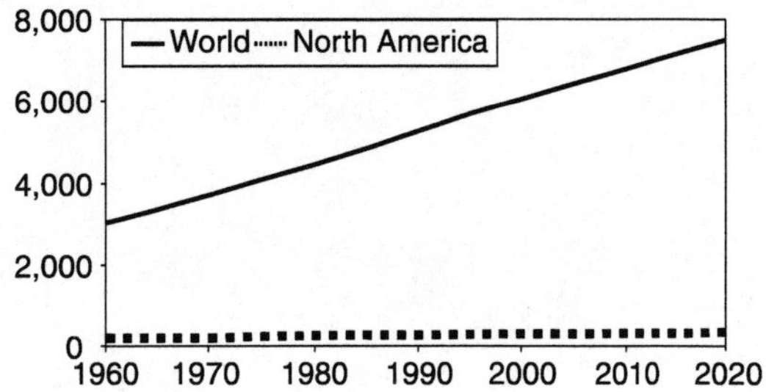

Figure 8. Global and North American populations (millions).

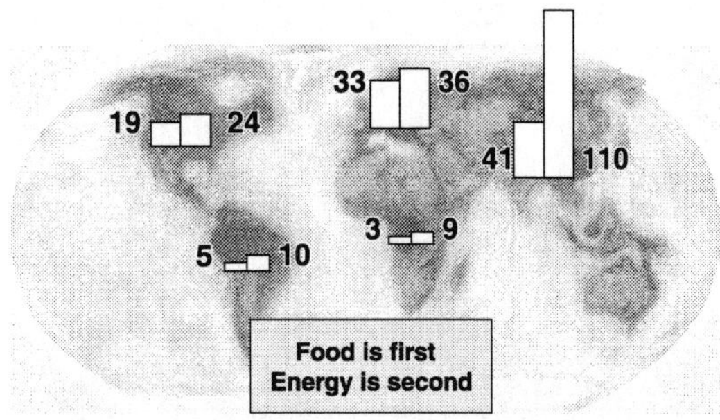

Figure 9. N mobilization rates (Tg N/yr) for the continents. Values for 1990 (left) and estimated for 2020 (right).

here to be talking about nitrogen in the United States and sulfur and the linked impacts. As a global community of scientists, policy makers and citizens, we would be remiss if we did not put the United States in the context of the global world.

If you are a biogeochemist, go to the Pacific Ocean and see what is coming off the Asian continent.

If you are a policy maker, I would spend some time in China and India, learning from them and perhaps showing them what we have done right and wrong in this country to regulate the emissions of nitrogen.

Asia is the driver, because food is first. You have to feed.

There are currently 1.2 billion people in China. Food is not a luxury there like it is here.

This is what the atmospheric deposition of nitrogen is going to look like. This is the result of global output by Frank Detner of Europe (Figure 10). These are the 1990 depositions of nitrogen, wet and dry, ammonia and NO_X, and NOI.

You can see the growth of deposition in Asia. There are some slight changes in North America and Europe but, again, if you are concerned about

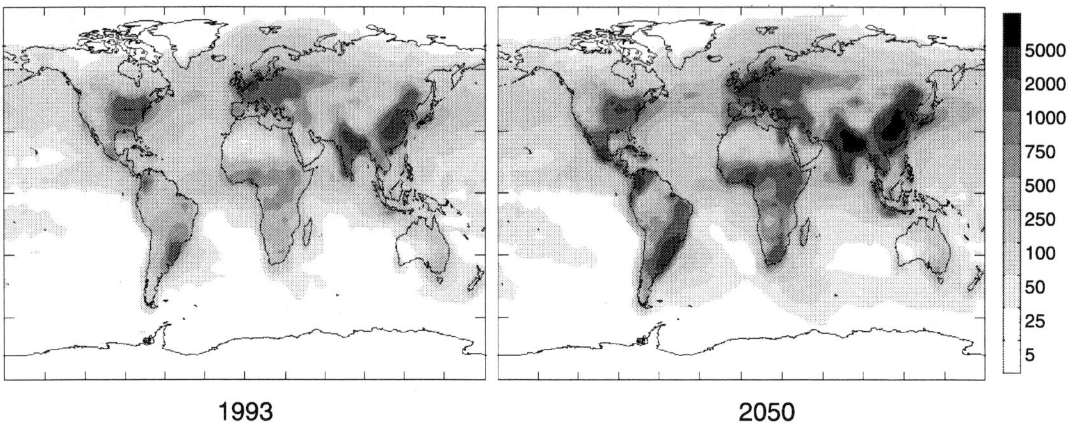

Figure 10. Nitrogen deposition - present and future (mg N/m²/yr) (Dentener, 2000).

nitrogen coming out of the atmosphere, if you are concerned about sulfur coming out of the atmosphere, if you are concerned about ozone forming in the atmosphere, particulate matter forming in the atmosphere, what we are going to see in the United States is probably pretty much business as usual, the same thing in Northern Europe, but Asia is where you are going to see significant changes.

Summing up:

- The production of food and energy have increased the concentrations of S, N, O_3, Hg and particulate matter in the atmosphere.
- The increased concentrations have significant health and ecosystem effects that are linked, just as their sources are linked.
- The linkages can occur in series, especially those involving nitrogen.
- The forecast for the future includes increased environmental accumulation of Hg, S and N species in the United States and significant increases in emissions and accumulation in the world, with concurrent detrimental health and environmental effects.
- Linkages among effects require a multi-pollutant strategy for control.

Linkages among effects require a multi-pollutant strategy for control. We can't just focus on ozone by reducing NO_X emissions in the summer. We can't just focus on SO_2 by reducing emissions to 40 percent of what it was in 1990. We can't just focus on mercury alone. We have to use a multi-pollutant approach to our science and policy. Thank you very much.

QUESTIONS

MR. KLEIMAN: My name is Gary Kleiman. I am from NESCAUM. I had a question regarding ozone's role in the oxidizing capacity of the atmosphere.

I am wondering, in your opinion, would it be a legitimate justification for controlling NO_X to say that it leads to ozone formation which, in turn, would increase the oxidizing capacity and accelerate the formation of sulfate and secondary organic aerosols.

MR. GALLOWAY: Yes, again, I would say that focus is too narrow about ozone, NO_X, oxidizing capacity. You have to, I think, once you put NO in the atmosphere, look to where it goes.

That is really the big message I am trying to get across here that NO in the atmosphere, after it does the ozone thing, is going to cascade through the environment and eventually end up in those coastal ecosystems.

MR. BULGER: Art Bulger, at the University of Virginia. I just want to make sure that I got this straight. You are attributing the quadrupling of the human population in one human lifetime to the contribution of fertilizer from the atmosphere.

MR. GALLOWAY: Yes, Bob Scheer(?) has just written a wonderful book, *Enriching the Earth, the*

Haber Process. It went to press in the year 2000 or 2001. In that book, the statement is that 40 percent of the world's population would not be here if fertilizer wasn't made.

He went on to say that in Asia that people actually need fertilizer for survival. In the western world, the fertilizer is really not that needed because our production is so much greater than our consumption.

MR. VALENTINETTI: Dick Valentinetti, New England State Directors. This is a question, I guess, for both speakers. Ammonia seems to be a real big player and our inventory of information is terrible. We really haven't focused on it.

How do we end up trying to focus on, besides putting out a lot of money, how do we end up focusing on inventories for ammonia and activity levels of things that produce ammonia?

MR. GALLOWAY: I don't know how you do that. That seems to me to be a resource allocation issue that a government has to make about gathering the basic data on ammonia emissions over a very complex landscape, natural and managed and unmanaged.

What my role, as a biogeochemist is, and I think the role of the atmospheric chemist is, is say, this is what ammonia does in the atmosphere and this is what ammonia does after it leaves the atmosphere, and this is how it relates to sulfates and to nitrates and to ozone, mercury, et cetera.

MS. MIDDLETON: I think there are going to have to be equally creative approaches to this food thing. I thought you did an excellent job of drawing that connection between the food and the energy, because we have got the ammonia coming from the food side.

There is where a lot of very creative thinking is going to have to occur, especially given the global situation.

MR. KAHL: Steve Kahl, University of Maine. How would you suggest the next clean air act be phrased relative to the balance of production between S and N?

MR. GALLOWAY: First of all, all sources of nitrogen have to be included. You have to have that. Then you have to take into account this cascade of effects for sulfur and for nitrogen.

Of course, there is a primary concern which is human health, but let's not forget about the secondary concerns, especially as nitrogen and sulfur accumulate in our ecosystems.

Then, after that, I am not sure what you do. If you are thinking percentages, what needs to be reduced the most, I think sulfur emissions – from a scientific point of view, if you reduce SO_2 emissions, you are going to see some recovery of acidified soils and lakes.

There is some recovery now. I am sure as Charlie told you yesterday it is not complete and in some cases it doesn't exist, it is just that if more of it goes up, more comes down. If you reduce what goes up in the atmosphere you are going to reduce what comes out of the atmosphere.

For sure, if you reduce sulfur emissions we will have ecosystem recovery, some. Of course, it depends on how much.

For nitrogen, you have to pay a lot more attention to ammonia emissions and this relationship between NO emissions from food production and NO emissions from combustion, especially if you take a more global view and look at Asia.

China is concerned with the increased emissions of NO from very heavy use of fertilizer and from increasing use of fossil fuels, making the ozone concentrations so large that their crop productivity is being decreased.

There, food, as I have said before, food production is much more important than here.

MR. TANG: Don Tang from the Center for Research, EPA. I think this is the most interesting presentation in this two day conference, but since you emphasized the food production, I have two small questions to ask.

One is, you emphasized the food production so much. The food industry is a major polluter. The other question is, you showed in your figure, 1,000 years of experience in production of nitrogen. I think we have more than that. One more thing – China's population is 1.3 billion now.

MR. GALLOWAY: Two very good questions.

The first question is, is the food production industry the world's largest producer. Is that right? That is a good question.

I would be very remiss if I didn't say that the food production industry is vital to our survival. Food is first. We have to have food.

We just had a workshop in Woods Hole last week in support of the second international nitrogen conference, which has already been mentioned as happening in October at the Bolger Center, and everybody please come.

The fertilizer industry was there. People from the food production sector were there. Their key message was, hey, we are providing a tremendous service to society by producing food very efficiently in this country. They said, we didn't realize that when we tell a farmer to put on so much fertilizer that a lot of it is going elsewhere. Our main concern is getting in the crop, and that is about 50 percent efficient.

So, yes, the food production industry is a polluter. We are a user, however. The reason that there is so much extra put on the crops in the United States has to do with our eating habits, first of all. I didn't eat meat for breakfast, but I had it last night for dinner and it was wonderful. That is a real driver for nitrogen globalization.

I don't want you to leave this room thinking that I am down on the food production industry. I appreciate them.

MR. LEWIS: Preston Lewis from New York State. The question I have is, why aren't we doing more modeling? We seem to be putting our finger on nitrogen and CO_2, but why aren't we doing more complete modeling of the system, all the aspects of the system, the food production, the people, the energy, and the world.

It would seem to me that mathematical modeling and simulation should be done here.

MR. GALLOWAY: That is a very good question. Let me give you a short illustration. Up until the work of Frank Detner, I showed you his nitrogen deposition figure now and in the year 2050. That was work about eight years ago, the atmospheric nitrogen models were of two types, those that dealt with NO and those that dealt with ammonia.

Yet, they weren't coupled. Part of the reason is because of what the scientists have done, but in part they have done it because of where we can get money. That goes back to the government.

Well, what do you want the money for? Well, we want it for regulations of NO. They want it for regulation of ammonia. They want it for regulation of SO_2.

I think one of the key things for this conference is that all these things are linked and we have to pay attention to them together. Thank you very much.

SESSION V: North American Regional Impacts Panel

Ellis Cowling – North Carolina State University

Our most important objective is to present and discuss information about pollutants and linkages and effects. What we are going to do in the next panel discussion is have an introduction into the nature of their effects on a regional basis.

Northeast

Richard Poirot – Conference of New England Governors and Eastern Canadian Premiers

I first became interested in and concerned about acid rain about 25 years ago, as a brand new employee for the Vermont air pollution control program. My first assignment was to write a state implementation plan revision to assure attainment of air quality standards for particulate matter in Vermont.

In those days we were regulating a pollutant called total suspended particulate or TSP which, as Dr. Lippmann explained yesterday, was a fairly primitive concept of particulate matter. Nonetheless, it was the state of the knowledge at the time, at least from a regulatory perspective. We had a bit of non-attainment in Vermont in those days (actually, it was the last non-attainment that we had), and it was measured at an urban monitor next to a heavily traveled city street.

Every spring, as the snow banks would melt away from the sides of the road, the sand that we had put down all winter long would blow into the monitor and we would get high TSP concentrations. We could examine the high concentration filters under an ordinary light microscope and actually see the huge grains of sand on filters, and we could analyze the filters chemically and find lots of silicon and aluminum and iron and calcium (we even had calcium in those days). We fixed that problem in the usual way – by moving the monitor a little bit further back from the roadway – and our air quality improved substantially.

But as we also got into this business of analyzing some of our filters, we realized that we had another period of high concentrations in the summer. When we analyzed those filters, there wasn't much sand on them, but instead we found an exotic new pollutant called "sulfate". When the concentrations were high in our largest urban areas, we found that they were high everywhere across the state, and were sometimes even higher at our most remote background sites.

This led us to start to look into the historical literature and as it turns out, we actually knew quite a bit about sulfate aerosols, acid deposition and so forth decades ago. We just keep forgetting and re-discovering this information.

These figures (Figure 1) showing health and visibility effects of sulfate aerosols, were taken from a federal criteria document on sulfur oxides

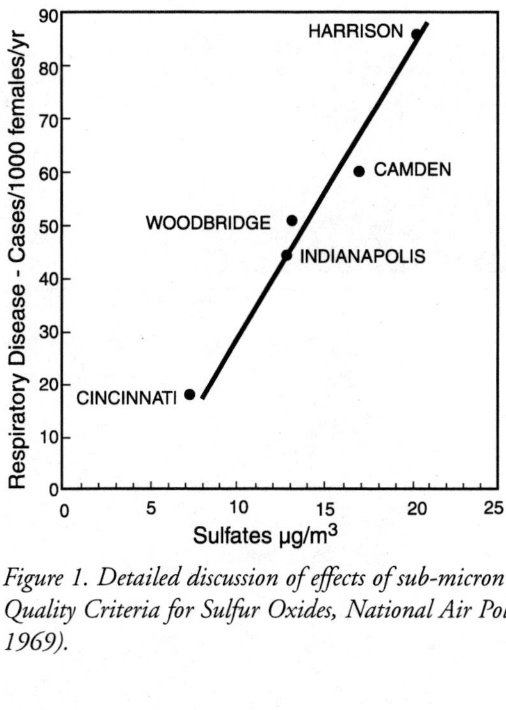

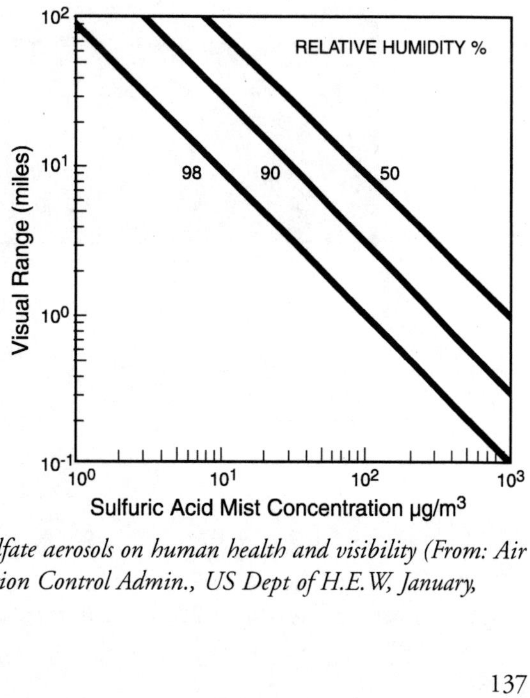

Figure 1. Detailed discussion of effects of sub-micron sulfate aerosols on human health and visibility (From: Air Quality Criteria for Sulfur Oxides, National Air Pollution Control Admin., US Dept of H.E.W, January, 1969).

published by the US Dept. of Health, Education and Welfare in 1969 – a year before the US Environmental Protection Agency was established. This 1969 document also included extensive discussions of the effects of sulfuric acid mist on metals, stonework and masonry (not as detailed as Mary Striegel described yesterday, but nevertheless indicating a thoroughly evolved scientific understanding of materials damage functions), as well as detailed discussion of the effects of sulfuric acid on vegetation.

The estimated health effects vs. sulfate relationship in 1969 was not exactly the same as the ones that Dr. Lippmann was showing you yesterday, but actually it is not all that dissimilar. The plot of sulfate vs. visual range, including effects of a hygroscopic aerosol growth function with increasing relative humidity, is conceptually very similar to the "modern" equations used to calculate "reconstructed light extinction" (from sulfate and relative humidity) which have been recently included in new EPA regulations to reduce regional haze in National Parks and Wilderness areas. Congress called for these visibility regulations in the 1970s; the scientific understanding of sulfate/visibility effects was reasonably clear in the 1950s; and now in 2001, we finally have a regulatory program in place – although these regulations will not be fully implemented for another 60 years.

As I began to discover a large body of historical scientific literature indicating that the aerosol conversion products of sulfur dioxide emissions had a wide variety of health, environmental and welfare effects, I was encouraged to learn that EPA, in 1973 Congressional hearings, indicated that they were working on an ambient aerosol sulfate standard, and that this would take some three to five years to develop. So when I started work for the State of Vermont in the late 1970s, I figured that any day now we were going to see a strong federal regulatory program to control these sulfur emissions. Just a matter of time.

The next implementation plan revisions that we had to develop in Vermont were in response to the visibility protection requirements of the 1977 Clean Air Act. In 1985 we tried to sneak in an acid rain control program as our State Implementation Plan to protect visibility in the Lye Brook Wilderness area. This cover of this 1985 document (Figure 2) illustrates a sulfate episode that occurred on July 17, 1982, with sulfate concentrations increasing from 1 to 7 to 14 to 35 micrograms per cubic meter over a 4-day period, and visibility declining to less than two miles. We had modeled this sulfate haze event using seven different

Figure 2. An early and unsuccessful attempt to protect visibility with an Acid Rain Control Plan hidden in a SIP Revision. VT's Proposed "Long-Term Strategy" to reduce visibility impairment was the 1985 "New England Eastern Canada Sulfur Dioxide Emissions Reduction Plan."

techniques that clearly indicated it was a midwestern sulfate transport episode. We proposed in this visibility plan, a VT annual average state sulfate standard of two micrograms per cubic meter, to be attained over a 10 year period, which would have achieved a roughly 50% improvement over early-1980's levels. We argued that this would represent "reasonable progress" toward protecting visibility as required under the Clean Air Act, and would have the "ancillary benefits" of reducing regional haze, fine particle concentrations and acid deposition throughout the eastern US and Canada. That is, given the large regional scale of the problem, improvements could not be seen in Vermont without improvements throughout most of eastern North America.

We also included as our long-term visibility strategy an international acid deposition control plan that had been developed by the New England Governors and Eastern Canadian Premiers in 1985. A key element of that 1985 international agreement called for national SO_2 reductions in the range of 50 percent from 1980 levels, by 1994, in both the United States and Canada.

Another key element of this international agreement was that it merged the "effects-based" approach that the Canadians have used, with emphasis on controlling specific sources to maximizing the benefits on acid sensitive regions, with the U.S. "emissions-based" approach, which is more of a broad based roll-back of utility sulfur emissions in the most cost efficient way (allowing a maximum SO_2 emissions rate of 1.2 pounds per million BTUs of heat input). A very significant aspect, in my view, of this early governors/premiers plan was not only proposals for upwind reduction but also commitments to reduce our own emissions, clean up our own house, put our money where our mouth was – regardless of what happened in upwind jurisdictions. Five years later, under the United States – Canada Air Quality Agreement, both countries did commit to significant national SO_2 control programs, which were not directly based upon, but were actually very similar to those proposed by the New England Governors and Eastern Canadian Premiers in 1985.

These national programs have, to a very large extent, run their course, causing the New England Governors and Eastern Premiers to pick up the issue again and take stock of how far have we gone, how far do we need to go. This led to a recognition that, while we made some progress, we were not seeing the kind of ecological recovery that we were interested in, and a number of additional issues had emerged.

In 1998, the Governors and Premiers signed another acid rain action plan calling for yet another round of reductions of at least 50 percent sulfur dioxide and annual NO_X reductions of 20 to 30 percent in both countries (Figure 3). As with their 1984 plan, this new agreement also included

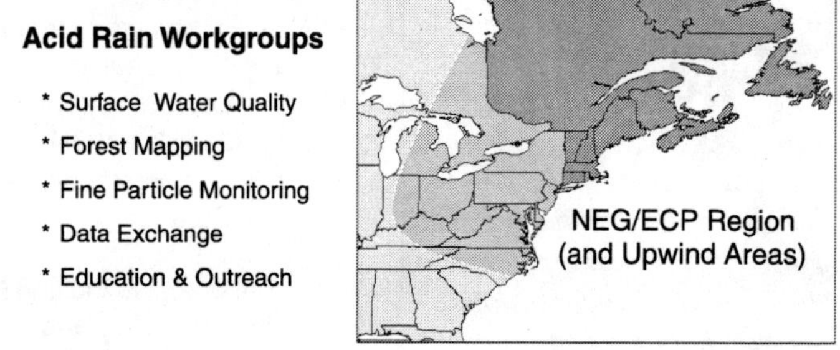

Figure 3. NEG/ECP Acid Rain Action Plan.

commitments to reduce our own emissions within this 11-jurisdiction transboundary region, regardless of what happens upwind. In addition, a number of technical work groups were established that are developing, and in some cases repairing or rediscovering, some of the historical data sets, and monitoring and research activities, that were much more prominent (and more adequately funded) during the "acid rain decade" of the 1980s.

One area of emphasis in this 1998 plan, and I think we have heard it emphasized by almost all the speakers so far, is the need to really look at the big picture and understand the multiple effects that are associated with these pollutants. You can jump into the issue any which way – call it acid rain if you like – but really, it is not very different from the visibility issue, the fine particle health issue, the effects on ozone and eutrophication of coastal estuaries and so on and so forth. So the plan specifically defines its focus as on "the multiple effects associated with emissions of sulfur and nitrogen oxides" or conversely, "the multiple benefits that would accrue with further reductions of sulfur and nitrogen oxide emissions" (Figure 4).

Sulfur and nitrogen oxides are certainly not the only atmospheric pollutants of concern, but from a northeastern perspective, they are sufficiently important, their effects are sufficiently adverse,

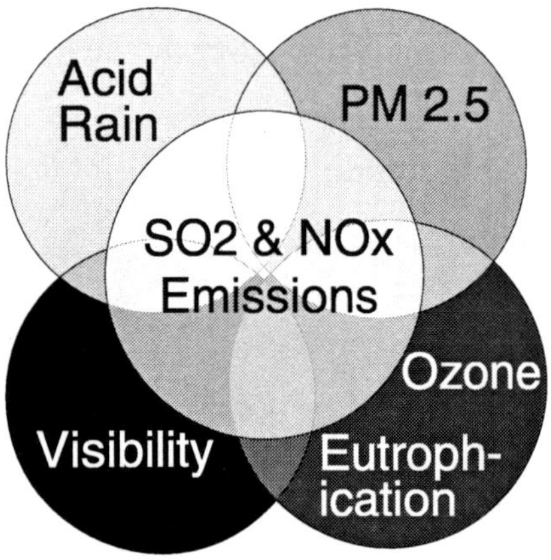

Figure 4. "Ecosystem sensitivity" refers to all the environmental effects associated with the emissions of sulfur and nitrogen oxides which cause acid rain... Or conversely, a consideration of all of the environmental benefits that would result from further reductions in sulfur and nitrogen oxide emissions.

continuing, cumulative and spatially widespread, and their control costs are sufficiently modest, that additional future emissions should be initiated without further delay. Within the New England/ Eastern Canada Region a number of substantial intra-jurisdictional emissions reductions commitments have recently been achieved (Table 1). Generally, most of these include commitments of additional sulfur reductions of 50 percent or more. Some of these programs include commitments for NO_X reductions, some include mercury and some include sulfur, nitrogen, mercury and CO_2 as well. There are also similar emission control commitments that are currently in the works or in legislation in areas directly upwind of us in New York and Ontario.

In addition to the need to extend these regional control programs to an international scale in the US and Canada, other areas where coordinated bi-national initiatives would be especially helpful include the development of emission inventories that are consistent across the United States and Canada, and that are consistent over time. I think this is kind of an international embarrassment that we really do not have consis-

Table 1.

NEG/ECP Unilateral SO_2 Emissions Reductions Commitments:

- Connecticut: 40% SO_2 by 2002 (0.5 lbs/ MMBtu), > 50% by 2003 (0.3 lbs MMBtu = 0.3% S in fuel)
- Quebec: 40% by 2002, 50% by 2010
- Massachusetts: 53% – 74% by 2008 (6 lbs/ MWH by 2004 or 2006; 3 lbs/MWH by 2006 or 2008, later dates for re-powering)
- New Brunswick: 30% by 2005; 50% by 2010
- New Hampshire: (multi-pollutant Bill currently in Legislature)
- Nova Scotia: (currently in Negotiation, proposed 50% SO_2)
- Similar commitments in progress from NY & Ontario

All of above also include NO_x reductions, some include Hg, CO_2 and a few include emissions/allowance trading.

tent emission inventories that are more modern than 1985 NAPAP. We also need to have some much better agreement and coordination on what do we consider to be reasonable US and Canadian emissions reduction scenarios for the future. How do we allocate future emissions reductions so that we can evaluate them in our modeling assessments? Depending on which country is doing the modeling, one country tends to make really silly assumptions about what might be happening in the other country for future emissions. We also need these models to be comprehensive in nature to evaluate the multiple effects and multiple benefits in not only deposition but aerosol concentration, fine particles and so forth. These models can't continue to focus on just a few short-term events (although we need those kinds of sophisticated models), we also need to look at more long-term concentrations of multiple pollutants and deposition over at least continental-scale domains. We would also want to consider alternative controls to maximize reductions of mercury and carbon dioxide, but these emerging issues should not be used as an excuse to further delay emission reductions of sulfur and nitrogen.

To illustrate a few of these points with bits of regional data, here is some good news in the historical precipitation record of sulfate deposition from the Bennington, VT NADP site (Figure 5). In the most recent year, we have received the lowest deposition rate and lowest concentration in this long-term record! We are concerned, however, that proposed budget cuts at USGS may jeopardize some of the water monitoring and research activities there, possibly reducing future support for this kind of monitoring effort. The critical importance of monitoring programs is a hard thing to sell politically, but it is absolutely critical if we are going to move forward.

In parallel to the modest regional improvements in sulfate deposition, we have seen similar reductions in aerosol sulfate concentrations, as evident in this long-term data from Acadia National Park in Maine from the national IMPROVE network (Figure 6). The blue line is a long-term average, about a one-year running average and the red dots are the actual daily concentrations. The trend line indicates a fairly steady long-term improvement. From the right hand scale (in nanograms per cubic meter), you may note that we are actually getting down close to the annual average of 2 micrograms per cubic meter, that we proposed 20 years ago as a visibility standard in Vermont. We have made some progress.

On the other hand, take a look at the actual daily concentrations. The episodes are as high as they've ever been, and July 17, 1999 was the highest sulfate level ever recorded at Acadia. In Vermont we had even higher sulfate concentrations that were up in the range of 35 micrograms per cubic meter, tying the all-time record for aerosol sulfate that was last measured exactly 17 years ago to the day earlier on July 17, 1982 (the day on the cover of our old visibility plan).

This was a whopping big sulfate episode, and resulted in exceptionally high fine particle concentrations extending from the Ohio River Valley well

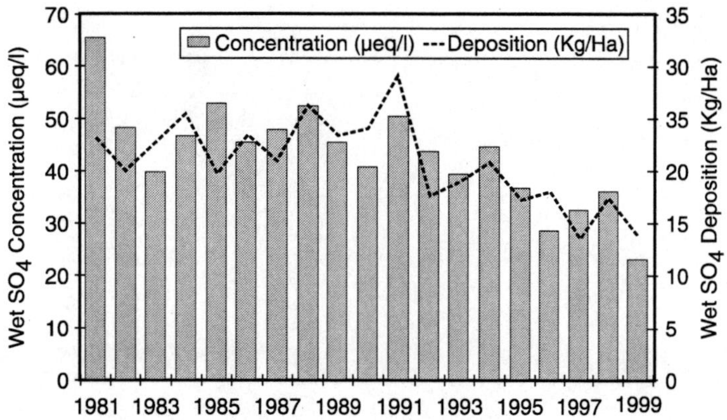

Figure 5. Wet sulfate deposition in Bennington, VT (NADP).

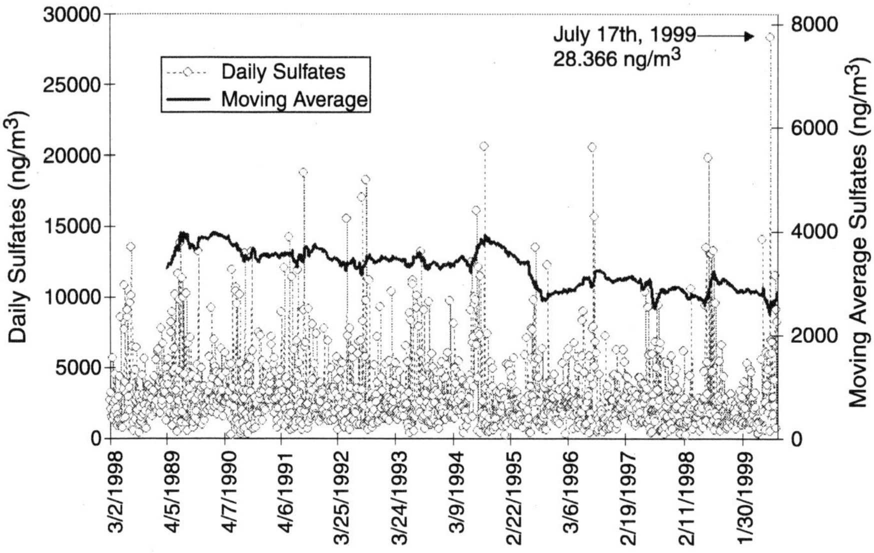

Figure 6. Daily and "annual average" sulfate at Acadia National Park, 1988-1999 (sulfate expressed at $[NH_4]_2\ SO_4$; 100 sample mixing average).

up into northern New England, and of course into eastern Canada as well, although we always tend to cut off our data at the borders. These concentrations, by the way, were well below – you might say "safely" below – the U.S. fine particle short term standard. On the other hand, they were well above the Canadian short term standard (Figure 7).

Healthy or not (and we suspect not), one thing that's clear is that you can't see through air which contains this kind of acidic sulfate pollution. The chart on the lower left (Figure 8) shows that hourly fine particle concentrations in Roxbury, MA were approaching 100 micrograms per cubic meter. In the upper left panel we see that these fine particle concentrations coincided with light extinction measurements (the inverse of visual range) that approached 1000 inverse megameters at sites in Vermont and New Hampshire. The panels on the right show the view of Boston on a clear day (top) and on July 16, 1999 (bottom) in the midst of this regional episode. Four hours after that picture was taken, John F. Kennedy's plane went down over extremely hazy skies over Martha's Vineyard. In a National Transportation Safety Board report on that accident, the word haze is mentioned 23 times and the word pollution is not mentioned once.

So, even though we have made progress, I think that we still have a long ways to go. Thank you.

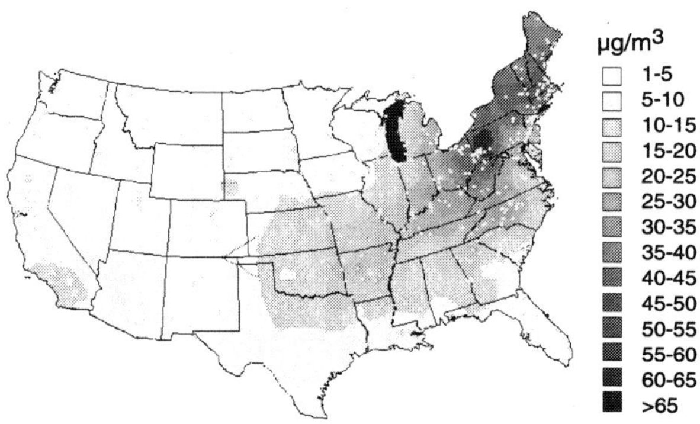

Figure 7. $PM_{2.5}$ concentrations on 7/17/99. "Safely" below US short-term $PM_{2.5}$ standards in New England but well above Canadian standards throughout Northeast and Eastern Canada.

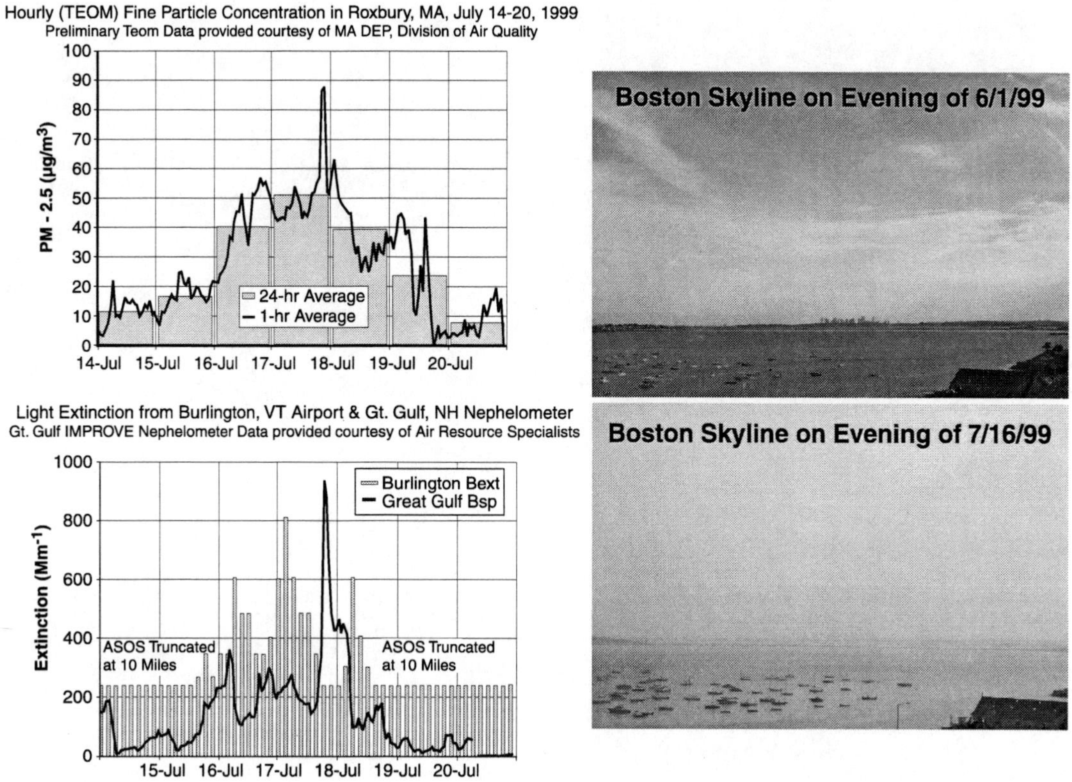

Figure 8.

Questions

MR. FRANKEL: Frankel, GAO. With regard to the loss of data if monitoring money is reduced, there is a tradition of getting rid of the messenger.

I wonder in a similar vein if all the dealings of the New England Governors and Canadian Premiers have ever been subject to scrutiny under the trading with the enemy act.

MR. POIROT: I don't know that act. I guess we would invoke the free trade agreement.

MR. LEWIS: Preston Lewis, New York State Department of Environmental Conservation. I guess my favorite topic is modeling and you mentioned it. I am curious to know who is going to do this comprehensive modeling. Who would you suggest?

MR. POIROT: I think there is a tendency for the modelers to keep thinking of new sophisticated tweaks to add to their models that almost exactly eats up all the advances in computer speed, such that we can never afford to model more than a few short episodes with these state-of-the-art models. While these continuing advances in model complexity are necessary as our knowledge increases, I think its also essential to keep working on developing the best multi-pollutant models we can afford that can be run to cover multi-year periods and large, continental-scale spatial domains. In both cases, I think the federal governments (perhaps in a collaborative effort between the US and Canada) are logical places for the development and deployment of these large regional models, and also for the detailed (and expensive) emissions inventory and meteorological information needed to run the models

As these regional models are developed and evaluated at the federal level, I think its also important that they become "transferable" to States, Provinces and other users. We each have specific pollutant mixes, time periods, and sensitive resources of local interest, and we also likely have slightly different ideas of what would be "reasonable" future control (or growth) strategies to consider.

Mid-West

Douglas Knauer – Wisconsin Department of Natural Resources

One of the things I am going to do is first describe the resources that we have in the midwest. These resources are quiet resources. These resources are basically different than what we see in the New England states. We have great changes as a result of the lowering of the SO_2 emissions.

Basically, what we are concerned about for aquatic resources in the upper midwest are seepage lakes. I want to tell you a little bit about seepage lakes and how they work.

This is typically a seepage lake in the upper midwest (Figure 1). You can guess the name of it, Round Lake. Basically, what we see is no surface inputs. It is basically what falls out of the sky from a hydrologic point of view and what comes in from the groundwater system (Figure 2).

We selected a few of these seepage lakes and modeled them extensively.

Figure 1. Round Lake.

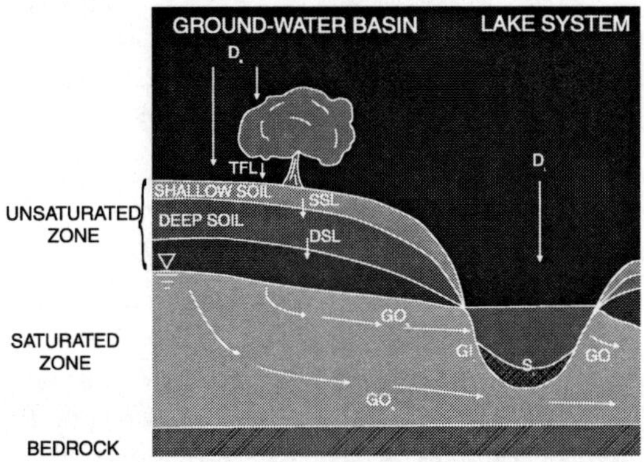

Figure 2. Hydrology of a seepage lake system.

This is one of the modeling efforts, Round Lake, where we took a look at what falls out of the sky, what goes through the trees, down into the soils, into the groundwater system, and the processes in the lake itself, to try to figure out the mechanics of acid rain.

In New England the soils have been damaged to the point where you don't have any weatherable materials left, your base saturation is pretty low. In Wisconsin, that is not necessarily true. We do have sandy, loamy soils. Take a look at this table (Table 1), it will give you an idea.

The base saturation index is about 30 percent. The Adirondacks is around 10. So, we have some base cations left, and we are still holding our own as far as the soils go, although we don't want to go any lower than this, if we can help it. The pH of the soil solutions down around eight centimeters in Wisconsin are running about 5.6 to 6.1. Down about 60 centimeters, 6.5 to 6.7. Compare that to the Adirondacks, where you see that the pH of the soil solution is much lower, much more acidic.

One of the biggest features with these seepage lakes, is the graphic watershed which, in this case, is the dashed line around the outside of Round Lake (Figure 3). It does not influence the chemical constituents coming into the lake.

The sandy soils would absorb any water. We had no surface run off coming into these lakes. It was the groundwater contributing basin. You can see the arrows coming in. They defy the concept that what falls in the topographic watershed is the only important issue here. You have to go beyond that. It is what falls into the groundwater contributing area that really counts. So, groundwater becomes an important issue here.

Table 1. Wisconsin soils.

Omega Loamy Sand

BSI ~ 30% <10%Adirondacks

Soil Solution pH

	Wisconsin	Adirondacks
8 cm	5.6 - 6.1	3.6 - 3.9
60 cm	6.5 - 6.7	4.7 - 4.9

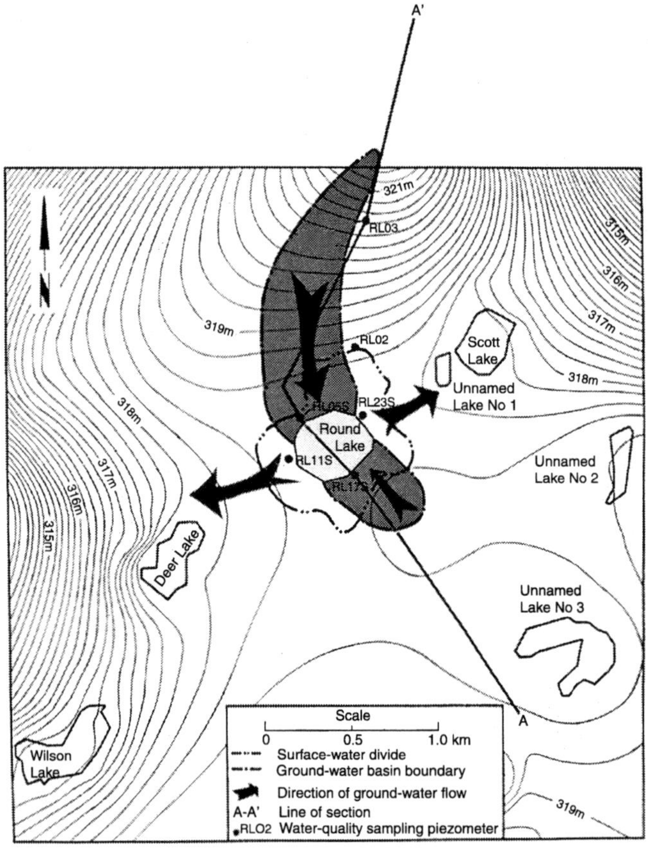

Figure 3. Groundwater flow in and out of Round Lake.

We did some model runs. This is with an EPRI funded project.

This is another lake that has an ANC of around 25, Vandercook Lake (Figure 4). You see in the solid black line what you would normally predict, what we would see from the present situation. In the dashed line is basically what we called the sensitivity. We made a sensitivity run with the groundwater out. You can see that with the groundwater out of this situation, the lake would go acidic.

I don't always believe in models until we verify them. So, we went to one of our acidic lakes and put wells around it to show that there was no groundwater moving in.

We put a well down, a high capacity well, and we pumped water in (Figure 5). We can control the volume, we can control the chemical inputs into the lake and we will see if our model really holds.

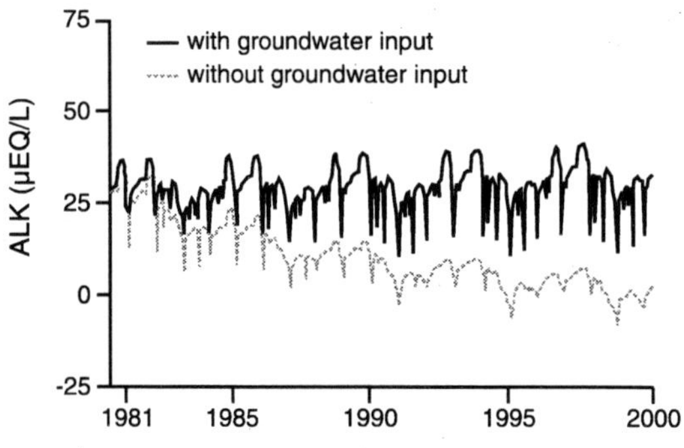

Figure 4. Vandercook Lake alkalinity with and without groundwater input.

Basically, we pumped the groundwater in with sprinklers. The groundwater is colder in the summertime than the surface waters. We didn't want it to just sink to the bottom. So, we put out sprinklers around the lake, basically, not very large volumes.

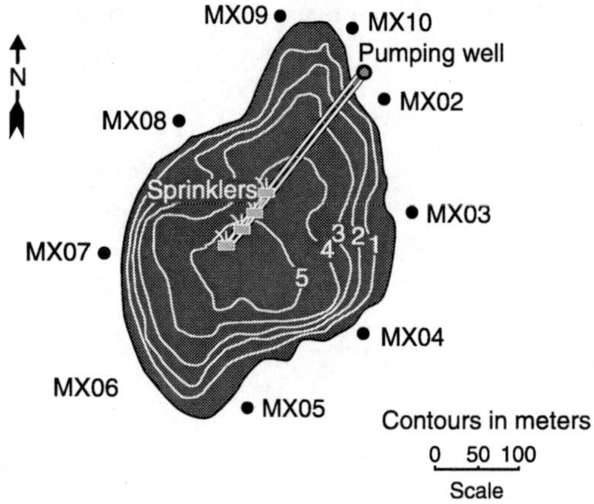

Figure 5. Location of groundwater well around Max Lake.

The background, the pH of the lake was 5.1 and the alkalinity, A and C, was negative.

As we put more and more groundwater in, of course, we changed the pH and the alkalinity, as you might expect (Figure 6). So, once again, it confirmed the model. In these lakes in the midwest that we are concerned about, groundwater is extremely important as far as buffering capacity goes.

Morgan Lake, is a lake that sits above the ground water table, a mounted situation, we call it (Figure 7). These are the ones that we are concerned about. We have no groundwater inputs to them. Basically, hydraulically, they receive everything out of the sky. These are the ones we find that are acidic in the state, and in the upper peninsula of Michigan.

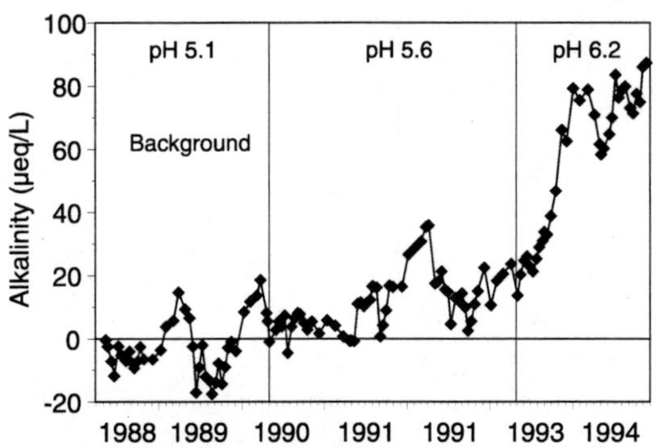

Figure 6. Alkalization of Max Lake.

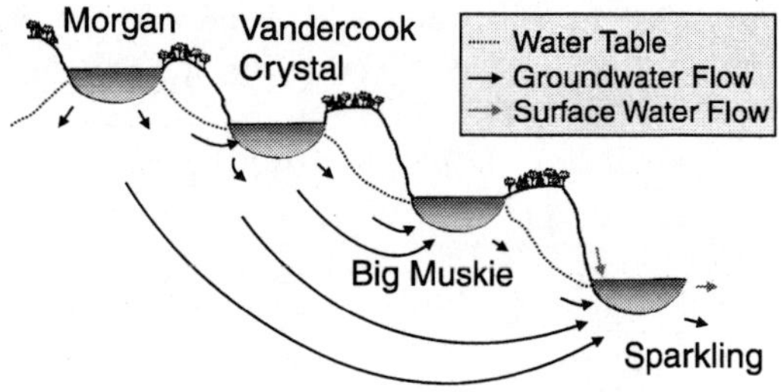

Figure 7. Relative landscape position of Wisconsin Lakes.

As you get more and more groundwater, like Vandercook, Crystal, Big Muskie, and work our way down to Trout Lake, the pH goes down and the alkalinity goes up, all down that chain.

Finally, we start getting surface water inputs like at Trout Lake, a pretty hard water lake that we don't have to worry about. So, we are concentrating basically on the impacts on lakes like Morgan Lake.

In our state in 1986, we had legislation passed, a reduction in SO_2 that was mandated by 1993, a 50 percent reduction in our SO_2 emissions, going from 650,000 tons of SO_2 down to 325,000.

The utilities were at about 500,000 and they reduced to 250,000 tons of SO_2 . As a result of that, and also with national legislation that came along in the 1990s, we did achieve by 1993, a 50 percent reduction and we have maintained that as far as SO_2 reduction goes in the state.

Now we will take a look at the impacts (Figure 8). If you take a look at the upper left hand corner, that is the sulfate deposition, kilograms per hectare per year, with time along the bottom. You can see that there is a tremendous reduction. We are down to levels now that are less than 10. We are talking about eight kilograms per hectare per year. We used to figure 10 was the magic number and if we got below that, we would start saving our resources.

In contrast to that, I was looking at the NADP stuff for the northeast, and they are still going from 40 to 30. So, there is still a long way to go. We do have much lower loading levels.

There have been some slight increases in the pH .

Now, also looking at the resources, I will give you a summary (Figure 9). Looking at box plots here, 1988 the sulfate on the left-hand side, versus 1997. You can see there was a decline in these lakes, basically in the sulfate.

If you look at the pH, there is definitely an increase in the pH of these clear water sensitive lakes.

This basically summarizes significant trends (Figure 10). Again, it is just reiterating what I just told you.

The last figure (Figure 11), there are other things that helped control our acidity in these lakes

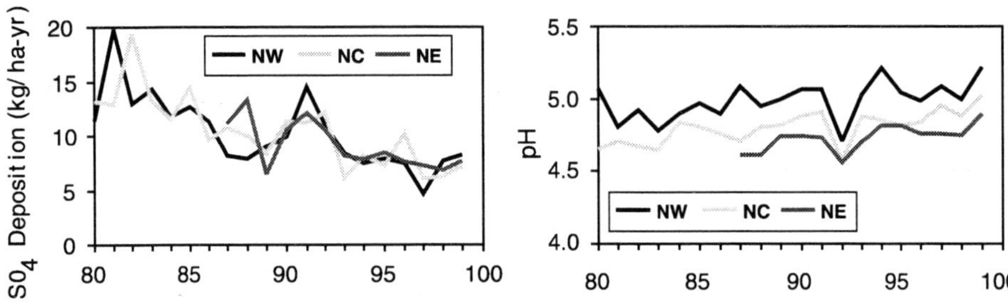

Figure 8. Acid deposition in Northern Wisconsin. Sulfate deposition has declined by ~50% since 1980. There has been an ~0.3 unit increase in pH since 1980. NADP data.

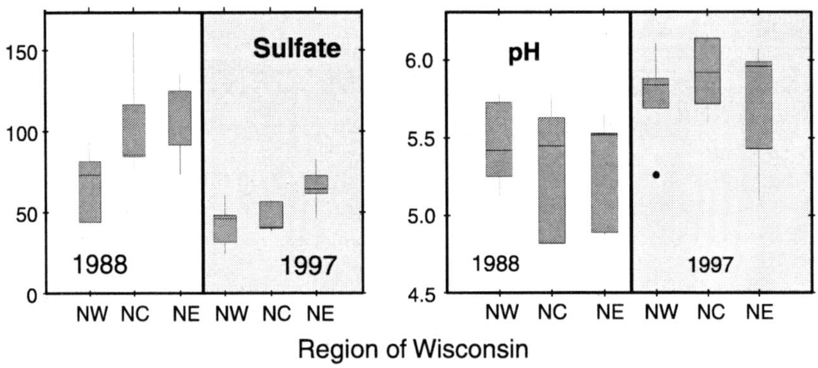

Figure 9. Decadal responses of clear-water lakes across Northern Wisconsin to decreased deposition. In 1988, lakes of similar pH had higher [SO4] in NC and NE. Over 10 years, SO4 decreased and pH increased in all regions. The largest declines in SO4 have been found to the east.

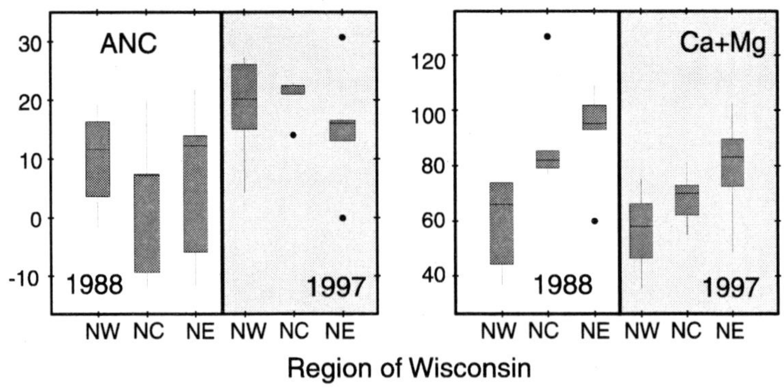

Figure 10. Delayed recovery due to cation decreases. There have been only small increases in ANC over the decade. Cations have decreased over the same period, particularly in NC and NE. Like with SO_4, cation concentrations increase from west to east.

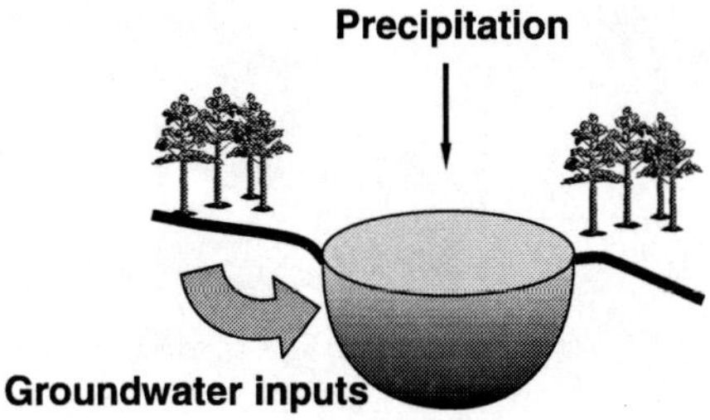

Figure 11. Sources of alkalinity and cations to seepage lakes. Other Issues. Climate change could reduce groundwater inputs of cations, causing lakes to become more sensitive to acid deposition.

that we have to be concerned about, and one of them is climate change.

If we get a change in the amount of precipitation that falls in the groundwater contributed areas of these lakes, and we either increase or decrease the groundwater inputs, it makes a difference in what the alkalinity is going to be in these lakes.

If we have a drought, which we had in some of these situations, we decoupled the groundwater from the lakes. We have actually seen the alkalinity, A and C, decline considerably.

We have great concern over climate changes and how this will impact our lakes also. So, there are a lot of features that come into play, although I think most importantly is the sulfate deposition.

Questions

MR. COWLING: Here is a state that put its money ahead of where its mouth was, I guess you could say, in the sense that you have committed yourself to decrease of emissions well in advance of the federal requirement to do so.

MR. KNAUER: Of course, the concern was how much falls locally versus how much comes in on a national basis. I think the position we are sitting in as far as the midwest goes, most of the high levels of sulfur dioxide emissions, of course, are going up into the northeast, thank goodness.

MR. SCHULER: John Schuler , American (sic) Department of Natural Resources. I was wondering if you looked at mercury in any of those lakes?

MR. KNAUER: Yes. We have been looking at mercury since 1988. If you want to get into that, that is another whole area. There is a relationship between the pH and the mercury you see in the fish. We first discovered that when we looked at Little Rock Lake as an acidification project. We actually divided the lake in half and acidified one side slowly over a number of years, six years. We did notice at that point in time when we did that, that the mercury on the acidified side versus the reference base was definitely significantly elevated. Is that because the pH itself is controlling that?

The sulfate side of it, the sulfate reducing bacteria are the methylators. This is methyl mercury, of course, that is in the fish so it has to be methylated first.

There is that definite relationship that is going on with acidity. If you increase the pH, where you start to change is maybe the methylation rate – not the methylation rate so much as probably the evasion.

If you lower the pH the Hg_2 that comes in has more opportunity to bump into the bacteria to be methylated. That is basically what we found in the Little Rock Lake experiment.

Atmospheric Deposition and Its Effects in the Intermountain West

Donald H. Campbell – U.S. Geological Survey

The Intermountain region of the western United States includes vast arid and semi-arid lands that would likely benefit from acid rain or any other type of rain they could get. But it also includes high-elevation alpine and subalpine ecosystems that are among the most sensitive in the world to harmful effects of atmospheric deposition.

These sensitive areas include some of the crown jewels of the National Parks and National Forest Wilderness areas. Many are Class 1 areas protected by the Clean Air Act Amendments from any degradation of air-quality-related values, including degradation that might be caused by atmospheric deposition. This discussion focuses on these sensitive areas by examining spatial and temporal trends in atmospheric deposition, ecosystem sensitivity and response to deposition, and concerns for protection of these ecosystems in the future.

In the Intermountain region, elevational gradients determine regional patterns in precipitation, with higher elevations receiving greater precipitation. Variability in precipitation amounts is generally greater than variability in concentrations of pollutants, so deposition is usually greater at higher elevations as well. Mountain ecosystems are sensitive to atmospheric deposition because they often have bedrock that is resistant to weathering, which results in poorly buffered soils, ground water, and surface water. Cold temperatures and a seasonal snowpack that persists for 6-10 months limit the growing season and the ability of ecosystems to assimilate nitrogen in atmospheric deposition. Much of the annual load of pollutants in atmospheric deposition is delivered to the ecosystem during a few weeks of spring snowmelt. Thus, in the Intermountain region high-elevation areas that receive the greatest amounts of atmospheric deposition also contain ecosystems that are most vulnerable to impacts from that deposition.

Spatial and Temporal Trends in Atmospheric Deposition of Sulfur and Nitrogen

Total loadings (deposition) of pollutants are important for assessing ecosystem effects, however, spatial and temporal trends in deposition are confounded by natural variability in precipitation amounts. Trends are discussed here in terms of concentrations in order to minimize the effects of climate and focus on variability caused by anthropogenic emissions. In order to better capture the spatial variability in amounts and sources of atmospheric deposition to the Rocky Mountains, concentrations and isotopic signatures of pollutants are measured in the seasonal snowpack at maximum accumulation during the spring [Turk et al. 2001; Mast et al. 2001; Campbell et al. 2002].

Sulfate concentrations in the snowpack generally increase from north to south along the Continental Divide, and both concentrations and isotopic signatures of sulfate are related to anthropogenic sources of sulfur dioxide emissions [Turk et al. 2001; Mast et al. 2001]. Spatial trends for hydrogen ion concentration (acidity) are similar to those for sulfate [Turk et al. 2001]. Greatest acidity and sulfate concentrations in the snowpack have been found in the Park Range in northwestern Colorado (Figure 1), an area downwind of coal-fired power plants that have recently begun installing emissions controls as a result of a legal settlement regarding visibility impairment to Class 1 Wilderness Areas [Turk and Campbell 1997]. Elevated concentrations of sulfate are also found in other areas near emissions sources such as northcentral Colorado, southeastern Wyoming, and northern New Mexico. Concentrations of sulfate in wet deposition were either unchanged or decreasing from 1981–1998 at all sites in the Intermountain region, which is similar to trends

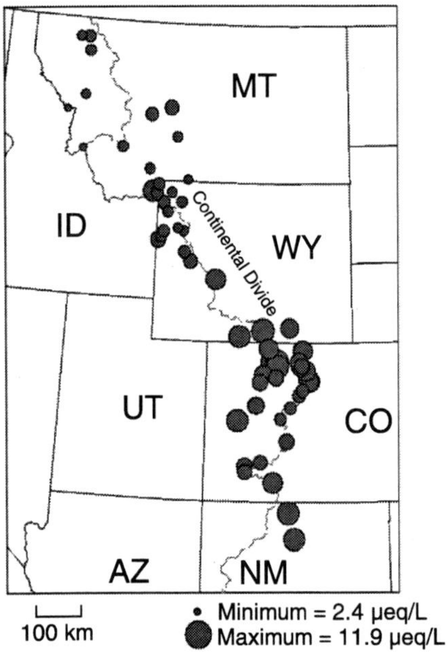

Figure 1. Average sulfate concentrations in snowpack for 1993-2000 [after Turk et al. 2001].

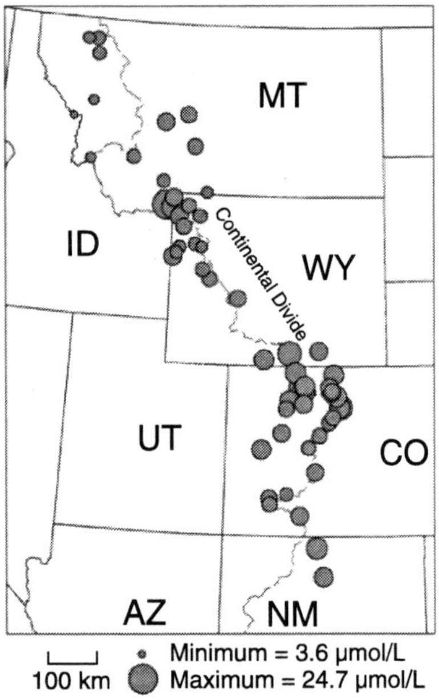

Figure 2. Average dissolved inorganic nitrogen concentrations in snowpack for 1993 – 2000 [after Turk et al. 2001. Ecosystem Response to Deposition of Acids and Nitrogen

resulting from decreased sulfur dioxide emissions in other parts of the country [Nilles and Conley 2001].

Spatial trends in snowpack concentrations of dissolved inorganic nitrogen (DIN) are similar to those for sulfate and acidity, except that the local areas of high deposition in northern Colorado are more widespread than they are for sulfate, and nitrogen deposition in northwestern Wyoming also is high (Figure 2) [Turk et al. 2001]. Ammonia makes up about one-third of the total nitrogen deposition at many sites, and is important for its contribution to fertilization of ecosystems as well as its potential acidifying effect as it is assimilated or nitrified in the ecosystem. In contrast to sulfate, concentrations of nitrate and ammonia were either level or increasing from 1981–1998 at all sites in the Intermountain region [Nilles and Conley 2001].

The snowpack concentration data can be combined with wet deposition measurements collected at the National Atmospheric Deposition Program / National Trends Network (NADP/NTN) sites [*National Atmospheric Deposition Program / National Trends Network*, 2001] and high-resolution precipitation maps [Daly et al. 1994] to produce high-resolution maps of deposition that better incorporate precipitation amount gradients than maps using the NADP/NTN data alone [Nanus et al. 2000]. These deposition maps illustrate regional trends in atmospheric deposition caused by precipitation amount and regional emissions, as well as highlight areas that are affected by local sources of pollutants.

The Intermountain region is fortunate to have deposition rates of acidity, sulfate, and nitrogen that are moderate compared to some parts of the world. Chronic acidification from atmospheric deposition has not been documented in any lakes or streams that do not have acid rock drainage, in the western United States. Episodic acidification is more difficult to evaluate because few sites have sufficient data, but its effects are likely found in only the most sensitive high-elevation lakes and streams. A recent study in the Park Range, however, indicates that many ephemeral ponds are much more sensitive to acidification than nearby lakes, with acid-neutralizing capacities reaching less than 20 ueq L-1 during parts of the summer [Muths et al., 2002].

Nitrogen deposition contributes to acidification of ecosystems in advanced stages of nitrogen saturation, but, in earlier stages, there may be other ecosystem effects that are more subtle. Elevated amounts of nitrogen in soils may disrupt soil nutrient cycling processes before inorganic nitrogen appears in surface waters. Advanced symptoms of watershed nitrogen saturation have resulted from atmospheric nitrogen deposition levels of 3-5 kg ha-1 y-1 in high-elevation ecosystems of the Colorado Front Range [Williams et al. 1996]. Front Range ecosystems have greater nitrogen content in soil and vegetation and higher potential net mineralization rates in soil compared to other areas in Colorado with lower nitrogen deposition rates [Baron et al. 2000]. Even in areas with little vegetation and soil, isotopic signatures of nitrate in surface and ground water indicate that most of the atmospherically-deposited nitrogen has been biologically cycled [Campbell et al. 2002], underscoring the importance of soil microbial processes in all landscapes. Export of nitrate from the terrestrial ecosystem results in higher concentrations of nitrate in lakes [Baron et al. 2000; Clow et al. 2002], and changes in community composition and biovolume of diatoms in lakes [Baron et al. 2000; Wolfe et al. 2001].

Although there has been substantial research on ecosystem response to atmospheric nitrogen deposition in the Front Range of Colorado and nearby areas, little or no data are available regarding the status of many high-elevation ecosystems in other areas of the Rocky Mountains. A survey of 370 lakes in and near National Forest Class 1 Wilderness Areas provides a snapshot of other high-elevation areas that are potentially sensitive to atmospheric nitrogen deposition (Figure 3). Most of these wilderness areas have nitrogen deposition rates of 2 kg ha-1 or less (about one-half of that found in the Front Range). Yet there were lakes in almost every area that had measurable nitrate concentrations during late summer, when ecosystem nutrient demand causes nitrate concentrations to reach their annual minimum. This leakage of nitrogen from the ecosystem indicates watershed

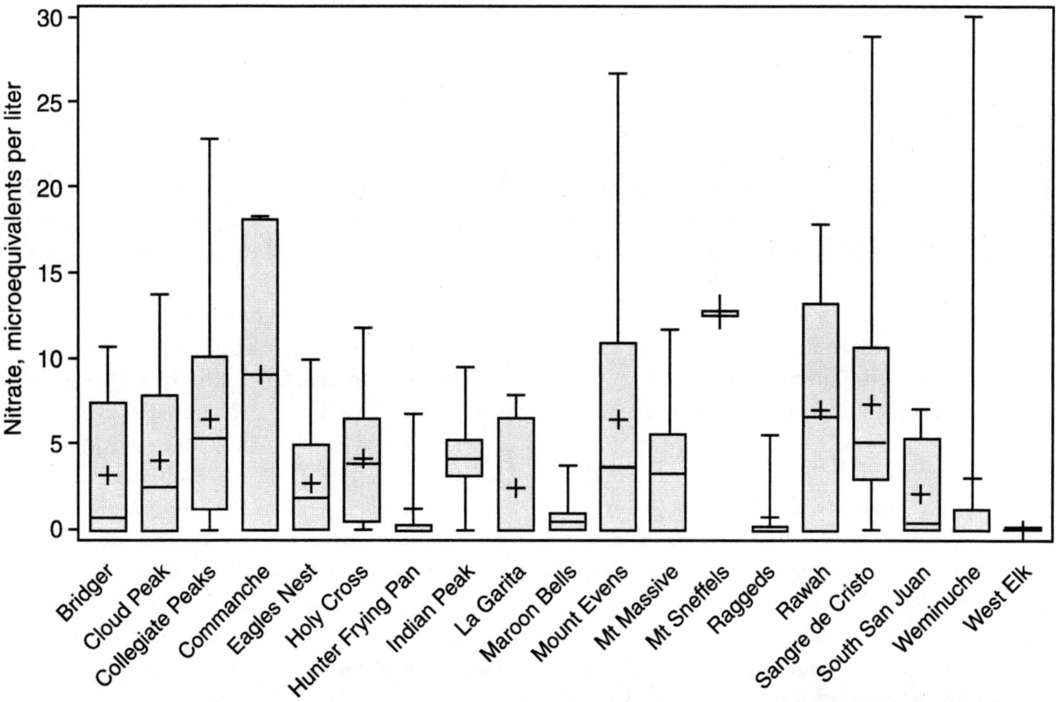

Figure 3. Nitrate concentrations in National Forest Lakes of Wyoming and Colorado, by Wilderness Area. Total number of lakes is 331.

nitrogen saturation at levels of nitrogen deposition that were not previously thought to cause ecosystem effects. Previous studies based on the Front Range data suggested a critical threshold of 3-6 kg ha-1 before advanced effects of nitrogen saturation such as nitrate leakage would be seen [Williams and Tonnessen 2000; Burns 2002]. These results indicate a need for better understanding of ecosystem sensitivity to nitrogen deposition, and re-examination of critical thresholds of nitrogen in atmospheric deposition to undisturbed high-elevation ecosystems. The role of organic nitrogen in nitrogen cycling and transport also needs to be examined in ecosystems with both moderate and high levels of nitrogen deposition [Williams et al. 2001].

Trends in Emissions

The primary source of sulfate and related acidity in atmospheric deposition of the Intermountain region is coal combustion for electric power generation. Decreased emissions of sulfur dioxide from some existing plants is expected as improved emissions controls are installed. However, there are also new coal-fired power plants planned or under construction, and development of new sources of energy such as oil shale could also increase regional emissions.

Emissions of nitrogen compounds come from a wider range of sources. Fossil-fuel combustion in motor vehicles contributes a large amount of nitrogen oxides, and increases in both population and per-capita vehicle miles are expected to continue. The extent to which this trend could be offset by improvements in vehicle emissions controls is uncertain. Electric power generation from combustion of both coal and natural gas also contributes substantial emissions of nitrogen oxides. As with sulfur dioxide emissions, some power plant emissions will be reduced by installation of emissions controls, but new coal and natural gas fired power plants are planned or under construction. In addition, further development and production of fossil fuels reserves such as natural gas, coal-bed methane, and oil shale will also add to emissions of nitrogen oxides. Because many of these reserves are located near sensitive high-elevation ecosystems in the region, these emissions may have acute local effects as well as contribute to regional emissions.

Emissions of ammonia come primarily from agricultural sources such as animal waste and volatilization of fertilizer applied to crops, but may also come from motor vehicles and industrial sources. Fertilizer use has increased rapidly during the last 50 years in the Intermountain region as it has worldwide [Baron et al. 2000], and livestock production has increased in some areas such as northeastern Colorado where large confined animal feeding operations exist.

Toxic Pollutants in Atmospheric Deposition

The abundance and effects of toxic pollutants in atmospheric deposition in the Intermountain region are poorly understood. The only high-elevation site where mercury deposition has been monitored in the Rocky Mountains is Buffalo Pass, in the Park Range of Colorado [National Atmospheric Deposition Network Mercury Deposition Network, http://nadp.sws.uiuc.edu/mdn/]. Mercury deposition at Buffalo Pass averaged 9 µg m-2y-1 in 2000, which is comparable to mercury deposition at sites in the Midwestern and Northeastern United States, where fish consumption advisories are common. Significant atmospheric mercury deposition to Rocky Mountain National Park also is indicated from sediment cores collected at Mills Lake [Krabbenhoft et al. 2002].

Persistent organochlorine compounds and current-use pesticides in atmospheric deposition are another concern. Concentrations of total DDT (DDT + DDE + DDD) as high as 14 µg kg-1 have been measured in lake sediments in Rocky Mountain National Park [P. Van Metre, U. S. Geological Survey, written communication]. In the mountains of western Canada, deposition of organochlorine compounds was shown to increase with altitude, and concentrations of more volatile compounds were enhanced by cold-condensation [Blais et al.1998].

Conclusions

Chronic acidification of surface waters from atmospheric deposition is not a widespread problem in the Intermountain region. Episodic acidification is a concern in some high-elevation headwaters and in some ephemeral ponds. Recent trends of decreasing sulfate in deposition suggest that acidification from sulfur compounds is not getting worse; however, it is unclear whether these trends will persist in the future, and they may be offset by increasing trends in deposition of nitrogen compounds.

Fertilization effects of nitrogen deposition on high-elevation ecosystems are currently the greatest concern related to atmospheric deposition. Both terrestrial and aquatic effects have been demonstrated at levels of nitrogen deposition that are now occurring in some parts of the region. Recent trends and future projections suggest that nitrogen deposition is likely to continue increasing, which would intensify ecosystem effects and expand their areal extent.

In order to protect sensitive ecosystems while managing growth and energy development in the Intermountain region, better understanding of source-receptor relations for specific pollutants and ecosystem responses to changes in atmospheric deposition and climate is needed. More monitoring and research is also needed in areas that currently are showing ecosystem responses to moderate levels of atmospheric nitrogen deposition. Finally, there is a need for monitoring and research on the distribution and effects of atmospherically-deposited toxic compounds such as mercury and pesticides.

References

Baron, J.S., H.M. Rueth, A.M. Wolfe, K.R. Nydick, E.A. Alstott, J.T. Minear, and B. Moraska, 2000. Ecosystem response to nitrogen deposition in the Colorado Front Range. Ecosystems 3: 352-368.

Blais, J.M., D.W. Shindler, D.C.G. Muir, L.E. Kimpe, D.B. Donald, and B. Rosenberg, 1998. Accumulation of persistent organochlorine compounds in mountains of western Canada. Nature 395: 585-588.

Burns, D.A., 2002, The effects of atmospheric nitrogen deposition in the Rocky Mountains of Colorado and southern Wyoming—a synthesis and critical assessment of published results, U.S. Geological Survey Water-Resources Investigations Report 02-4066, 41 p.

Campbell, D.H., J.S. Baron, K.A. Tonnessen, P.D. Brooks, and P.F. Schuster. 2000. Controls on nitrogen flux in alpine/subalpine watersheds of Colorado. Water Resources Research. 36: 37–47.

Campbell, D.H., Kendall, C., Chang, C.C.Y., Silva, S.R., and Tonnessen, K.A., 2002, Pathways for nitrate release from an alpine watershed: determination using $d^{15}N$ and $d^{18}O$, Water Resources Research 38 (5) 10.1029/2001WR000294.

Clow, D.W., G.P. Ingersoll, M.A. Mast, J.T. Turk, and D.H. Campbell, 2002, Comparison of snowpack and winter wet-deposition chemistry in the Rocky Mountains, USA: Implications for winter dry deposition, Atmospheric Environment 36: 2337-2348.

Clow, D.W., Striegl, R.G., Nanus, L., Mast, M.A., Campbell, D.H., Krabbenhoft, D.P., 2002. Chemistry of selected high-elevation lakes in seven national parks in the western United States. Water, Air, and Soil Pollution, (in press).

Daly, C., R.P. Neilson, and D.L. Phillips, 1994. A statistical-topographic model for mapping climatological precipitation over mountainous terrain. Journal of Applied Meteorology. 33, 140-158.

Krabbenhoft, D.P., Olson, M.L., Dewild, J.F., Clow, D.W., Striegl, R.G., Dornblaser, M.M., and VanMetre, P., 2002. Mercury loading and methylmercury production and cycling in high-altitude lakes from the western United States. Water, Air, and Soil Pollution, (in press).

Mast, M.A., Turk, J.T., Ingersoll, G.P., and Clow, D.W., and Kester, C.L., 2001. Use of stable sulfur isotopes to identify sources of sulfate in the Rocky Mountain Snowpacks. Atmospheric Environment 35: 3303-3313.

Muths, E., D.H. Campbell, and P.S. Corn, Hatching success in salamanders and chorus frogs at two sites in Colorado, USA: effects of acidic deposition and climate, Amphibia-Reptilia (in press).

National Atmospheric Deposition Program / National Trends Network, 2001. NADP/NTN Coordination Office, Illinois State Water Survey, Urbana, Illinois, http://nadp.sws.uiuc.edu/.

Nanus, L. Campbell, D.H., Ingersoll, G.P., Clow, D.W., 2000. Atmospheric Deposition Maps for the Rocky Mountains. EOS Transactions of the American Geophysical Union 81:48.F527

Nilles, M.A., and B.E. Conley, 2001. Chages in the chemistry of precipitation in the United States, 1981-1998. Water, Air, and Soil Pollution 130: 409-414.

Turk, J.T., and Campbell, D.H., 1997. Are aquatic resources of the Mt. Zirkel Wilderness Area affected by acid deposition and what will emissions reductions at the local power plants do? U.S. Geological Survey Fact Sheet FS-043-97, 4 p.

Turk, J.T., Taylor, H.E., Ingersoll, G.P., Tonnessen, K.A., Clow, D.W., Mast, M.A., Campbell, D.H., Melack, J.M., 2001. Major ion chemistry of the Rocky Mountain snowpack. Atmospheric Environment: 35, 3957-3966.

Williams, M.W., J.S. Baron, N. Caine, R. Sommerfeld, and R. Sanford Jr., 1996. Nitrogen saturation in the Rocky Mountains. Environmental Science and Technology, 30:640-646.

Williams, M.W., E. Hood, and N. Caine, 2001. Role of organic nitrogen in the nitrogen cycle of a high-elevation catchment, Colorado Front Range. Water Resources Research 37:2569-2581.

Williams, M.W., and K.A. Tonnessen, 2000. Critical loads for inorganic nitrogen deposition in the Colorado Front Range, USA. Ecological Applications 10: 1648-1665.

Wolfe, A.P., J.S. Baron, and R. J. Cornett, 2001. Anthropogenic nitrogen deposition induces rapid ecological changes in alpine lakes of the Colorado Front Range (USA). Journal of Paleolimnology 25: 1-7.

Southwest

Brent Takemoto – California Air Resources Board

I want to thank the Center for inviting someone to speak on conditions in the southwest. As Don Campbell has shown, there are quite considerable differences in the chemistry of acidic deposition in the west versus conditions in the east and the midwest.

I would also like to begin by thanking my colleagues at the Air Resources Board for sticking with the acid rain program when it wasn't apparent that we would be setting a standard for either acidic pollution or nitrogen or sulfur deposition.

In my opinion, California's acidic deposition is dominated by nitrogenous inputs stemming from our heavy use of motor vehicles. In most of our urban and rural settings, most of the nitrogen is largely deposited as part of gases during what is called the dry half of our year.

This is a map of California (Figure1). It is the third largest state with a land area of about 164,000 square miles .Because of our geographic and elevational ranges, the total range in deposition varies quite considerably across the state . There are roughly about 1,300 miles of coastline and east-west differences range from about 150 to 250 miles along longitudinal lines.

California's climatology is classified as Mediterranean. What this means is that we have very wet winters and dry summers. In most parts of the state, about 80 percent of our annual precipitation falls between the period from October to March.

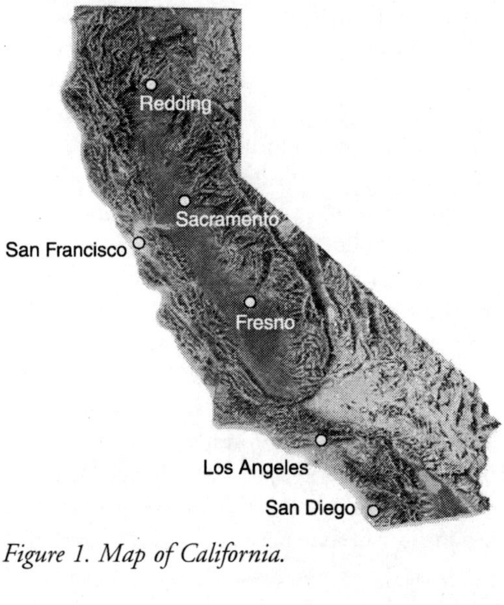

Figure 1. Map of California.

In terms of ecological receptors, there are 10 climates zones, ranging from the deserts in the southeastern part of the state where considerable invasion of grass species is occurring in the deserts. The researchers there are attributing that for now, anyway, to increases in nitrogen inputs due to primarily motor vehicle use.

Our other ecosystem types range from hardwood, mixed conifer, sub-alpine, alpine transition zones in the Sierra Nevada which extends over about two thirds of the state.

In terms of air quality, California's 34 or 35 million residents drive about 800 million miles per day in the state's 15 air basins. This clearly is at the heart of our nitrogen-dominated acidic deposition concerns.

Here is a summary of our statewide emissions inventory for the calendar year 2000 (Table 1). In rough terms, there is about an 11-fold higher amount of NO_X emitted than sulfur dioxide. On a mass basis, the same relationship holds for emissions of gaseous nitrogen and gaseous sulfur.

By source category, the emissions of NO_X are about 80 percent from motor vehicles and about 20 percent from stationary sources and area sources. For sulfur dioxide, about 55 percent comes from mobile sources and about 45 percent from area and stationary sources.

Through our acid deposition monitoring program – and at this point we only have about five wet deposition stations still collecting data –

Table 1.

California Emissions: CY 2000

Pollutant	TPD	TPY
ROG	3,300	-----
NO_x	3,600	600,000 (of N)
SO_2	300	55,000 (of S)
PM_{10}	2,400	-----

here is some of the data showing the basis for our concerns with respect to acidic nitrogen (Table 2).

The data in this table are from 10 sites, three urban, four what I have called mixed – and I will explain that in a minute – and three remote forest sites.

At the urban sites, the third line down shows the ratio of total nitrogen to total sulfur deposition. The number for urban systems, 9.1, approximates, roughly speaking, the statewide emission ratio for NO_X and sulfur dioxide of about 11.

Coming next to the mix sites, which are a collection of moderate to small urban areas like Sacramento and Fremont in the San Francisco Bay area, the total nitrogen to sulfur deposition ratio drops to about 5.3, and then it falls another 50 percent to about 2.5 at our remote forest sites.

I want you to take note that, at these three classes of receptors, total nitrogen drops from about 23 to 3 kilograms of nitrogen per hectare per year, while sulfur drops from about 2.5 to 1.5.

Another point I would like to make is on the fourth line, the importance of dry nitrogen deposition. It accounts for anywhere from 80 to 90 percent of total nitrogen deposition in our urban and mixed sites, only about 40 percent in our forest sites.

With state funding, we were able to conduct an effects research program in five areas. Our human health research focused on the effects of acidic fog and nitric acid vapor alone and in combination with other pollutants in both human and animal systems.

Our aquatics program focused on wet deposition effects to high elevation systems in the Sierra Nevada. That includes an intensive study site at Emerald Lake in Sequoia National Park.

Table 2.

CADMP Multi-year Data

	Urban	Mixed	Remote
Total N	23.0	9.2	2.9
Total S	2.5	1.7	1.5
N:S	9.1	5.3	2.5
% Dry N	92	84	40
% Dry S	63	54	20

Forest program studies were conducted to assess the combined effects of ozone and atmospheric nitrogen on a mixed conifer forest in Southern California and, for agricultural crops, we did acidic fog studies to characterize structural and physiological impacts.

In our manmade materials program, we looked at the effects of ambient Southern California air on deterioration rates in a range of building materials.

For acidic fog, acute exposure studies were conducted. The work of Shepherd and others in the University of California system indicated that a lot of the observed changes in lung function were largely associated with the application of the water droplets alone. Acute exposure studies using nitric acid vapor indicated slight changes in airway reactivity and inflammation when using above ambient exposures to nitric acid vapor.

In terms of chronic effects, an important finding was recently reported as a result of the airborne children's health study. This is an epidemiological study being conducted in 12 communities in Southern California. Researchers from the University of Southern California have found an association between ambient concentrations of NO_X, PM_{10} and nitric acid vapor and reductions in what is known as lung function growth, which is a physiological response in children living in the most polluted communities. This suggests that there is a potential for increased risks of chronic respiratory illness in adulthood, which stems from a reduction in the maximum lung function that can occur as a young adult.

In terms of air pollution, as was mentioned earlier, studies conducted since the 1950s have shown that ambient ozone has an adverse effect on pines in Southern California. In our forest program, research looked at the combined effects of ozone and nitrogen deposition, as ozone is an over-arching pollutant stress, and reports of nitrogen saturation were quite prominent in the literature.

Researchers from the U.S. Forest Service in Riverside conducted a three-year study at Barton Flats, which is in the San Bernardino National Forest. At this particular forest site, the daytime ozone level is about 60 parts per billion and the nitrogen deposition rate is about five to nine kilograms per hectare per year. This is what they considered to be a mid-range forest site along this west to east ozone gradient that occurs in the San Bernardino Mountains.

The high side at the western end is a place called Camp Pifica. At Camp Pifica, the daytime ozone levels are about 90 ppb and nitrogen deposition is about 30 to 40 kilograms per hectare per year.

Using the nutrient cycling model developed for the integrated forest study, this group found that the potential for chronic soil acidification and aluminum toxicity was fairly low, but that rates of soil nitrogen emissions were comparable to those that you can measure in agricultural fields. As a follow on to these studies, these researchers also measured nitrate levels in streams from the watersheds. The highest nitrate levels were observed at rates exceeding the highest atmospheric nitrogen loads.

Our studies on aquatic systems looked at the effects of high elevation links in the Sierra Nevada above 2400 meters elevation.

Researchers from the University of California, Santa Barbara, documented that snow is the main source of water and analytes to these systems, but summer rains could also be a significant source for both acidity and nitrogen.

Although many lakes were surveyed, there was no indication, as Don Campbell indicated, of chronic acidification. Episodic acidification does occur pretty much every year during snow melt in the spring. In terms of biological effects, episodic acidification was linked to reductions in the populations of some invertebrates as well as some spring spawning fish species.

In keeping with the theme of the conference, is California's acidic deposition problems solved? In some ways, yes, and in some ways no.

I think yes in terms of, there is a commitment to pursue further emission reductions of both criteria in toxic air pollutants, and no in terms of not being able to fully appreciate what the extent of chronic health and ecological effects may be.

Our air quality data indicate that acid precursor emissions are headed down. By 2010, ambient air quality will be improving in urban areas of the state.

Without improving revisions to our state implementation plan, more measures will be adopted to reduce NO_X emissions to control ozone, PM_{10} and acid deposition as a secondary or contributing stress to both human and forest health.

By focusing our control programs on ozone and PM_{10}, it is anticipated that there will be concurrent reductions in any potential effects that could occur from acidic pollutants.

The phrase, uncontrollable externalities, comes from my economic friends. Given our current energy crisis in California, some concessions may have to be made in terms of air quality to meet the greater good of California.

I guess over the next 50 years or so, only time will tell if we are embarking on the right kind of program. Thank you.

Questions

DR. COWLING: Brent, you have a huge concentration of animals in the lower part of California, the central valley. I wonder if the state has a good ammonia emissions inventory.

MR. TAKEMOTO: As of about three years ago, there has been a very concerted effort to at least identify where all the feedlots and dairy facilities are located. As you mentioned, in the San Joaquin Valley, there is an increase in their production. It has been a three-year work in progress, but that was certainly one of the issues, and that was driven by particulate matter concerns of ammonium nitrate.

MR. COWLING: We also heard a very interesting report on ammonia emissions from automobiles. I wonder if you could comment on your sense of the inventory development attributing ammonia emissions to animal agriculture and automobiles, on the other hand?

MR. TAKEMOTO: I am not sure exactly who did that work. It might have been either Cal Tech or Desert Research, I am not sure. There are significant, or at least measurable, ammonia emissions from cars.

We have one of the most complex models for looking at vehicle emissions, and that is also being worked into the process as well. What the nitrogen outputs from cars, ammonia, derived from ammonia will be, and how that could possibly affect our strategies for PM_{10} will be studied.

Canada

Guy Fenech – Environment Canada

I will give you a brief update of what we are doing in Canada to address the acid rain problem, as well as give you information about one test case related to the changes in soils, forests and how the trees are reacting.

Acid rain in Canada has certainly had a long history. After several years of scientific concerns, in 1976 the government established the integrated program for long-range transport of air pollutants. Eight years later, we had the first control programs where the seven easternmost provinces agreed to reduce their emissions of SO_2 by 50 percent by their allowable levels by 1994.

Emissions from the United States are certainly a big influence in Canada, and one very important date for us has been the signature 10 years ago last March by President George Bush and Prime Minister Mulroney of the Canada/US Air Quality Agreement, whereby both countries committed to reduce their emissions of SO_2. More recently, our ministers have agreed to update the strategy to address acid rain.

Like everywhere else, our emissions of SO_2 have declined fairly rapidly (Figure 1). The bar graph is the emissions in eastern Canada, the seven easternmost provinces, where the problem resides. The other line is the total emissions for the rest of the country, or for the country as a whole.

NO_X emissions have stayed just about constant, no surprise there (Figure 2).

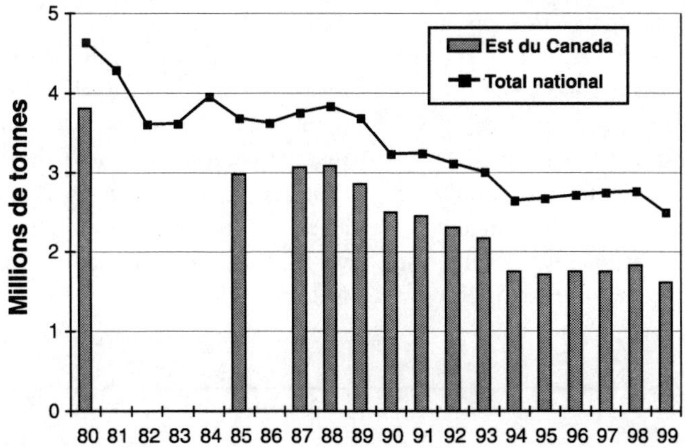

Figure 1. SO_2 emissions.

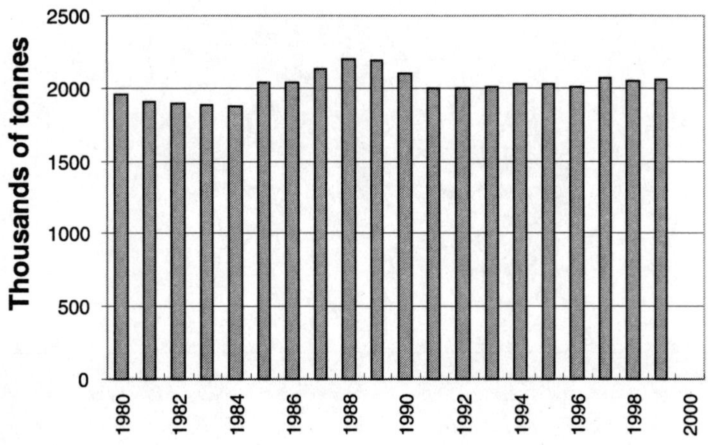

Figure 2. NO_X emissions.

In terms of ecological response first some very good news. There has been some very significant improvement in the area around Sudbury. The deposition in those lakes has been reduced by 80 to 90 percent during the period. There has been chemical as well as biological recovery. Anybody who would be interested in doing recovery work, we have a wonderful lab for you. Just get in contact with me or John Gunn at Laurentian University. We would be delighted to have you there to work on recovery.

The improvements elsewhere have been rather modest. In fact, some lakes and rivers in the easternmost part of the country are still acidifying mainly because they are very sensitive to rain.

For example, the Atlantic salmon rivers of Nova Scotia: Nova Scotia used to have 63 salmon rivers. The salmon runs have decreased in all those rivers. Several of those rivers do not support salmon any more, and predictions are that only two of the rivers may have a chance to support salmon in the long run.

We are extremely concerned about the chemical alterations of forest soils, and I will give you an example of what this means for us.

As other speakers have pointed out, acidifying pollutants do affect human health, human air quality, and contribute to regional haze problems.

Acid rain continues to be a sizeable problem in Canada. What we have done since 1990, in fact, is to evaluate or quantify the sensitivity of the terrain or the sensitivity of the lakes, or the tolerance of the lakes to receive sulfate deposition and we compared the actual deposition to the critical load or the tolerance level, to determine areas where this level is exceeded and by how much (Figure 3).

We have estimated that in order to get rid of this problem, we would need reductions in SO_2 emissions from Canada as well as from the eastern United States on the order of about 75 percent from Phase II of the program.

The Canada-Wide Acid Rain Strategy was adopted by ministers in 1998. Its long-term goal is to meet the critical loads. It commits the provinces to set new eastern Canada SO_2 emission reduction targets. It also commits the federal government to pursue further U.S. reductions. The provinces that do not have an identified acid rain problem have committed to keep their area clean by preventing pollution.

The strategy commits the governments to review the compliance with international commitments on an annual basis. It commits the governments to maintaining an active role in science and monitoring, and for the next assessment report, to assess the role of nitrogen in acidification. Finally, we have to report annually to ministers as well as to the public.

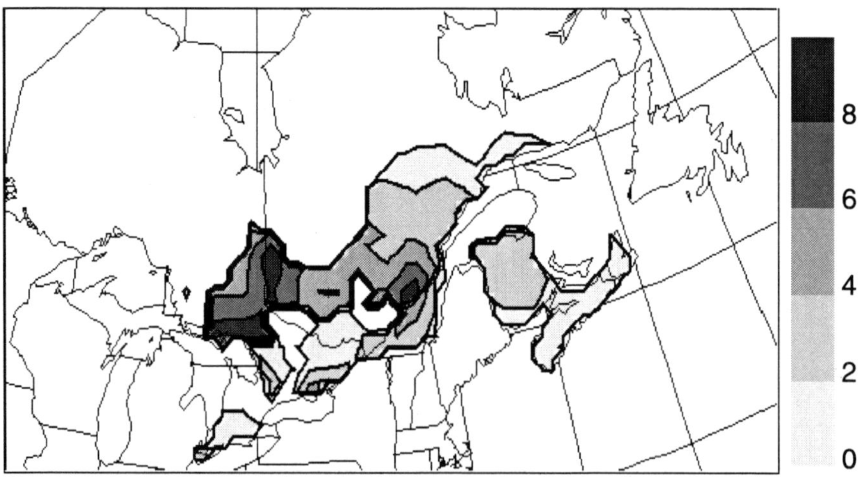

Figure 3. "Exceedance" loads. The area of eastern Canada expected to receive SO_4 deposition above critical loads without further controls beyond provisions in the 1991 Canada - U.S. Air Quality Agreement (in kg SO_4 per ha per year).

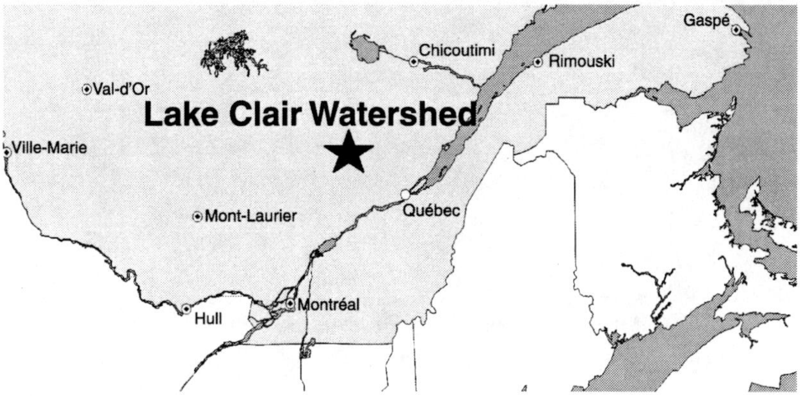

Figure 4. Lake Clair Watershed.

Now I would like to tell you about one test case about the impact of acid rain on soils and forest. I will talk about the Lake Clair Watershed (Figure 4).

What the Quebec Ministry of Natural Resources has done is to calculate a calcium budget for that watershed, taking into consideration the deposition of calcium, the immobilization, the wearing, the calcium coming from the bedrock and how much is going out from the watershed (Figure 5).

The budget shows that, on average, per year the watershed is losing approximately five kilograms of calcium per hectare per year.

Yesterday, Dick Valentinetti was referring to taking money away from the bank without knowing exactly how much is left in the bank. We have done those calculations now for Quebec, for that watershed. The reservoir of available calcium is estimated to be on the order of 200 kilograms per hectare. If you lose about five kilograms per year, it means that basically you have about four decades left before you run into serious calcium problems.

The Quebec Ministry of Natural Resources took a look at the dendrogeochemistry of a few maples (Figure 6). This is fairly typical here. The x axis goes back to 1920, but from 1920 to the mid-1950s or so, there was no change in the concentration of iron or aluminum in the wood, which probably means, to my simple mind, that the soils were probably quite capable of buffering the incoming acidity. Then, in the second part of the century, the toxic aluminum started to go up. These are the red dots, and the iron also was going up in the wood.

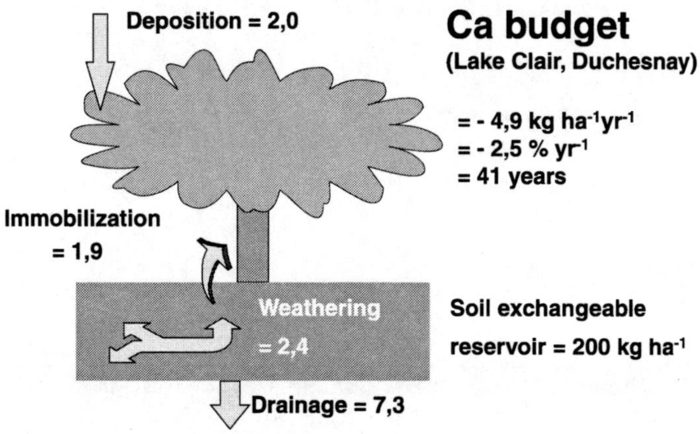

Figure 5. Calcium budget for the Lake Clair Watershed.

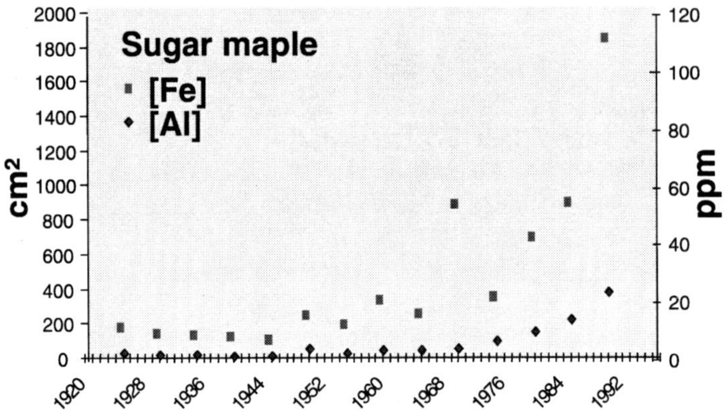

Figure 6. Dendrogeochemistry

In terms of the effects on the tree growth itself, a liming experiment suggests that calcium is the limiting growth factor (Figure 7). Basically they looked at four plots, one reference plot and three other plots to which they applied lime. They compared the growth index for each plot. Before the application of the lime, the four plots were behaving more or less the same. After the application of the lime in 1994, the growth of the reference plot continued with a slight decrease while the other plots showed significant growth and, the more lime was added to the plot, the more the tree responded.

To my mind that tells me that the trees were really starved for calcium.

There has been significant progress in addressing the acid rain problem in both countries, but much more is needed.

Much more is needed in terms of significant additional emission reductions. There will be some co-benefits. The benefits will not be only to the ecosystems, but they will be also related to better air quality.

Another point that is very important is that monitoring must continue to verify that the programs are having the desired effects.

In the future we will formalize the new targets that deal with the provinces. We will complete the targets for the eastern provinces and there will be some activities to keep the areas clean.

And, we will reassess the situation in 2004. Thank you.

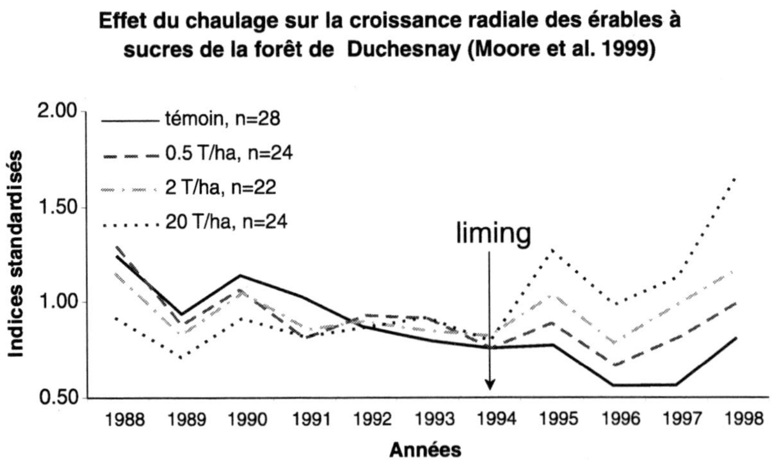

Figure 7. Tree growth response to liming suggests that the calcium is the limiting growth factor.

QUESTIONS

MR. LEWIS: My name is Preston Lewis. I am with the New York State Department of Environmental Conservation. One of the things that is very helpful to me in the work that I do is the scoreboard, the EPA scoreboard that is on the internet. Does Canada have anything that is comparable?

MR. FENECH: I am not familiar with the EPA scoreboard.

MR. LEWIS: It is basically the Phase I and Phase II units in the Clean Air Act, and what their emissions are annually.

MR. FENECH: I know we report emissions on a regular basis, an annual basis. I am not 100 percent sure about the media on which it is reported. I know it is reported in an air quality report that we do every second year, and it is probably available on the internet also, but I would have to check on that one.

MS. BAUM: I am Ellen Baum with the Clean Air Task Force. I was curious about the map on critical limits that you showed, given that you have begun your talk talking about Nova Scotia rivers having such a problem with salmon reproduction, yet Nova Scotia didn't show up as exceeding critical limits.

MR. FENECH: The map is for the problem as it is in 2010, after all programs are in place, both in Canada and in the United States.

Nova Scotia rivers are exceeding the critical loads now. However, the exceedence will be reduced over the years. Even after the implementation of Phase II, Nova Scotia will continue to exceed critical loads, not by much, between zero and two kilograms per hectare for sulfate.

That is not much, but they are in exceedence. Those rivers are expected to continue to acidify and not recover.

MR. FRANKLIN: I know that national salmon stocks are having problems with interference from farm salmon. Is there any possibility that some of the problem in Nova Scotia may be interference from farm salmon, or is it clear that it is purely ecological effects on the rivers?

MR. FENECH: There are certainly two components. One of the components is related to the salmon habitat which is acidified, which has caused a decline in previous years.

There is something that is relatively new, which is the return rate of the salmon. The return rate has dropped significantly in the past few years. The reason for the reduced return rate is unknown at the moment. I don't know whether it is competition by other genetically modified salmons, or it could be also because that salmons which reach the ocean are not in enough good shape to survive the trip.

Southeast Regional Impacts

Ellis Cowling – North Carolina State University

[Editor Note: The photos referred to in the text were not available]

I am just going to show a very few slides that will simply illustrate that the southern United States has yet continuing concern about visibility effects. We have continuing concerns about acidification effects. We have a growing concern about eutrophication effects. These effects are leading to both industrial and public interest in these matters.

This is a picture taken from the summit of Mt. Mitchell, which one might envision was one of the more remote parts, since it is the highest elevation location in the eastern part of our continent.

The chemistry of the air that you can see at the top there, where it is brown, this is near the Blue Ridge Parkway, but we might reconsider its name these days. It is more commonly brown.

The chemistry of that layer of the air is roughly what you might find over Philadelphia. That is not a natural circumstance.

This is a condition in Raleigh, North Carolina, where white pines continue to suffer from ozone exposure. My wife told me last night that we have a serious ozone problem in the central part of North Carolina yesterday, and I expect we have one today as well.

This is what the forest looks like near the summit of Mt. Mitchell. It is not true that all of that is due to acid deposition or to acid water, but is from a combination of stresses that include a nematode that is unique to these high elevation forests. It includes very cold circumstances. It includes wind damage and ice freezing and so on.

Among the contributing factors are continuing high exposures to ozone concentrations and, at this elevation, those occur 24 hours a day under appropriate summer conditions.

This is the picture of our state flower. The dogwood is a much appreciated part of our landscape. The four blossoms that you see to the right received acid precipitation before being infected with an anthracnose organism. Those at the left did not receive simulated acid precipitation. This is from the work of Robert Anderson.

Rich showed some pictures of what Boston looks like and I would like to show you some pictures from the Blue Ridge Parkway.

This is the sign of the Blue Ridge Parkway. Actually, many of you may have seen these slides. They were given to me by John Skilly(sic).

The Peaks of Otter is a very special place along the Blue Ridge. The Park Service has somewhat exaggerated the steepness of the slopes of the Peaks of Otter, but that is not what I am wishing to call to your attention.

This is what they really look like when there are two parts per hundred billion in the atmosphere. This is what they look like when they are 10. This is what they look like when there are 16 parts per hundred billion of oxidants in the atmosphere.

It may not matter if you are there on your honeymoon, but on the other hand, if you go back later at other times, indeed, it may matter to you more.

The south is not out of the woods yet. I should tell an interesting story. Many of you know that I lead the southern oxidants study, and my predecessor in the leadership of that organization was a man named C.S. Cleone (sic).

Among the very, very important things that he tried to do was increase concern about acid deposition and ozone exposures and particulate matter exposures in this part of our region.

I don't know how many of you know C.S. Cleone, but he came to this country in 1957, and he likes to tell this story.

I came to this country in 1957 and I heard about this civil war. I also heard that it ended in 1865. I looked up recently the amounts of federal investment in scientific research in the different states and I found the state of New York, the State of California and the state of Massachusetts each were receiving more than the combined investment of the Federal Government in research in the 14 southern states. I wondered if the war was over.

I would like to end with the suggestion that all of us in this room are recognizing that we have a long ways to go if we are going to manage air quality in this country or in Canada or, for that matter, learn enough to help others in other parts of the world to manage it more wisely.

We must continue to make significant investments in research, both in those 14 southern states, and in the rest of the country, even in New York and California and Massachusetts.

QUESTIONS

AUDIENCE: I come from Virginia myself in the mountains, and I am curious to know how much of that is long-range transport and how much of that is coming from downwind, upwind, whatever, nearby states.

MR. COWLING: It comes from west of us, of course, but we are a substantial contributor to our own pollution. We had some discussion earlier about ammonia emissions. The ammonia emissions of the state of North Carolina now exceed the NO_X emissions of the state of North Carolina.

That is based on an assumption that cars produce only six percent. The current best evidence would suggest that it is probably 13 percent of the total ammonia emissions are from automobiles in North Carolina, instead of the six that the state has included.

I think we certainly are a downwind state from those west, and those of you who have looked up the emission statistics will recognize that Tennessee is a big emitter state and not so big as North Carolina.

We are all in this together, and I think that is one of the most important things we must understand. No state can do an effective job of managing its own air quality. No province in Canada can do it either.

We will have to undertake to consider the values our society wants to have from our landscapes and from our scenic vistas and from our trout streams and so on, and we will have to make more effective use of science in making those decisions.

I think North Carolina must do its part, together with many other states, if we are to improve air quality in my state or in other parts of the southern United States.

Question: (Inaudible)

MR. COWLING: I will repeat the question. To what extent does acid rain predispose flowering dogwood to anthracnose.

I think the experiment that Bob Anderson has done, he has done under controlled circumstances. At least under those simulated conditions, it is pretty clear we have a modest epidemic of dogwood anthracnose developing in our state.

We presume that the acidifying effects are a part of the predisposition that may lead to this episode. We do not know how much genetic variability we have in the populations of dogwood.

Among the things that we would like to know is not that the whole species is threatened, but perhaps some are, and we will need to do further investigations if we are to protect our state flower.

Remarks of Congressman John E. Sweeney

Thank you for the invitation to participate in the Center for Environmental Information's conference on one of the most urgent environmental problems facing my state of New York and the nation – the issue of acid rain.

I am pleased you are all here to combine your experiences and expertise in an effort to highlight the scope of the acid rain problem facing our nation.

Acid rain is truly a national problem, and the participation of all of you concerned citizens hailing from dozens of states not only proves this point, but will help make the case throughout our country. Thank you all for your contributions.

As a New Yorker, it is good to see Adirondack Park Agency Chairman Richard Lefebvre, representatives of New York State DEC, and my good friends at the Adirondack Council, Adirondack Nature Conservancy, and Adirondack Mountain Club.

I have seen firsthand the devastation that has slowly and steadily eroded the ecological health of the region.

Since coming to Congress, I have been working to stop acid rain damage to the Adirondacks on two fronts: emissions legislation and funding for acid rain research. As the lead sponsor of H.R. 25, the Acid Rain Control Act in House of Representatives, I am working hard to advocate for further emissions reductions using the framework of the current acid rain program.

The bill would provide cuts to sulfur dioxide and nitrogen oxides to stop the damage of acid rain to our precious Adirondack Park; would require that the EPA keep its commitment to regulate mercury from power plants in a timely manner.

Introduced the first day of the new Congressional Session, H.R. 25 – so designated as a sign of its importance – currently has 26 cosponsors. We now have the opportunity to reach out to more members from many more states and seek their support for this legislation – and I ask for your help in doing just that.

We also have a unique opportunity to work with President Bush and EPA Administrator Whitman. Both have made strong public commitments to further reductions in Sulfur, Nitrogen and Mercury, and I am working hard to transform those commitments into legislative action.

Five years ago, H.R. 25 was a radical concept. I am confident that we have a real opportunity to make our vision a reality.

I am also pleased, as a member of Appropriations Committee, to be working on an equally important front – to ensure adequate funding for acid rain monitoring and other basic research.

Information derived from acid rain research lays the groundwork for the evaluation of our current programs and for the advancement of acid rain policy initiatives. I am a firm believer in the idea that sound science drives good policy.

I would like to specifically commend New York State scientists present at this conference: Charlie Driscoll of Syracuse University and Greg Lawrence of RPI and US Geological Survey. Your work in New York has been crucial.

Although I won't go into too much detail, it is imperative that we continue the highly successful acid rain monitoring programs – such as the Adirondack Lake Survey Corporation, which does water quality sampling of 52 lakes in the Adirondack Park, and the cloud water monitoring program at Whiteface Mountain. I will continue to work to ensure that these programs receive full funding.

Again, these research efforts – which you are highlighting here this week – pay dividends by providing a greater understanding of the impacts of acid rain. Let me give some examples.

Another Adirondack scourge, black flies, is being compounded by acid rain. Because black flies thrive in acidic waters the acid rain problem has exacerbated the black fly problem in the Adirondacks.

Loons, which are the symbol of a healthy Adirondack wilderness, are adversely affected by higher levels of mercury, one of the deadliest toxic metals associated with acid rain; Cranberry Lake and Stillwater Reservoir, located in the Adirondack Five Pond Wilderness Area, have substantial mercury contamination.

Research has shown that our spruce forests are

dying because the trees, made weak by acid rain, are threatened by insect infestation and are susceptible to winter injury.

Adirondack lakes, such as Honnendaga Lake, that are dead appear to be clear and beautiful. It is ironic that these lakes, which are so aesthetically pleasing, are devoid of life.

Lakes are an eerie shade of blue because acid rain has killed off plankton and other microscopic life that gives a healthy lake its aqua-marine color.

Amid all of this bad news does come a bit of good. Data now shows that while acid deposition continues to do damage as evidenced above, the current acid rain program, based on the 1990 Clean Air Act Amendments, has shown measurable success and is putting us within reach of a recovery.

This data confirms that the current acid rain program can be an effective model for instituting greater emissions controls.

Without good science, we could not make such a determination, nor could we understand the full extent of the acid rain problem.

Your work and commitment to this issue are therefore crucial to our efforts to form good public policy. I would like to closely identify with your work and again thank you for the opportunity to speak with you today.

Again, we have a legitimate opportunity for progress in this Congress. This is a fight we can't afford to lose.

SESSION VI: Economics of Acid Rain

John D. Kinsman – Edison Electric Institute

I want to give you two or three minutes of introduction. I will lead off with a question or two.

How many of you know that, after the implementation of NO_X controls by 2004 these emissions from the electric utility sector will be less than 20 percent of total U.S. man-made NO_X emissions?

How many of you know that, when Title IV is completely done around 2010, that actual U.S. emissions of SO_2 will be at their lowest mark in a century, largely due to the reductions from utilities, except for two years during the Great Depression?

I will give you a little information from DOE and EPA. You see in Figure 1 that over the last 30 years, roughly, coal use and coal electricity are approaching a tripling. You also see that the emission rate – pounds per unit of energy – for NO_X is down 50 percent, SO_2 is down 70 percent, PM_{10} is down 94 percent.

Absolute tons of emissions from electricity are on their way to about 50% reductions (Figure 2). This is all the more significant, of course, in the face of this tremendous increase in electricity.

Why is this? Because we have substantial regulatory requirements on electric power production, not only currently but in the future. We have national ambient air quality standards for ozone, SO_2, nitrogen dioxide and PM, which led to reductions in emissions even before the acid rain program. We have the two phases of the acid rain program, reducing NO_X and SO_2. Don't forget the NO_X SIP Call in 2004 and related ozone reductions in the Northeast Ozone Transport Region. Mercury reductions are required at the end of 2007. We have fine particle and 8-hour ozone regulations toward the end of this decade and regional haze, essentially taking emissions down to zero over 60 years.

People tend to ignore all these reductions that have taken place or are in the pipeline. I will conclude by noting that we have been having discussions about what we call a comprehensive approach or integrated strategy to take these reductions a step further. We have been talking, for several years now, with the administration, the environmental groups and Congress about an approach whereby we could eliminate these many steps and craft a more streamlined approach which would allow us to eliminate duplication, minimize confusion and reduce uncertainty. We call this the comprehensive approach or integrated strategy. Congress is very interested in this. The President is very interested in this, and so are the utilities.

With that, I would like to introduce our first speaker, who is Jeremy Platt, the Electric Power Research Institute.

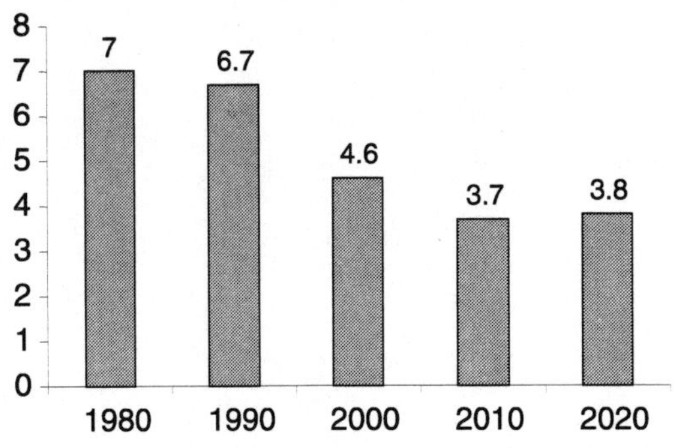

Figure 1. Nitrogen oxides emissions - fossil fuel electric generation (millions of tons). Emissions in 2010 and beyond will be substantially lower than shown here, due to new EPA regulations on particulate matter, ozone, and regional haze.

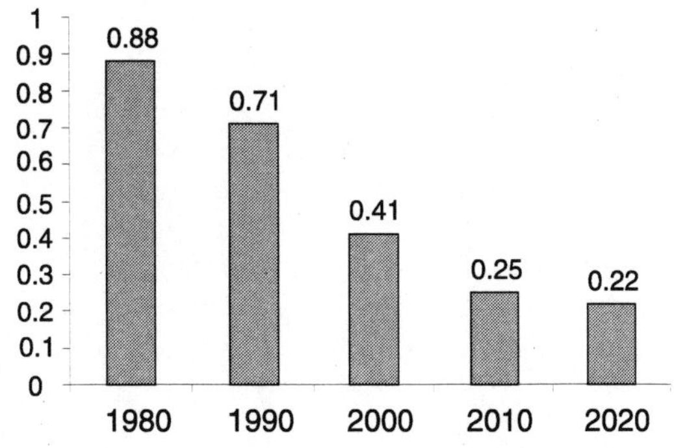

Figure 2. Nitrogen oxides emissions rate - fossil fuel electric generation (pounds per million BTU). Emissions in 2010 and beyond will be substantially lower than shown here due to new EPA regulations on particulate matter, ozone, and regional haze.

SESSION VI. Economics of Acid Rain

Compliance Costs and Markets

Jeremy Platt – EPRI

Acid rain has been an issue for over two decades. Federal action came about in the 1990s. The clock started ticking in 1995 – well, actually a little bit before that. This process, we have been talking about it for almost 20 years now. Maybe that is where some of the impatience comes about.

I was going to touch on the framework of regulation and what people actually did in Phase I, discuss some sources of error, or so-called forecast error that the industry is often taken to task for. I will speak a little bit about maybe our inability to look ahead in these markets, and particularly the SO_2 market, and give you some perspectives on cost, although there are experts here on the panel also who can answer questions about cost better than I.

I wanted to acknowledge some of the people who have assisted me, or references that have been helpful. Most of all, I thank Denny Ellerman, because it was his inability to be here that eventually led to the recommendation that I try to stand in and address some of these industry cost issues. I have worked with Denny for quite a while. He and a bunch of other authors at MIT have come out with this book called *Markets for Clean Air*. It was really, I think, widely accepted as the authoritative review of what is going on in the SO_2 business. I really recommend that to you very highly. It is fairly readable and parts of it are pretty inscrutable. You will find it a great background. It leaves almost nothing untouched.

Also, John Dean, with J.D. Energy, for some perspectives on markets, Tom Houston of UVA on the FTD data I will be sharing with you.

Here is a picture, and I really want to qualify this picture (Figure 1). I was laboring over it this morning when I realized that the scale looks very unfamiliar to all the other estimates that you have seen in the handouts. This is for only the Phase I affected units. That is why the scale looks like it does. So, keep that in mind.

What you see in red are the actuals, what is going on. You see an estimate of how those Phase I units might have behaved in the dashed blue line at the top. Then you see what happened when 1995 came along. The cap was imposed. It took a little bit of a step jump down, because there were incentives to add scrubbers and so on. There were higher emissions allowed for companies that were scrubbing in these years. Eventually by 1997, we

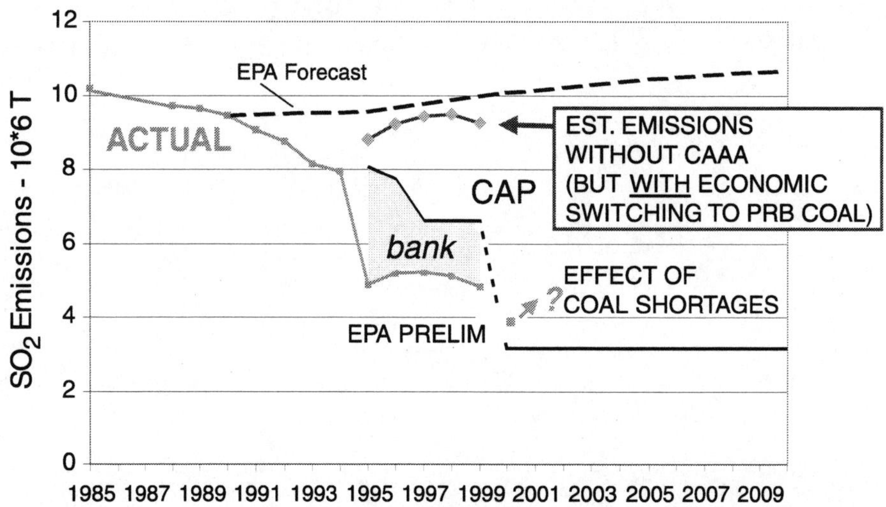

Figure 1. SO_2 allowance market: Title IV in historical perspective. Source: MIT 2000, updated 4-00; 2000 actual McLean 3-28-01.

reached the cap. Throughout that time, the industry was emitting well below the cap. That was probably, the most singular feature of Phase I. It was not entirely unexpected but it was interesting to see that build.

I am taking some liberties with some of these data as well (Figure 2). In this case, this is a view of where we might have been without the law, but primarily influenced by the role of Powder River Basin coal, that helped drive down allowances. Even if there had been no Act at all, we would have seen some dropping off in SO_2 to a lower level At the end of March 2002, EPA announced that we came down a million tons from the last year.

One thing we can assume is that right now we are facing coal shortages. The price of coal has run up dramatically. You hear about natural gas prices running up and electric prices running up. Well, who ever thought it would happen to coal? No one in the last 25 years. In the last couple of quarters we have had a boom in the price for every type of coal, reflecting the shortage conditions totally around the country, low inventories, and inability to get hands on enough supply. That means, in order to get their lights on, companies have to burn anything that they can get, regardless of the sulfur content.

What that means is that this year we are going to find that probably emissions will be up a little bit, because we are burning allowances to allow the industry to burn this range of coals that they weren't expecting to have to use so intensely. That is probably also one of the most singular features of this type of trading program. It allows the industry the flexibility.

Normally we would have thought of flexibility on a week-to-week basis. Now we are looking at almost a pervasive or seasonal or multi-month shortage, for which having a bank is going to make a tremendous difference. Having this kind of legislation to deal with these changes in market fundamentals is probably one of the most stellar features of this type of regulation.

On the SO_2 allowance prices, (Figure 3) I think it is important just to point out that it really does behave like a market. This is another whole thing about flexibility. There are all sorts of questions about not predicting what happened to these prices and I think costs were quite a bit higher than all these prices.

I think the costs for switching were on the order of \$150 a ton and above, and some of the scrubbing costs were on the order of \$230 plus on average, but in many cases, up around \$300 and \$400 and so on. That is what they incurred in their investments but we see the price behaving quite differently for a variety of reasons that I can't go through entirely.

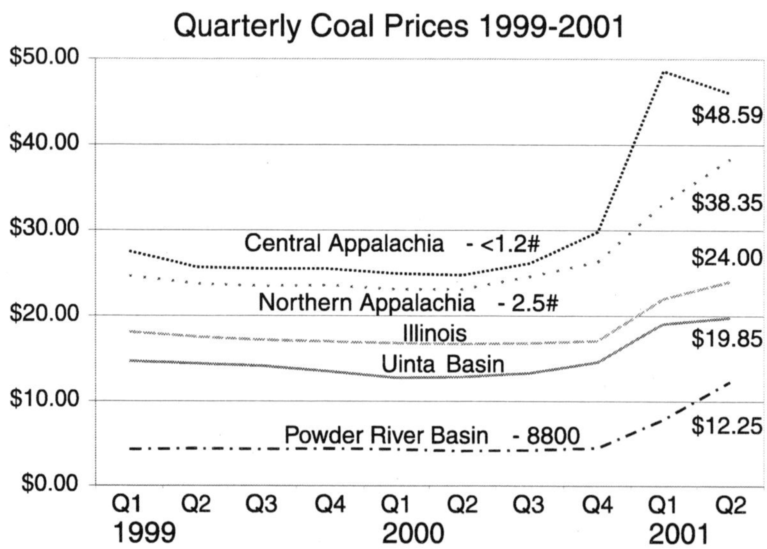

Figure 2. 2001 surprise - coal price spikes. Source: JD Energy.

We tested the variable costs for running a scrubber .You have some cost in your equipment and you are not recovering any fixed costs of scrubbing, but you are running very, very low. Then we found out that costs jumped up to around $200 first in 1998 and now again. For example, these prices at a fairly high discount rate are equivalent to 2010 prices of maybe $350 to $500 a ton. That is in the range of what people have been projecting for marginal costs of compliance in 2010.

We have been jumping all around. There is a price collapse that occurred, when the NSR cases, the new source review cases, were first announced. That really drove down prices because people thought there might be forced scrubbing across the industry and there wouldn't be anyone to buy an allowance and it would erode the value of whatever stock of allowances that they had.

When I think back over some of the elements of regulation, first of all, I say banking and trading were widely embraced, it permitted internal efficiencies which are very important within units under ownership of any one organization. It provided operating flexibility, as I mentioned. In particular, it had a big impact on the design of the scrubber itself, because you could build a single module scrubber.

You didn't have to have a spare as in the old rules when if any part broke, you needed to run a spare and make sure you stayed in compliance. Now you could build a more cost effective larger vessel with greater through-puts of gas and so on and achieve true engineering savings as well, not just modest savings. Finally, it provides intricate intertemporal flexibility, which I just mentioned seems to be quite important right now.

The cost of compliance did fall with a proliferation primarily of what I call unexpected switching options. At the two-and-a-half pound level, a number of coals came in. People thought that central Appalachian coal was going to drive the cost of compliance. It turned out that people got by without using much central Appalachian coal after all. They found some coal that was below two-and-a-half pounds out in Indiana. They used that. Then there was a tremendous effect from the Powder River Basin coal that came in, greater than people had anticipated. There were changes in infrastructure that permitted more coals to come to bear and more blends to take a role that gave companies options to switch.

There are some people who have said there has been too much over-compliance, if the utilities were really smart, they wouldn't have built so many scrubbers. I call that a red herring because it is just a question of when they might have built those scrubbers. If you build the scrubbers you will have them when needed; within a couple of years you would have built them anyway.

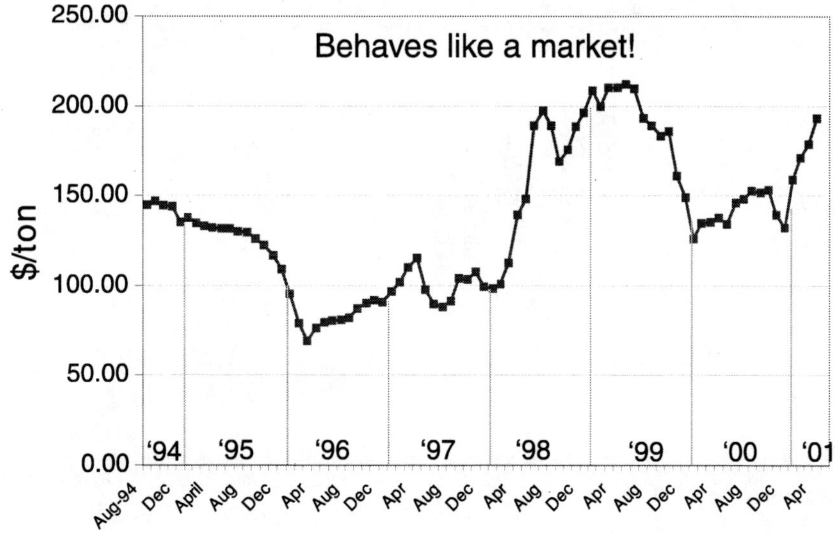

Figure 3. SO_2 allowance prices. Source: TradeSpark MPI.

You can offset your costs in overspending early by how you use allowances. So, in some sense, that whole thing comes out more or less in a wash.

There are no ill environmental effects from the trading and I will try to point that out. I would say the importance of the stringency of the two-and-a-half pound cap has probably been an underlying element in why all these competitive options were able to jockey against one another and help to drive down prices. I think that is the one thing that was not necessarily given as much explicit attention in the MIT book, but it is mentioned by Ann Smith in her paper.

Just looking at the geography of reductions, this is a quick view of what people did do. This ranks states from east to west (Figure 4). Here are the big states, Pennsylvania, West Virginia, Ohio, Kentucky, Tennessee, Indiana and Illinois and then Missouri and Georgia. So, we are moving more or less eastward.

You see about half the reductions were taken up by scrubbing and about half by switching. You see the tremendous reductions taking place in what I call elephant country, meaning that is where the emissions were. That is where the reductions took place.

I think it is interesting to compare what was done to what was required (Figure 5). What we found here is, looking at these different states, the required emission is in grey and the black is actual. Even in all the upwind states, they reduced their emissions. Even though there was trading going on, they ended up reducing their emissions considerably in the upwind states, the ones that I think are of considerable concern to legislators in the northeast.

I think I am probably going to try to be quick over this next figure (Figure 6). This is complicated, and we can go back to it maybe privately. What we are trying to show simply is that, relative to companies' allowances, they adopted quite a range of strategies. Most companies adopted to control their emissions on their own account and not to be net buyers. You can see that there have been some changes in that regard, meaning this is the allowance allocation, the thick black horizontal line, and the emissions relative to that were less than that allocation.

One exception is out there in Illinois. They were considered kind of a novel leader by buying allowances and being able to continue to use local coal. Now they are going to switch to Powder River Basin coal. They are going to go way down in emissions in that location.

There has been an increase in trading, as you see, in 2000, and that is recorded for both internal trades and external trades between utilities and with brokers (Figure 7).

If we summarize Phase I, we see the history of these cost estimates at around half a billion to $1.3

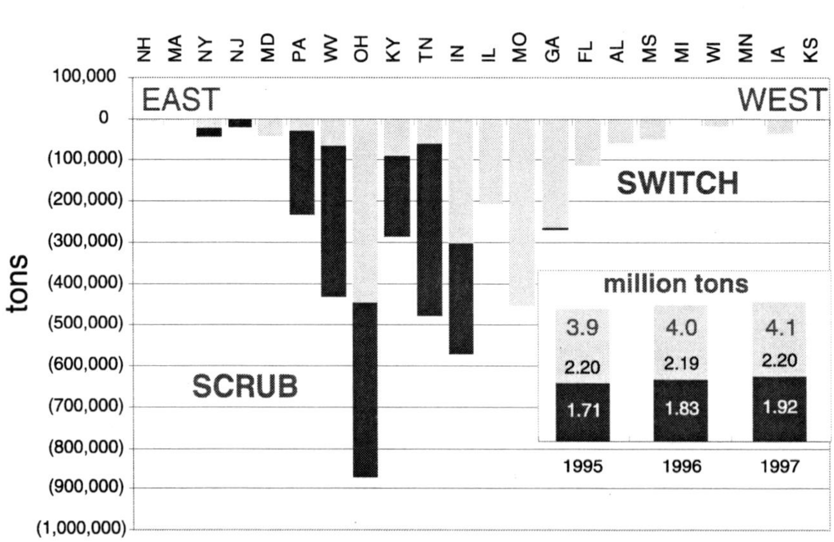

Figure 4. Geography of reductions (1997). Source: MIT 2000, updated MIT CEEPER 4-00.

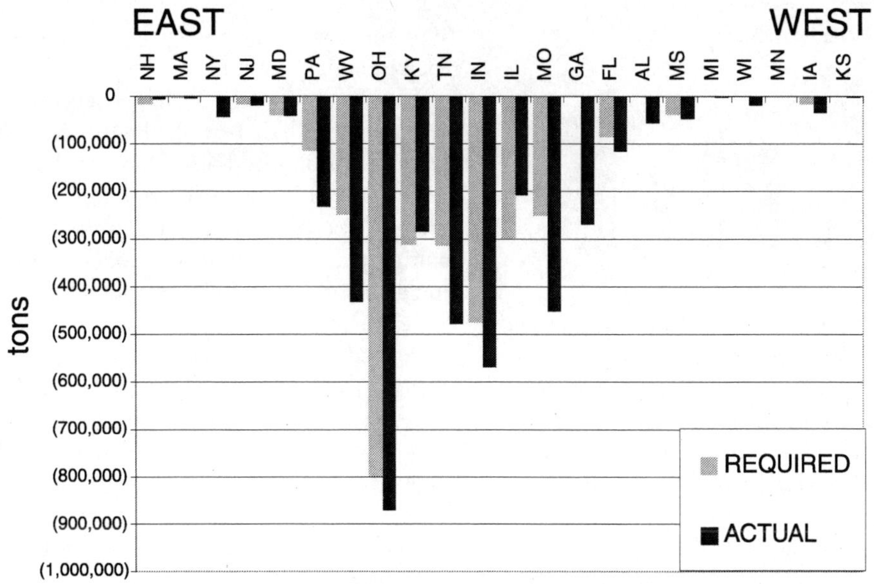

Figure 5. Impact of trading on reductions (1997). Source MIT 2000, updated MIT CEEPER 4-00.

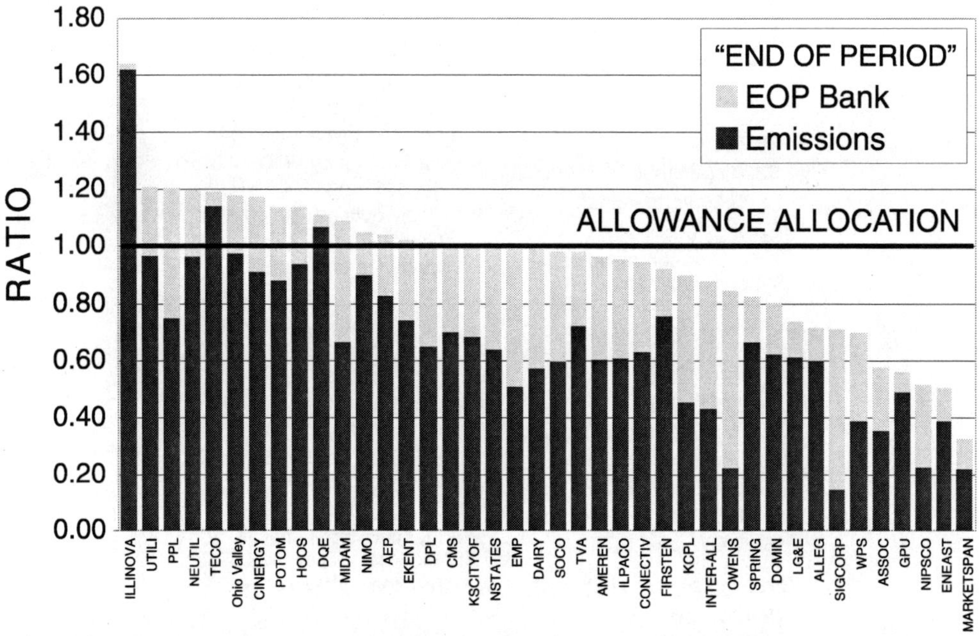

Figure 6. Phase I retrospective emissions, banks and market activity relative to allocation. Source: MIT CEEPER 4-00.

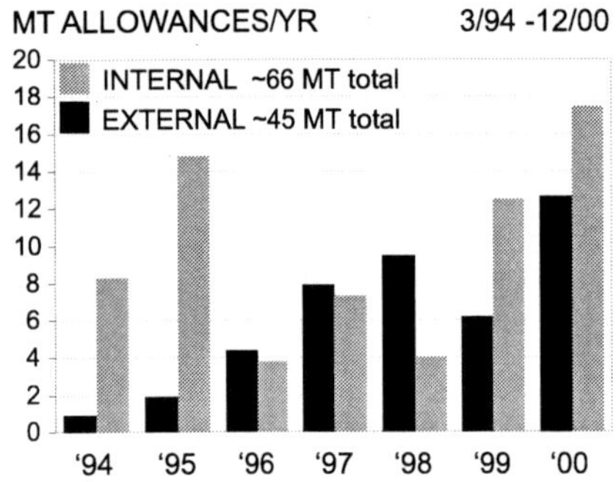

Figure 7. Trading activity is increasing - recorded on the EPA Allowance Tracking System. External transfers are with brokers, between utilities, with fuel companies ... 2000: ~50% of external trades were utility-utility. Source: http://www.epa.gov/airmarkets/trading/so2market/cumchart.html.

billion a year for the industry (Figure 8). It turns out that it came in at about $730 million a year, according to MIT's tally. Overall, they have studied what kind of savings have come out of all this and projected that to Phase II and they suggest there is going to be about a two-to-one ratio, that with trading, the savings are about two-to-one, not 10 to one, not 20 to one, which have also been reported in the literature.

Some of the factors that have happened in driving it down have been the Powder River Basin coal phenomenon, additional fuel switching, and alternatives that I mentioned. You realize that variable costs are quite low. Finally, there are multiple options of heightened competition, and I would stress the importance of the stringency of that experience.

As we look ahead, here is another set of cost estimates looking ahead (Figure 9). We don't have the retrospective on Phase II, plus the rules are changing. So, it is really anyone's guess.

By the mid-1990s, I would say those numbers had come down to about $1.7 billion and then that includes, say, the $700 million that I just mentioned.

We see where we are going. From 15 to 20 gigawatts could have been added with scrubbers. Then we are saying, well, what would happen as

$$$

Cost Estimates

$ 0.5-1.3 B/yr

$150-300/t (average)

Actual

$0.73 B/yr

$200/t (average)

$300-800/t (marginal)

Savings (Projected Ph. 1 and 2)

~ 50% vs. command and control *(MIT 2000)*

Factors

- PRB coal phenomenon
- Additional fuel switch alternatives (greater midwest, less Central App.)
- Lower FGD O&M and greater removals (>95%)
- Cost floor set by FGD (notion of "irreversibility")
- Multiple options - heightened competition
- Importance of stringency

Figure 8. Phase I summary.

$$$

Cost estimates

$1.5-6.5 B/yr => $1.7 B

$225-500/t => $200 ave.

=> <$435 to >$500 marg'l

($275 w/ lower coal use)

15-20 GW add'l FGD

Half-cap scenario (50% cut)

$2-4 B/yr add'l (& new gen)

$350-500/t ave.

($800-1,000+ marginal cost)

60-100 GW add'l FGD

Factors

- New Source Review -- forced scrubbing?
- Questions of "sterilization" of allowances, and when:

  ~0.5 M t/yr added to supply?
- Further cuts: additional 25-75% for acid rain or PM2.5?

  $100-400/t price range
- Regional haze - moving East?
- Mercury rules -- MACT

Figure 9. Phase 2 look ahead.

we go ahead. One is the half cap view, the 50 percent cut in SO_2. Sixty to 100 gigawatts of scrubbers could be required in addition to what might have otherwise been required for Phase II, and that is going to cost some real money. I am not saying too much or anything like that. I am just saying that is the consequence of that level of stringency.

We find that key factors include such things as the new source review. Are we going to force scrubbing for a while or will we sterilize the allowances that the new scrubbers might be generating? There could be an oversupply of allowances in that case or not. There are some regulatory and interpretative uncertainty about how the market might be affected by those NSR cases. Right now people are saying the price range in Phase II, is on the order of $100 to $400 per ton, for a more updated picture.

Here, I am saying there has been steeper discounting of prices in the out-years (Figure 10). This is a measure of that price curve from before, but I have shown some of the auction points.

Every March the EPA has held an auction. What they have done for the current year auction, in black diamonds, and the grey circles, the seven year auction, which is a measure of what they are willing to pay for an allowance, what they think it might be worth to pay now for something they could use seven years in the future.

Those things were fairly closely aligned for quite a while but now you see they have really come to split in the last two years, suggesting that there is a fairly steep discount rate being placed on putting that cash down now for something that can only be used seven years from now.

That is what you see now, a tremendous discount. Like I said, at those kinds of discount rates, the $200 price that we have just come to will align with about $350 to $500 prices in 2010.

We see the next rate in sight, but it was not driven by acid rain (Figure 11). It is driven by all these other reasons, particularly regional haze and resource review and so on.

The newer units that have already been constructed earlier are shown as dark grey circles. That is not a complete picture of those. We also see haze units in light colored circles and they are in the $100 to $200 range.

When I come down to this I ask, how do we view all these different costs in context? Here is just one company's views of what they might view the slate of costs that they could be facing, looking at all of these things together (Figure 12).

It comes out in sort of increment dollars as they project the cost stream out to about $450 a

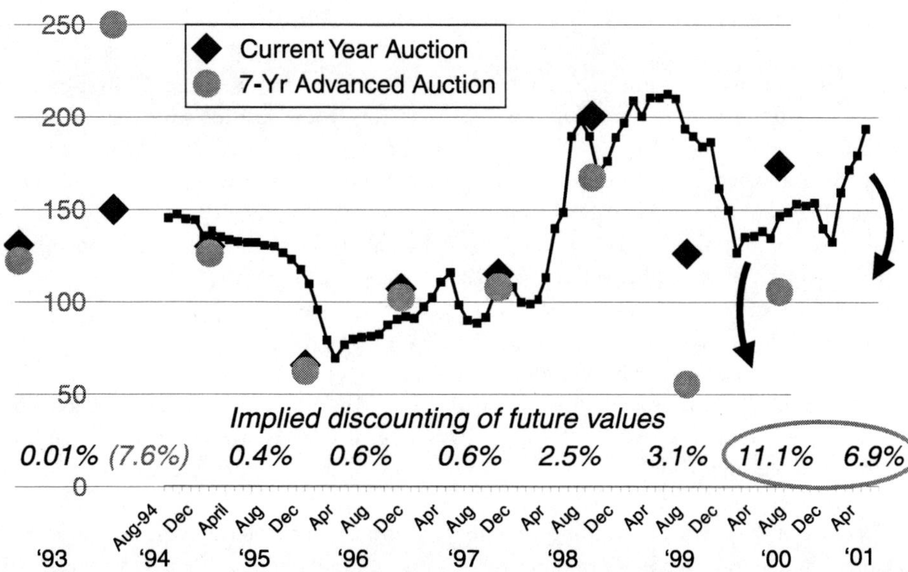

Figure 10. Steeper discounting of out years from EPA Auctions. Source: EPRI; EPA Auction Results; TradeSpark MPI.

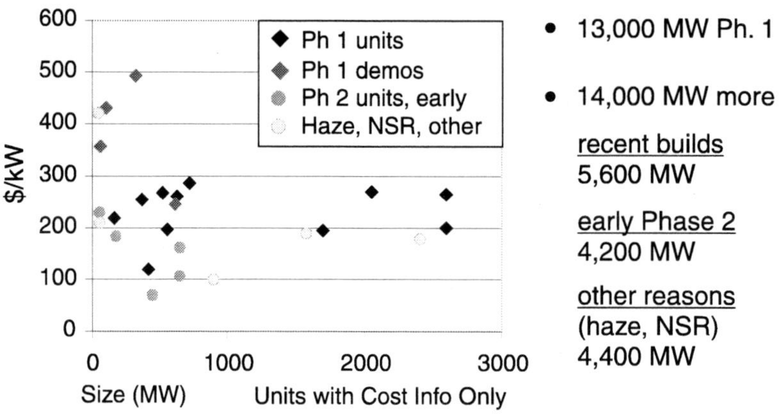

Figure 11. Next wave of FGD in sight. Source: EVA data.

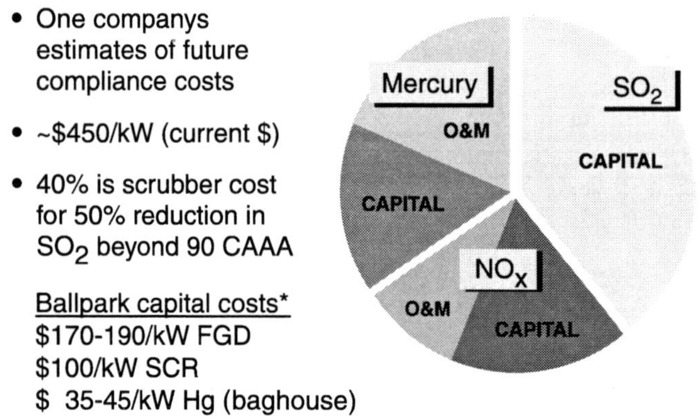

Figure 12. SO_2 control costs in context. Various sources: J Dean, EPRI, DOE.

kilowatt. A huge fraction of that is taken up by the SO_2 side of the story, and those scrubbers are the big capital cost.

You might ask, where is the O&M costs in this, because there are the O&M costs and other costs related to that and disposal costs and so on. They are being offset in the higher sulfur coal that they can use.

I think I have summarized my main points, that the costs have come down. The level of stringency, I would allege, is probably the principal agent in the Phase I experience.

The future looks highly uncertain, really driven by politics and regulations and stringency and so on. It is because of that, that I have come up with extrapolating the Phase I experience to other types of stringency levels.

You need some source of over-compliance at a relatively low cost to create a supply of allowances for trading or for consumption. So, where would the supply of allowances come from under strict NO_X or in a mercury case? You would expect it would be very difficult to generate excess allowances in those cases.

QUESTIONS

MR. MC CORMICK: I am John McCormick. I am with the Citizens Coal Council in Washington. I wanted to dispute some of your inferences that the field switch did not include as much central Appalachia as it did the northern plains.

I am looking at form 423 data. In 1985, in central Appalachia, compliance coal was about 33 million tons production. In 1999, it is 54 million tons. The non-compliance, or the 1.2 to 2, is mountain top mining.

I want to say that the Citizens Coal Council is 54 organizations in the coal-producing states. They are with you on the issue of reducing aerosols, but they realize that the central Appalachian Mountains have been sacrificed in large part to meet the goals of Phase I of the Clean Air Act.

They are not going to allow this to happen again. Further reductions of SO_2 emissions that are met by trading – i.e., fuel switching – that puts more pressure on the bituminous low sulfur coal deposits in central Appalachia. This will not be tolerated by the 54 organizations I speak for.

We need to come together because technology is part of the solution and not so much fuel switching, because this is the cost of fuel switching. If anyone would like to talk further about it, I would be more than happy to. Thank you.

MR. PLATT: One of my points was that the reliances on Appalachian coal turned out to be somewhat less than it was anticipated to be, but it has been very important.

I would say probably a lot of that strip mining in central Appalachia is because that is the economic way to produce coal for whatever reason, regardless of whether there had been a Clean Air Act. There would have been equal pressure to produce coal in that manner in that region.

MR. LEWIS: Preston Lewis, New York State Department of Environmental Conservation.

I know that EPRI has, for a long time, been the research arm for the utility industry. I am curious to know what kind of research is being done in reducing the sulfur dioxide emissions, particularly from existing units, perhaps competing with the scrubbers. Are there improvements in the scrubbers?

MR. PLATT: I don't think we have a very active program on scrubbing right now. I think we have cut back the intensity of that. Dick Rudy is the guy at EPRI who manages that program.

I know there are some advances being made, for example, whether it is by us or others. I hear about fertilizer by-products using an ammonia-based system rather than a limestone-based system that maybe produces a product that would have high value in the fertilizer industry. The problem is in permitting large tanks of ammonia on site.

I have heard about it for smaller plants. Maybe a circulating fluid bed might be one of the methods that might come into its own. This has just been demonstrated in Austria, I believe.

I say the commercialization and advance of technology is continuing progressively and incrementally on SO_2, aiming for higher removals, lower operating costs, better by-product utilization, that kind of thing.

In the old days we thought just about the biggest dirtiest units, wet limestone scrubbing or maybe magnesium enhanced limestone scrubbing, which turns out to have some real advantages.

SESSION VI. Economics of Acid Rain

Economic Benefits of Controls

Dallas Burtraw – Resources for the Future

We heard Representative Sweeney say today that one can perceive a trade-off between economics on the one hand and the environment on the other hand. He suggested that folks may be inclined to approach policy from one camp or the other camp.

The work that I am drawing on links these two areas: economics and the environment. This work is the attempt to estimate the economic cost of environmental degradation (Figure 1).

This body of work is at least two decades old, and to some the work in this area has been frustrating. Richard Ottinger wrote, in the early 1990s, that previous studies aimed at estimating external environmental costs failed to value such a significant portion of externalities as to render the values irrelevant. He went on to suggest that the very small values arrived at in these studies raise a serious question about whether it is worthwhile to pursue further the evaluation of environmental externalities. An environmental advocate may have a feeling that the environmental wars cannot be won on the basis of economic analysis; they can only be lost on the basis of the economic arguments. But that is not the case.

I want to draw on the main findings from the economics and integrated assessment literature about the performance of the acid rain program. This work has involved a large number of coauthors and has appeared in the *Journal of Political Economy, Environmental Science and Technology, Contemporary Economic Policy, Journal of Air & Waste Management*, plus some unpublished discussion papers. Some of the findings are encouraging, with respect to the success of the program, and some of them point the direction for ongoing research.

There are several findings that emerge from this review. One is that there is an adequate basis to value the effect of acid rain on the recreational and consumptive use of resources. This area of economic benefit is characterized as "use value." An unhappy finding from the literature that I will survey is the conclusion that improvements in use values alone do not justify the costs of Title IV emission reductions.

Another important category of value is characterized as "non-use value." These are values individuals hold for the resource even if the individuals do not intend to directly experience or consume the resource. The area of non-use value includes the estimation of broader ecological changes. A second finding in the literature is that the science and economics are still not adequate to value in economic terms most of the ecological effects that we have been hearing about at this

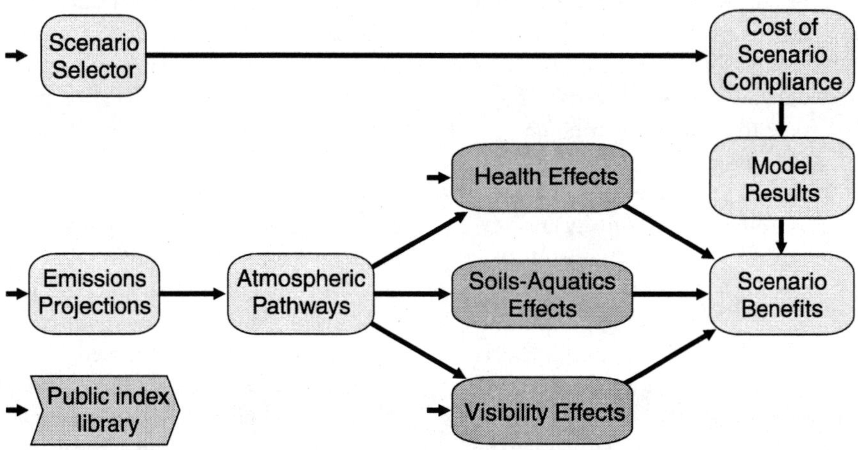

Figure 1. Drawing links between economics and the evnironment.

conference. However, there is evidence in the literature that suggests that the relative magnitude of non-use values and the value of ecological improvements viewed in this way could be very large.

A third and very important finding in the literature is that values associated with improvements in human health, which Dr. Lippmann talked about yesterday, provide the lion's share of quantifiable benefits. Taken alone, the value of improvement in human health is an order of magnitude greater than costs. These estimates, independent of the assessment of ecological improvements, appear to justify in and of themselves further reductions in the emissions cap.

A fourth finding that I will touch on is that trading of SO_2 emission allowances has not undercut the benefits of the emission reductions and may have added to those benefits slightly.

Finally, I want to look forward to a contemporary policy issue, which is the design for a regional NO_x trading program now expected to be implemented in 2004 in the eastern states.

In an important sense, my topic – the economic valuation of environmental improvements – has come late in the ball game. It has been considered largely after the SO_2 policy was designed and implemented. Indeed, the topic comes late in this panel and at the tail end of this conference. Rather, it is at the beginning when, as an economist, I might like to organize an assessment to identify where the greatest benefits are. It is also when one might want to use such an assessment to prioritize the expenditure of research dollars and public efforts to improve the environment. For obvious reasons, that is not the way that NAPAP unfolded in the 1980s. Extensive resources were spent to characterize airborne transport of emissions, and terrestrial and aquatic effects of acidification, but other processes omitted from this analysis proved to be equally critical drivers in calculating benefits and costs. We have heard about these other drivers in the last day and a half. They include the epidemiology of health effects, the economic valuation of health effects, the change in coal transportation costs (Carlson, et al. 2000; Burtraw et al. 1998), among other topics. These were not studied in adequate economic terms before implementation of the SO_2 trading program, and since the implementation of the program there have been significant contributions to the academic literature that improve the type of assessments that can be made. That is why, at the conclusion of my talk, I want to draw attention to the NO_x trading program as a contemporary policy issue that should be considered in light of what environmental economics can contribute to future policy.

I am going to begin by summarizing the results of an integrated assessment effort that involved 30 researchers from about a dozen different institutions. The result was a public domain integrated assessment model called the Tracking and Analysis Framework (TAF) (Bloyd et al., 1996). Figure 1 provides an illustration of the model, which is publicly available at www.lumina.com\taflist. One publication based on this effort appeared in *Contemporary Economic Policy* (Burtraw et al., 1998), and it was one of the few publications independent of federal agencies that attempted to estimate benefits and costs of the Title IV SO_2 and NO_x programs. The integrated assessment model linked changes in emissions to changes in transport, to changes in concentrations and deposition, to changes in a number of environmental end points, including lake acidity, and then to use values associated with lake acidities in the Adirondack Park region. The concentrations were estimated on a state-by-state basis across the nation with respect to changes in sulfates and nitrates. Human health as well as visibility effects were quantified and valued in monetary terms.

The midpoint estimates resulting from the study are reported in Table 1. I want to emphasize two important caveats. First, note that the numbers in the table are reported "per affected person." The populations affected by changes in the environment vary with the nature of the environmental change. In the case of health benefits, the population of the nation are included. In the case of recreation or visibility changes, only regional populations are included. Hence, the numbers in the table are not additive.

A second caveat is illustrated in Figure 2. The integrated assessment gave projections through 2030 for a number of environmental end points, as illustrated. Notice that the vertical axis displays the estimates on a log scale. This is appropriate when

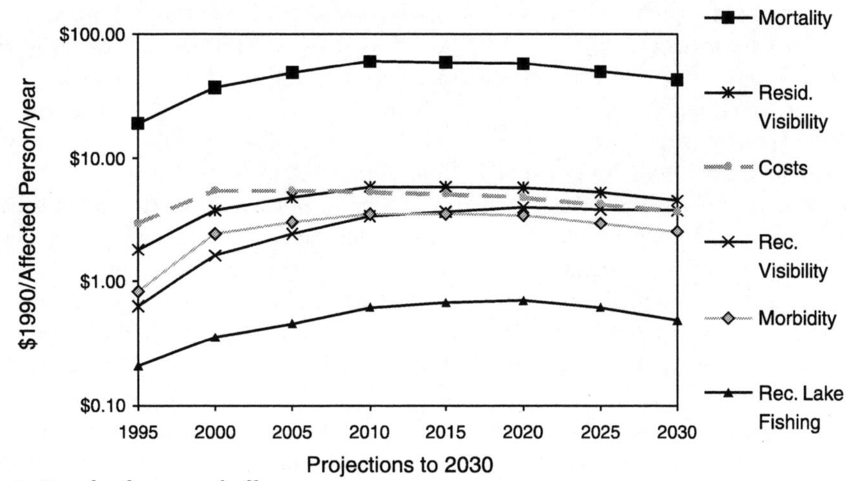

Figure 2. Benefits for assessed effects.

you look at the magnitude of uncertainty through the entire valuation pathways. The log scale is a useful visual tool for comparing the benefits and costs while retaining a flavor of the uncertainty associated with the midpoint estimates.

Costs are estimated to be about $6 per person for electricity consumers across the nation. The benefits of improvements in recreational lake fishing are substantially less than this cost (Table 1). This benefit estimate is achieved using a reduced form model that was calibrated to the Magic model and only considers effects in lakes in the Adirondack Park region. Although there are a lot of other lakes and streams that are going to be affected around the rest of the nation, there are also going to be different user populations that will be affected at these locations. So, on a per-affected capita basis, the estimate we obtained for the Adirondacks is a useful estimate of the benefits of

Table 1. Comparing benefits and costs.

Effect	$ per *affected* person from expected emissions reductions in 2010 (1995$)
Morbidity	4
Mortality	69
Lake Fishing	1
Recreational Visibility	4
Residential Visibility	7
Costs	*6*

environmental improvements pertaining to recreational fishing. What the study finds is that it would be difficult to justify the costs of the program from a benefit-cost perspective when considering improvements in recreational fishing alone.

Happily, one does not have to justify the cost of the program on the basis of any one of the environmental pathways taken alone.

A number of benefit categories individually weigh in at about the same level as cost, including recreational and residential visibility improvements per affected person, and morbidity health effects on a nationwide basis. However, mortality health effects on a nationwide basis weigh in at about an order of magnitude greater.

I want to focus a moment longer on the mortality estimate because of what Dr. Lippmann said yesterday. Part of the theme that has emerged in the last day and a half is what we have learned since 1990. Dr. Lippmann pointed out that as of 1990, the research was not firmly in place to convincingly establish the benefits associated with premature mortality from changes in nitrates and sulfates. Since that time, there has been a growth in the epidemiology literature. The integrated assessment estimates I have been citing relied on epidemiology in Schwartz and Dockery (1992). Subsequently, the Pope et al. (1995) study and the Krewski et al. (2000) re-analysis of this study suggest that the dose response coefficients are about three times what they were when we calculated the numbers based on Schwartz and Dockery in the work I have presented.

Another important development in environmental economics goes to valuing changes in premature mortality. The way valuation of mortality effects was done circa 1990 was to look at compensating wage differentials in the work place, for example the wage premium for coal miners or for police officers, as compensation for the extra risk associated with those professions. That approach is the basis for the value of statistical life that has been the underpinning of benefit-cost analysis for most of this decade.

There are philosophical and conceptual problems with the compensating wage approach. So, in the middle part of the decade there began efforts to try to, on an ad hoc basis, see if we could improve on this, paying attention to factors such as evidence that infirm or older populations are perhaps most at risk from poor air quality. Yesterday, Dr. Lippmann mentioned the attempt to look at life years lost as being sort of an ad hoc adjustment. Those ad hoc adjustments were already reflected in the integrated assessment estimates I reported.

More recently, however, there has been a series of studies that have tried to match the nature of the change in risk from PM exposure to the valuation of that risk (Krupnick et al., 2000; Mrozek and Taylor, 2002). This work suggests changes in values that go in the opposite direction of those from the new epidemiology, reducing the estimate of benefits from reduced mortality, and they largely off-set the changes in the epidemiology since the benefit estimates are the product of these two values.

There could be two or three reasons that the valuation of premature mortality associated with PM exposure might be less than suggested by the compensating wage approach. One reason that is often suggested, and which I mentioned previously, has to do with the age of individuals who are exposed. However, the literature does not find this to be statistically important. Instead, what does have an effect on the estimate is the delay between exposure and response. A subject's willingness to pay at the time of exposure is higher when the response is immediate than when it is postponed (that is, when the exposure shortens one's life because of changes in health status when one becomes elderly). The willingness to pay to avoid such exposures differs when health effects are realized immediately versus when there are changes in life expectancy that are realized 20 or 30 years from now.

In summary, considering changes in the health epidemiology and the valuation of changes in health status in the last decade, these changes head in offsetting directions with respect to the economic valuation of mortality health effects. So, developments in the 1990s suggest the health benefits from the reductions in emissions achieved due to the 1990 Clean Air Act Amendments are an order of magnitude greater than the cost of the program, as suggested originally by the studies done in the middle of the decade.

Another question of interest is the regional effect of emission allowance trading. We have used a simulation model that embodies the economic logic of trading with plants trading on the basis of relative marginal costs of emission reductions. This is used to find the expected location of changes in emissions, which is then coupled with other aspects of the integrated assessment framework to evaluate those changes (Burtraw and Mansur, 1999). Figure 3 illustrates the finding that trading leads to an increase in emissions in the Midwest and a decrease in the East and Northeast. However, what follows from this change in the location of emissions is that health status in the aggregate is

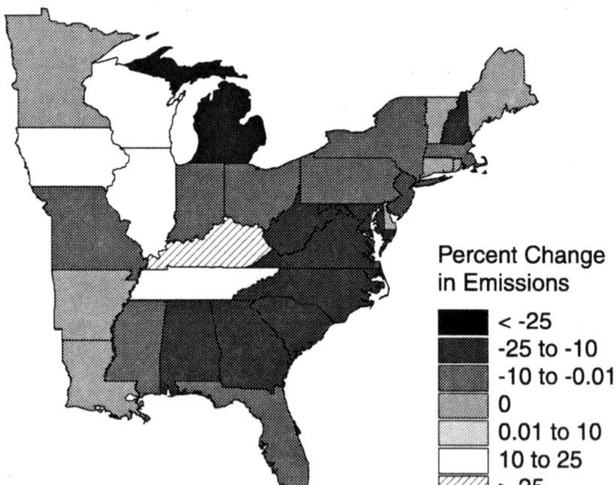

Figure 3. Effect of trading on emissions. Percent change in Title IV baseline utility emissions attributable to trading for 2005.

not degraded by trading, and may actually be improved, as illustrated in Figure 4. Many observers have objected to the fact that the northeastern states are affected by long-range transport of emissions from the Midwest. This is true, but this effect is overshadowed by the fact that residents of northeastern states suffer less from a unit of emission in the Midwest than when the emission is more proximate to where they live. The change in health benefits is most unfavorable in the regions where emissions increased. This view runs contrary to the intuition formed from studies about acidic deposition, which falls, of course, at the end of the atmospheric transport pathway, presumably many hundred miles from the emission source. Changes in concentration occur all along the transport pathway and are greater relatively close to the emission source rather than at the end of the transport pathway. It is in the Midwest that there are the greatest changes in health status as a consequence of the slightly greater emissions there as a result of the trading process.

We find that health effects vary slightly due to emission trading, but it is so trivial in magnitude that one could probably say the effect is zero. However, where there is an effect with respect to changes in health status or with respect to deposition, we find an improvement in aggregate public health due to the trading process. In some states there is an increase in exposure and in some states there is a decrease in exposure due to trading, but on net the effect of trading is to improve the aggregate measure of health status. The map in Figure 4 illustrates the expected percent change in monetized health benefits as a consequence of the geographic pattern of trading. Let me emphasize that this illustrates changes that come in addition to the large changes resulting from the decrease in aggregate emissions. The map indicates that the whole eastern seaboard will realize a slight improvement in health status as a result of the pattern of trading that we identify in the model.

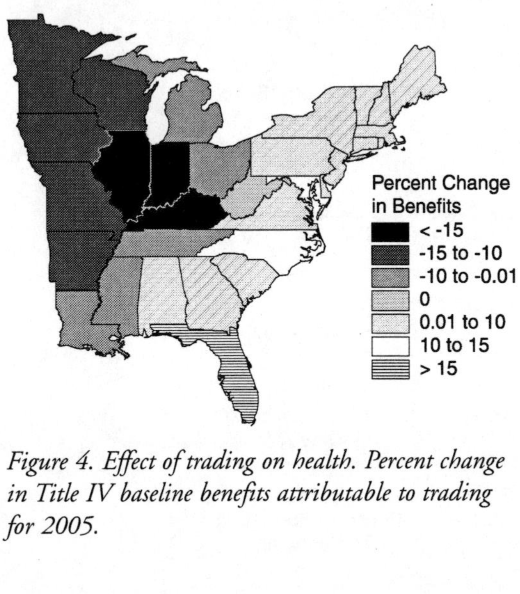

Figure 4. Effect of trading on health. Percent change in Title IV baseline benefits attributable to trading for 2005.

The measure of improvement in deposition that we have estimated is focused on the northeastern states including the State of New York. The improvement from the trading process is small and not important. Again, however, it comes on top of the large and very important improvement that results from the aggregate reduction in emissions under the emissions cap.

There has also been work by others that has examined the actual location of emissions in the first few years of the program, compared to the location of allocation of emission allowances. Yesterday we heard about the EPA's look at current allowance usage and where allowances were allocated compared to where allowances have been surrendered for compliance. That is a problematic measure because firms can bank an allowance, firms can buy one from Peter and buy one from Paul and use one and not use the other. So, it is not entirely meaningful to track the origin of an allowance used for compliance under the program. Nonetheless the EPA empirical assessment (EPA, 1998) and others (Swift, 2000; Solomon and Lee, 2000) are pretty much the same as the findings from the integrated assessment simulation model. The pattern of trading has had very little effect on the ultimate location of emissions.

Nonetheless, I want to validate the concern that there could be a geographic effect due to trading. The idea that the geographic pattern of trading could be undoing the benefits to the environment or human health, or in some other sense be fundamentally unfair, might strike one as an exciting potential piece of investigative journalism (Figure 4). It is a legitimate concern, and indeed it is found throughout the underpinnings of the economics literature about trading. To address this issue in public policy involves potentially complex institutions that are designed to accommodate concerns about the geographic or temporal patterns of emission trading. As the EPA escorts us towards a new NO_X trading program in the eastern 19 states, the same question may be relevant to the design of the NO_X trading program.

The institutional design of a trading program that would be necessary to accommodate this concern would come at a significant cost, which would be realized through greater transaction costs, less trading and greater costs of emission reductions in the aggregate. The simplicity, transparency and flexibility of the SO_2 trading program is one of the primary reasons for why costs have been low and it has been such a success. These virtues would be sacrificed if we had to design a complex program to accommodate the geographic concerns. Doing so would raise the cost of a trading program substantially, so it should only be done when evidence suggests that these conceptual possible problems are likely to be manifest in an important way.

In the case of the SO_2 trading program, it is serendipity that the evidence indicates the pattern of trading has not harmed health or the environment, and it may actually have contributed to an improvement. The concern in New York State is to the contrary. I want to validate that concern as a conceptual point, but in the case of the SO_2 trading program I cannot support it as an empirical point.

Perhaps the overarching point from the perspective of New York State, however, comes out of the concern about where the emissions cap is set and whether or not we are achieving sufficient reductions. That is a very good question, and I encourage that as a focus for ongoing dialogue.

In conclusion, the effects of trading are very small compared to the effects of the overall emission reductions, as summarized in Table 2 and Table 3. Yes, there are positive health benefits in the aggregate associated with trading according to our modeling, but they are only about $100 million a year, compared to health benefits due to the overall reduction in SO_2 emissions of over $17 billion per year, in the year 2005. I am comparing

Table 2. Effects of Title IV are small.

	1995	2005
Change in SO_2 Emissions	-152 thousand tons Nationally	-53 thousand tons Nationally
Change in Sulfur Deposition	-0.1 kilograms/hectare New York	-0.2 kilograms/hectare New York
Change in Health Benefits	566 million (1995 $) Nationally	124 million (1995 $) Nationally

Table 3. Effects of Title IV are large.

	1995	2005
Change in SO_2 Emmissions	-1287 thousand tons Nationally	-3213 thousand tons Nationally
Change in Sulfur Deposition	-1.3 kilograms/hectar New York	-3.1 kilograms/hectar New York
Change in Health Benefits	5998 million (1995 $) Nationally	17,400 million (1995 $) Nationally

$100 million versus $17 billion when I conclude that the main point is that the overall cap is what was really important in achieving the benefits that have resulted from the program. Trading underneath that cap has been important to realizing cost savings, flexibility for the industry, and it has had close to a trivial effect in terms of the benefit estimations.

As a final observation about the SO_2 trading program, I want to report on an effort to identify the weak links in the existing science and economics as identified by the integrative assessment. We did a number of value of information and value of uncertainty analyses. We extended the quantitative analysis through an expert solicitation process to obtain judgments about the value of additional information for a number of environmental and health end points, including health, visibility, cultural material uses, and non-use values. As I indicated previously, use values stem from changes in well being through direct exposure or experience with an environmental end point, such as swimming, boating, fishing, angling, bird watching or what have you.

An idea emerged in the economics literature 25 or 30 years ago, at first heretical, but one that now has gained widespread acceptance and legitimacy. This is the idea that individuals may be willing to pay for environmental improvement, even if the individual never plans to visit the resource, never plans on having his or her children visit the environmental resource, etc. Even in this case, it is possible that the individual is willing to sacrifice something in terms of monetary well being in order to see that resource preserved. This willingness to sacrifice forms a basis for estimating what are called nonuse values. Today, even among the most conservative economists, the concept of non-use values has legitimate standing. It has been shown in some very careful studies to be sometimes quite significant. The issue that divides the economics profession is whether non-use values can be reliably and consistently estimated in a fashion sufficient for use in the policy process. It is all too easy to come up with totally fictional estimates by just asking your neighbors and friends what they would be willing to pay for something like healthier air. The scholarly research in estimating non-use values is much more sophisticated, and it has broken out of economics. It is now a subject of interdisciplinary studies linking social science, economics and physical sciences.

Anyway, the end points that we looked at are indicated in the rows in Table 4. We solicited judgments from experts about the questions indicated at the tops of the first four columns in

Table 4.

Categories ● high **O** high-mid ◒ mid ⊙ low-mid ○ low	1. Link Between Science and Economics: Are benefit endpoints well established? Does science provide infomation needed for economic analysis?	2. Economic Methods: Are economic methods adequately developed?	3. Data Availability: Is data available from science and from economics for an assessment of benefits?	4. Expected Benefit: Are expected benefits large?	5. Value of Additional Information: With the goal of improving benefit estimates, what is the relative short-term return on investment?
Health: Mortality	**O**	**O**	**O**	●	●
Health: Morbidity	**O**	**O**	**O**	**O**	**O**
Visibility	**O**	◒	⊙	**O**	**O**
Materials / Cultural	⊙	◒	○	**O**	**O**
Nonuse Value: Ecosystem	⊙	⊙	⊙	●	◒
Aquatics: Recreation	**O**	●	⊙	⊙	◒
Forests: Recreation	⊙	**O**	○	⊙	◒
Ag. / Comm. Forestry	**O**	●	◒	◒	⊙
Radiative Forcing	⊙	○	○	⊙	○

the table. Are the links between science and economics firm enough to support a benefit estimation? Are the economic methods in place? Is the data available on both the science and economic fronts? Are the expected benefits large? The last column of the table indicates the expected benefit, or in other words the value of additional information, in terms of qualitative order of magnitude from a short-term research agenda to improve the links between science and economics.

We find that health mortality is the driving question here, because of its overall magnitude of expected benefit. The other issue which has high expected benefits from additional research is non-use value for ecosystem health. The problem with that research area is that it is a long-term research project. I wouldn't be too discouraged about that, however. If we were in this room 25 years ago asking the same questions, we would have been talking about valuing health effects with comparable uncertainty and many at that time would have felt that the ability to do so was ill founded, not reliable, and certainly not robust to repeated inquiry coming to study the same question to see if they could come up with the same answer. Twenty-five years ago, the estimation of health effects was nowhere near where it is today. Now, in the 1990s, we find dozens of epidemiological and health valuation studies that are robust, and stand up to repeated inquiries by various scholars. I think we can be optimistic that although non-use values are probably a decade or two away from the same level of rigor, the research is well along the way in that regard. However, although research has started in the academic community, the funding for it, as funding for anything we have talked about in the last two days, is not necessarily evident.

As we look forward to issues of the greatest contemporary policy interest, the one that stands out is the proposed NO_XSIP Call and regional summertime NO_X emission trading program for the 19 eastern states and the District of Columbia. The program is expected to take effect in 2004.

We have heard that NO_X emissions contribute to multiple problems. However, the regulatory handle for the NO_XSIP Call cap and trade program is limited to non-attainment of the ozone health standard. Consequently the program is designed at achieving reductions in ozone, which is a summer seasonal problem, but other NO_X-related effects are realized throughout the year. Our recent research is motivated by the observation that about 75% or 80% of the costs of NO_X control that is going to be realized at these power plants are one-time capital costs, and a relatively small proportion is variable costs. Consequently, the overall magnitude of costs would not change by much if the emission reductions were achieved on an annual basis. How would the benefits change, and would it be cost-effective to reduce emissions throughout the year?

To investigate this we looked at two scenarios (Burtraw et al., 2001; Palmer et al. 2001). One was a summer cap in 19 states. The other is an annual cap, where the cap was determined by an average emission rate that was kept the same as for the summer cap scenario. We looked at two scenarios with respect to the status of restructuring in the electricity industry. Limited restructuring characterizes current commitments to implementing competitive pricing at the state level. National restructuring is a scenario in which the entire country is assumed to implement competitive pricing by 2008. The regulatory structure of the electricity industry has big effects because restructuring, we think, is going to lead to greater NO_X emissions in the baseline (that is, in the absence of the NO_X SIP Call cap and trade program).

The results are illustrated in Table 5 for the limited restructuring case (Burtraw et al., 2001). The SIP seasonal policy would cause a reduction of about a million short tons of NO_X from the power sector. However, if implemented on an annual basis, the policy would more than double the emission reductions. Paying attention to the particulate matter health pathway, we calculate more than a doubling of improvement in health benefits from particulate matter that would result, while the compliance costs increase only by roughly about 20 percent.

The good news from a political perspective is that, for reasons that are quite interesting to an economist and would bore the rest of you, the change in electricity prices could be expected to be less under a SIP annual policy than under a SIP seasonal policy. We find that extending the NO_x SIP Call trading program to an annual program would lead to $200 million to $600 million per

Table 5. Nationwide, annual particulate related health benefits and costs for 2008.

	Benefits/Costs in Aggregate (million 1997$)			(thousand short tons)	Benefits/Costs per Ton (1997$ per short ton)		
	Particulate Related Health Benefits Only (from Table 3)	**Post Combustion Control Costs** (from Table 5)	**(Partial) Net Benefits** (Particulate Health Benefits minus Costs)	**NO_X Emissions** (from Table 2)	**(Partial) Average Benefit**	**Average Cost**	**Marginal Cost**
Baseline	-	30	-	5,533	-	-	1,356
SIP Seasonal	+749	+2,146	-1,397	-992	755	2,163	3,401
SIP Annual	+1,777	+2,728	-951	-2,378	747	1,147	1,985
National Annual	+2,564	+4,434	-1,870	-3,962	647	1,119	3,884

year in additional net benefits (additional benefits less additional costs). This estimate is based only on the value of particulate-related health effects. But, the finding is robust to all the omitted benefits, including changes in ozone, which are the driver for the SIP Call policy in the first place.

Looking forward, our conclusion is that the EPA and the eastern states, many of whom are represented here today, should consider revising the SIP Call initiative to aim at annual reductions. Thank you.

REFERENCES:

Bloyd, Cary, et al. 1996. *Tracking and Analysis Framework (TAF) Model Documentation and User's Guide*, ANL/DIS/TM-36, Argonne National Laboratory, December.

Burtraw, Dallas, Alan Krupnick, Erin Mansur, David Austin, and Deirdre Farrell. 1998. "The Costs and Benefits of Reducing Acid Rain," *Contemporary Economic Policy*, vol. 16 (October), pp. 379-400.

Burtraw, Dallas and Erin Mansur, 1999. "The Environmental Effects of SO_2 Trading and Banking." *Environmental Science and Technology*, Vol. 33, No. 20, (October 15), 3489-3494.

Burtraw, Dallas, Karen Palmer, Ranjit Bharvirkar, and Anthony Paul), 2001. "Cost-Effective Reduction of NO_X Emissions from Electricity Generation," *Journal of Air & Waste Management*, Vol. 51, 1476-1489.

Carlson, Curtis, Dallas Burtraw, Maureen Cropper, and Karen Palmer. 2000. SO_2 Control by Electric Utilities: What are the Gains from Trade?" *Journal of Political Economy*, 108 (6), 1292-1326.

Environmental Protection Agency, 1998. "The Environmental Impacts of SO_2 Allowance Trading," Acid Rain Program, (December); http://www.epa.gov/acidrain/effects/tradefx.htm

Krewski, D., R.T. Burnett, M.S. Goldberg, K. Hoover, J. Siemiatycki, M. Jerrett, M. Abrahamowicz, and W.H. White. 2000. "Reanalysis of the Harvard Six Cities Study and the American Cancer Society Study of Particulate Air Pollution and Mortality. A Special Report of the Institute's Particle Epidemiology Reanalysis Project." Cambridge, MA: Health Effects Institute.

Krupnick, Alan, Anna Alberini, Maureen Cropper, Nathalie Simon, Bernie O'Brien, Ron Goeree, and Martin Heintzelman. 2000. "Age, Health, and the Willingness to Pay for Mortality Risk Reductions: A Contingent Valuation Survey of Ontario Residents," Discussion Paper 00–37. Washington, DC: Resources for the Future (September).

Mrozek, Janusz R., and Laura O. Taylor, 2002. "What Determines the Value of Live? A Meta-Analysis," *Journal of Policy Analysis and Management,* forthcoming.Palmer, Karen, Dallas Burtraw, Ranjit Bharvirkar, and Anthony Paul, 2001. "Restructuring and the Cost of Reducing NO_X Emissions in Electricity Generation," Resources for the Future Discussion Paper 01–10 (March).

Palmer, Karen, Dallas Burtraw, Ranjit Bharvirkar, and Anthony Paul, 2001. "Restructuring and the Cost of Reducing NO_X Emissions in Electricity Generation," Resources for the Future Discussion Paper 01–10 (March).

Pope, C. A. III, M. J. Thun, M. M. Namboodiri, D. W. Dockery, J. S. Evans, F. E. Speizer, and C. W. Heath, Jr. 1995. "Particulate Air Pollution as a Predictor of Mortality in a Prospective Study of US Adults," *American Journal of Respiratory and Critical Care Medicine*, vol. 151, pp. 669-674.

Schwartz, J., and D.W. Dockery. 1992. "Particulate Air Pollution and Daily Mortality in Steubenville, Ohio," *American Journal of Epidemiology*, vol. 135, no. 1, pp. 12–19.

Solomon, Barry D., and Russell Lee, 2000. "Emissions Trading systems and Environmental Justice," *Environment,* vol. 42, no. 8 (December), 32-45.

Swift, Byron, 2000. "Allowance Trading and SO_2 Hot Spots – Good News From the Acid Rain Program," *Environmental Reporter,* vol. 31, no. 19 (May 12), 954-959.

QUESTION

AUDIENCE: NO_X is a worse contributor to deposition and damage in the winter. I wanted to bring that up.

MR. BURTRAW: I'll accept that as an observation rather than a question. I agree, that if we look at the omitted benefits, an annual NO_X reduction program would dominate a seasonal NO_X reduction program, not just because of health benefits, but also because of the non-health benefits, which we did not model thoroughly.

Remarks of Senator Hillary Rodham Clinton

I am delighted to be here and so pleased that my schedule could be worked out. I am happy to be here with the hearty band that is here and committed to this issue and so pleased to learn, as I just did, how many people are here from other parts of the country, not just New York and New England.

I think that speaks very well to the efforts that have been undertaken by so many of you over the years to make this issue a front burner priority for our entire country.

I want to thank the Center for Environmental Information for sponsoring this conference, and all the other groups who are co-sponsoring it.

I want to thank Bernard and the Adirondack Council for providing me with the opportunity to speak but, more than that, for all of the hard work and dedication that has gone into years and years of better understanding and addressing the challenges posed by acid rain.

I am very honored to serve on the Environment and Public Works Committee, following in Senator Moynihan's footsteps. He was a great advocate and champion of what needed to be done on acid rain in the Senate.

Now, Senator Schumer and I have banded together to keep up that fight and hopefully find more and more allies throughout the Congress and the country.

I know that both Senator Schumer and Congressman Sweeney have been here and have addressed many of the issues that we are so concerned about in New York.

We know that the kind of continuing deterioration that we have seen means that, although there have been efforts over the past 20 years to address this issue, we haven't done anything near what we need to do.

We, for example – I am sure this has been said over and over again – already live with the fact that 41 percent of the lakes in the Adirondacks are affected in some way, some to the point of all the life being gone from them.

In 40 years, left unabated, we will have mostly dead lakes in the Adirondacks, other parts of New York, other parts of our country.

Someone mentioned to me as I walked in, Big Moose Lake, where I was privileged to go during the last year and speak about my commitment to ending acid rain.

The Hubbard Brook study that was released earlier this year is another loud alarm bell warning us what we need to do.

Here is what we are trying to do, and this is where we really need your help. Senator Schumer, Congressman Sweeney, Boehlert and McHugh in New York, have all joined together in sponsoring the acid rain control act. We need more sponsors. We need more people who view this as a priority.

Senator Schumer and I have also joined with a bipartisan group of our colleagues in the Senate in introducing the clean power act of 2001.

We are trying to have as many different approaches to this issue as we possibly can, to reduce NO_X, to reduce sulfur dioxide, to reduce mercury.

The clean power act also calls for reductions of carbon dioxide. We are hoping that the administration, despite their reversal on carbon dioxide, will find it appropriate and appealing to at least move on sulfur dioxide, nitrogen oxide and mercury.

I didn't get to hear the previous panel, but I know that focusing on the economic benefits of decreasing acid rain is one of the most important arguments we can make, and I commend those of you who are doing the work to quantify those benefits.

Clearly, there are benefits from tourism, there are benefits from energy savings, there are economic benefits of all kinds. We have to get that message out more clearly.

I hope that the emphasis that has finally come to this issue in the Congress means that, for example, on the Senate Environment and Public Works Committee, both the chairman – Chairman Smith of New Hampshire – and the ranking member, Harry Reid of Nevada, have said that they are willing to move a multi-pollutant bill in Congress.

We have already had three hearings on air issues in the committee, including one this week on global climate change.

We are working very hard to find common ground with our friends and colleagues in the midwest, who acknowledge that they are homes to the plants that do so much of the damage to our lakes in the Adirondacks and further north in New England.

N.Y. Attorney General, Elliot Spitzer, testified before our committee a few weeks ago on clean air and particularly what we needed to do to further reduce emissions.

He had quite an interchange with Senator Voinovich from Ohio. Senator Voinovich kept saying, yes, but Ohio meets the clean air standards.

Attorney General Spitzer said, yes, because you built smokestacks 700 feet into the air so that the soot and particles and emissions are carried far away from where you are.

We are looking for a kind of common ground and objective analysis that will give us the opportunity to move our colleagues, who think this is somebody else's problem, who don't see it as a national problem, in their interest to address.

We are very open to ideas about how we can provide some encouragement and even incentives so that other representatives and senators will join us in this crusade.

I am optimistic that, on a bipartisan basis, we can move legislation addressing acid rain and other air quality issues in the committee.

The brownfields bill, which just passed the Senate 99 to nothing is a perfect example. It has been tied up in the committee and in the senate, held hostage to superfund reform for several years.

Finally, a bipartisan group said, look, this has got to end. We have got to have a bill on brownfields. It won't be everything that all of us want, but it will get the process going again to get brownfields cleaned up, to get redevelopment going.

I was pleased that two amendments I offered about public participation and environmental health were accepted by Senator Smith as part of the manager's package.

We have got an example coming out of that committee of what was once a very contentious issue, that pitted regional interests against one another, sectoral interests against one another, and we now have a victory we can claim.

I am hoping we will have fast action on that bill in the House, so that we can point to a real environmental victory for America.

I would like to see our efforts on acid rain be the next victory. I would like to see if there is any way we can push it by the end of this year.

We are going to continue to hold hearings on air issues in the Environment and Public Works Committee. I am hoping that we will have a hearing on the acid rain problem in New York and have many of you who have been working on these issues for years be part of testifying and try to establish a public record that will impact on the Senate's deliberations.

You know, the same emissions from midwest power plants that cause acid rain in the northeast generally degrade the quality of our air.

You know, they contribute to the smog problem which, as we learned this week from the American Lung Association, is only getting worse in many areas of New York and states across the nation.

I think that my constituents in western New York, particularly in beautiful Chautauqua County, were surprised to learn how bad their air was.

That had never been an issue that they had focused on, and now there is a lot of publicity about the deteriorating air in places that always considered themselves kind of rural and off the beaten path.

We also know that if we are going to address air issues, we are going to have to be very vigilant about the energy policy that this administration puts forth.

I know there are people here from the EPA and from the administration. I would just make a heartfelt plea. We cannot address our energy problem at the expense of our environment.

We can have a balanced energy program that emphasizes both supply and demand, that will focus on conservation and the development and promotion of cleaner sources of energy.

What we are hearing is that the plan will be largely one of advocating increased production. The budget that has been proposed to the Congress cuts significantly those programs in EPA and the Department of Energy that try to provide credits and incentives for efficiency, for cleaner forms of energy.

I really would just hope that the administration

will think very, very hard about this issue. It will do us little good to try to move forward on an acid rain bill if an energy bill is pushed through the Congress to meet our short-term energy needs that exacerbates our environmental and air problems.

We cannot be short sighted about what we need to do. I hope that we will be able to make that case. Your meeting here today provides even greater impetus and more voices to the extraordinarily urgent message that we can and we must develop a clear set of standards to govern air emissions, deal with acid rain, and have a balanced approach to energy.

There are a lot of veterans in this room. Some of you have met with me on several occasions. A number of you came to a forum on the environment that Attorney General Spitzer and I hosted in Rensselaerville a few weeks ago.

I am absolutely committed to these issues and will do everything I can, working across the aisle and with both houses of Congress.

The time is really now. I heard reports that some of your meetings with unlikely members were quite productive, that there are those who have not been interested in, or supportive of, these efforts in the past who, for whatever reason – epiphanies, ideology, politics, whatever reasons – are now willing to sign on and be part of the solution, not the exacerbation of the problem.

We have to move quickly to take advantage of that opening, because it is, indeed, an opening, but I don't know how big it is or how long it will stay. Your meeting could not be more timely.

To those of you from New York, I thank you for your decades of effort and diligence, and for the support many of you gave to me.

For those of you from other parts of the country, I am so glad you are here. I am so pleased you are part of this effort.

I urge you to do everything you can, both with your elected members and colleagues back home, to sound the call to action, that we will and we must address acid rain this year, before it ends, before we cause any more damage, and it must be addressed in a bipartisan effective manner. Thank you all very much.

QUESTIONS

AUDIENCE: I hope I can ask this question without being too nervous. You made such an excellent point about the need to tie together clean air and clean energy policy and whatever.

What would you suggest that we in this room do tomorrow to do this, and then, how can we help you move forward with this kind of alignment?

SENATOR CLINTON: I think there is a lot that each one of us in this room can do from our various capacities. I would urge you to go home and get on local radio programs, write letters to the editor, hold press conferences with like-minded people, talk about what has been discussed and learned here at this conference.

Talk about the importance of combining clean energy and clean air and clean water issues together. You know, the environment has always been a kind of sleeper issue.

It breaks through public consciousness when people have to face hard choices, and that is where we are right now.

The more that all of us – I would particularly urge each of you as members of groups, as citizen activists, as officials, to really take every chance you have, both to speak to the public and to speak to your senators and representatives.

Write to the White House. I really believe – I have always believed this and now, as an elected official, I believe it even more, that communication matters.

If people believe that there is a groundswell of opinion and activism out in the country on behalf of clean energy, clean air and clean water, they will sit up and take notice.

Our problem often is that the numbers and the intensity are on the other side, and the arguments are made with so many resources behind them.

So, what we can't match necessarily for resources for ads on television, we have to match, as best we can, with every other medium available to us.

I know that Deb Callahan, who I spoke with a few days ago through the League of Conservation Voters, is getting very serious, which I applaud, as has the Sierra Club and others, to become major players in the political process. The environment has to have that kind of force behind it in order to be taken seriously.

I think we have wonderful openings, because a

lot of the issues that the administration has addressed in the last few months have really been outside the mainstream and outside people's understanding of what is right.

It doesn't appear that they share our values if they don't want to deal with arsenic in the water. That is something nearly everybody gets and understands.

It is a little troubling that a promise is broken. During a campaign we are going to see CO_2 addressed. Once the White House is attained, that promise goes out the window.

If you look at the progress that has been made, there was a front page article in one of the papers today about a very fuel-efficient automobile.

We have invested federal dollars the last eight years in new generation vehicles. It was a partnership with American auto makers to move more fuel efficiency forward. That is being cut back or even cut out in the Bush budget.

The standards for fuel efficiency in cars, they need to be raised. I joined Senator Schumer and Senator Feinstien with a bill to raise them. We need to get people to focus.

It is opportune, because people are worried about their electricity rates and supply. If they only knew what it would mean to increase fuel efficiency in cars or standards in refrigerators, and that energy efficiency is the second biggest source of energy we could tap into, as well as being an economic driver – upstate New York has a lot of jobs that could be created and built on if we moved toward more energy efficiency, like fuel cells technologies.

There are a lot of arguments we could make. I would just urge you to find every possible opportunity to do that.

Call those radio talk shows. Get yourselves booked as a guest. I am serious. You don't have to be an expert.

You know, look in the mirror, talk to a friend about how to make your points, but don't leave the field to people who are paid to put forth a different point of view.

AUDIENCE: I understand the Senate environment Democrats aren't willing to move a multi-pollutant bill that doesn't address less than four pollutants, without CO_2.

With the President opposing regulating CO_2, doesn't that mean that bill is sort of in a logjam right now, that can't really be addressed unless the President changes his mind?

SENATOR CLINTON: That is why I am on both bills. I think there are opportunities in pushing both bills forward at the same time.

We have a lot of very strong support on both sides of the aisle and in both houses of Congress for the clean power act, which is a four-pollutant bill.

I think we should keep pushing that. There are Republicans who stood with us when Joe Lieberman and Harry Reid and I and others introduced that bill, because they know that just because the President reverses himself and says CO_2 is not a problem doesn't mean that it is not a problem. We want to continue to push that.

At the same time, we want to have real progress on NO_X and SO_X and mercury. I think there is a deal in there. I don't think we get to where there is a deal unless we push as hard as we possibly can and get as much support behind both. I think we are in a stronger position.

I would hope that this President thinks that he needs an environmental victory. I would love it to be acid rain. That would be fine with me.

If we can't address CO_2 this session, CO_2 is not going anywhere. In fact, we are getting more of it every day. So, it will be around for us to deal with. But at least we can make some progress.

MS. LAMBERT: Thank you for coming, Senator. I am Kathy Lambert with the Hubbard Brook Research Foundation. I think as you know, the only way we can tell if we are solving this problem is to track it.

I was wondering if you could support a modest investment of about $15 million in scientific monitoring equipment that hasn't been updated in decades to help solve this problem.

SENATOR CLINTON: Absolutely. I would be happy to support that. That is part of a bigger problem, if I may, and I don't want to be partisan. I know this is a non-partisan group.

If you go back to the 1980s, the Reagan administration began to undermine our capacity to collect data across the board in everything, social policies, environmental policies, scientific efforts, began to dismantle a lot of the government programs that collected data.

We saw a further assault on data collection beginning in 1995 with the Republican-led Congress. That is extremely dangerous. I would just say that, no matter Republican, Democrat, independent, if we don't have good data, we don't make good policy.

If we don't have good data, we can't marshal the arguments about why we need to do what we need to do.

I am hearing increasingly from so many different constituents and parts of New York and parts of our economy about how crippled we are now because we don't have good data.

We haven't invested in upgrading our data collection in the public sector. We are far behind the private sector.

Often, sort of ruefully but jokingly, we say that Washington is sometimes the only evidence-free zone I am aware of, because when you come with evidence, and oftentimes it is very sound evidence but it is not politically correct or politically popular and so, if there are 100 people saying X and there are two people saying Y, it is more convenient to go with Y than X.

We have got to get better data. I would certainly be willing to support that effort, and any other kinds of information technology needs that the environmental community has are ones that I would be open to.

We don't want to be in a position where the information is largely held on one side of the political equation and is, therefore, easily manipulated for whatever political end is to be sought.

I will just end, because I know that was the last question. I think we have a wonderful opportunity. I really do.

I believe that, with coordination and focus on the part of our environmental groups, with setting some priorities and kind of going down the list saying we have focus here and then we are going to win one and then we are going to move on, I think the next two years could, in a kind of odd way, maybe, a kind of unpredictable way, be very good because they got off to such a bad start.

I think that is something that we can take advantage of and hold them accountable for. It means that all of us are going to have to do our part and be very vigilant and be willing to get out there and to make our case, and to make our case in language, in using examples that people who are not already immersed in the issue can understand.

I mean, I am guilty of policy wonkism from time to time. You know, we talk four pollutant and two pollutant. Most average people don't have any idea what it means.

When we say that, you know, the loons are gone from Loon Lake, and you don't see any more moose around Big Moose Lake because they can't drink the water, people kind of get that.

When you say that it doesn't make a whole lot of sense to keep 50 parts per billion arsenic in our water when all the scientific studies say we should drop it at least to 10, people kind of get that, too.

I think we have to make our arguments with clarity and strength, but we have to just be persistent and diligent.

I don't want to jinx us, but I think we can have some real progress on acid rain, and wouldn't that be great, and we could all celebrate because of your work over the years. Thank you very much.

House of Representatives Science Committee: Hearing on Acid Rain - The State of the Science and Research Needs for the Future
Opening Statement for Acid Rain Hearing May 3, 2001

Congressman Sherwood Boehlert (R-NY) - Chair of the House Science Committee

[Editor Note: The hearing of the House Science Committee on Acid Rain: The State of the Science and Research Needs for the Future was held at the end of the conference and was transferred to the Rayburn Building to enable Committee members to participate in key votes in the House of Representatives that afternoon.]

It's a pleasure to be here this afternoon at this important conference. We're holding this hearing because we want to be sure that Congress takes note of the important research findings that have been aired over the past two days. We will take pains to ensure that the information in today's hearing record is widely circulated and known.

That will be an especially pleasurable task for me because acid rain is one of the first issues I focused on when I first came to Congress back in 1983. To many back then, acid rain was just a theory – and not a very credible one at that. But up in the Adirondacks, in the northern part of my District and beyond, the evidence was accumulating that acid rain was all too real. The dead lakes and thinning trees told an alarming story that many people were not interested in hearing.

Back then, I used to frequently quote from a speech by Tom Kean, who was at that time the governor of New Jersey. Kean said that if all we did was study acid rain, "we'd end up with the best documented environmental disaster in history." It was a good line, and it sounds even more prescient now because of the way things have turned out. First, we heeded the warning, we overcame the naysayers, and we did indeed take action. We passed the Title IV of the Clean Air Act Amendments of 1990, and we launched the war on acid rain, and I'm quite proud of my role in helping to draft that legislation. And, even more encouragingly, as the latest research shows, that legislation has made a real – if insufficient – difference.

Second, the continuing research has, as we had expected, demonstrated that the problems caused by acid rain are truly severe, varied and widespread. We can now document the impacts of acid rain on lakes, aquatic species, trees and soil in ways that we could only guess at in the early '80s.

So what we've ended up with is genuine action, real improvement, and growing documentation that we need to do still more. And I'm trying to get the Congress to do more as an author of both H.R. 25, the Acid Rain Control Act, and H.R. 1256, the Clean Smokestacks Act. Those bills would require significant additional cuts in sulfur dioxide, nitrogen oxides, and new cuts in mercury.

But to make the case for those additional cuts we need to rely on the best, most recent, credible science. And that's why today's hearing is so important.

We'll hear this afternoon from the top scientists in the country studying acid deposition and related problems to get a clearer sense of what steps we need to take. And – and this is especially important from a Science Committee perspective – we'll hear what kind of research and what kind of monitoring networks the federal government needs to support to ensure that we continue to get the science we need to avert disaster.

The response to acid rain has thus far been a story of the system succeeding – a story of Congress acting on the basis of the best research, even in the face of uncertainty, and then trying to update policy on the basis of continuing research. That's a pattern we need to perpetuate. We need to keep funding the research and heeding its results. Today's hearing should further those goals.

Testimony by Charles T. Driscoll

University Professor of Environmental Systems Engineering
Syracuse University

I would like to briefly comment on three issues pertaining to effects of emissions of air pollutants associated with fossil fuel combustion (i.e., sulfur dioxide, nitrogen oxides and mercury) on sensitive watersheds in the northeastern U.S. These comments focus on: 1) effects of acidic deposition on soils, trees and surface waters, 2) effects of atmospheric mercury deposition on watersheds and lakes in remote regions and 3) monitoring programs to evaluate long-term trends of these pollutants.

Acidic Deposition

Acid rain, or acidic deposition, originates from emissions of sulfur dioxide and nitrogen oxides, largely derived from fossil fuel combustion and ammonia, largely released from agricultural activities. Effects of acidic deposition include: 1) accumulation of sulfur and nitrogen, depletion of exchangeable nutrient cations and mobilization of aluminum in soil, 2) stress to red spruce and sugar maple leading to death from climatic disturbance or insect defoliation, and 3) acidification of surface waters and decreases in species richness of aquatic biota. Since 1973 emissions of sulfur dioxide have declined about 35% due to controls on electric utilities in response to the 1970 and 1990 Amendments of the Clean Air Act. These reductions have decreased concentrations of sulfate in surface waters across the northeastern U.S., but with limited improvement in the acid-base status of surface waters. The computer model PnET-BGC has been used to project how an acid impacted forest ecosystem might respond to current proposed additional controls on emissions from electric utilities. Model calculations have indicated that additional controls in sulfur dioxide emissions beyond those required in the 1990 Amendments of the Clean Air Act will be necessary to accelerate chemical recovery of soils and surface waters.

Mercury

Many lakes in remote regions of eastern North America contain fish with elevated concentrations of mercury. This mercury is largely derived from atmospheric deposition and is predominantly in the form of ionic mercury. In wetlands and lakes, ionic mercury is converted to methyl mercury by bacteria. Methyl mercury is a powerful neurotoxin. Although concentrations of methyl mercury are very low in water, processes allow this substance to bioconcentrate through the aquatic food chain. Large older fish and piscivorus fish generally exhibit the highest concentrations of methyl mercury. Concentrations of methyl mercury are concentrated by a factor of a million to ten million from water to fish. Consumption of fish is the primary mechanism of human exposure to methyl mercury. Birds which consume fish as a large portion of their diet are also susceptible to mercury poisoning. Data from lake regions impacted by acidic deposition show that fish mercury concentrations increase with decreases in lake pH. Thus acidic deposition and acidification of surface waters exacerbates the problem of elevated mercury in fish tissue.

Data from sediment cores collected and dated in remote lakes in the Adirondack region of New York show a 3.5 fold increase in mercury deposition since 1850. Deposition of mercury increased markedly in the late 1800s and in most lake watersheds peaked approximately 20 years ago. There have been modest decreases in sediment mercury deposition in Adirondack lakes in recent years. Initially with the onset of increases in mercury deposition most of the mercury entered remote lakes from direct deposition to the lake surface. Mercury deposited on the land surface was largely retained in soil. Over time watershed retention of mercury has decreased. As a result, today much of the mercury entering many lakes is derived from mercury previously deposited on the

land surface. It is anticipated that if atmospheric inputs of mercury would decrease associated with emission controls, mercury that was previously deposited on the land would continue to be transported to lakes, thereby delaying recovery.

Monitoring

Environmental monitoring is critical to national environmental policy. Monitoring of atmospheric deposition and surface water chemistry provides the only quantitative means of assessing the efficacy of state and federal policy. There are several national monitoring networks that provide data to scientists and policymakers and need greater security and support. Five specific networks require increased federal funding to stabilize, expand and/or update the monitoring network.

1. National Atmospheric Deposition Program (NADP)

The NADP program is a successful inter-agency network that monitors wet deposition of sulfate and nitrate associated with fossil fuel emissions. The U.S. Geological Survey (USGS) is the lead federal agency and the Environmental Protection Agency (EPA) plays a strong supporting role. The coverage and baseline funding for this program are adequate to ensure a high-quality network. However, as the oldest network in the U.S., the system needs substantial modernization and a modest number of new sites. The NADP network currently relies on WWII vintage equipment.

Proposal:	Amount:	Agency:
Federal support for annual operating costs	$3.6 million	Inter-agency[1]
Modernization of existing 260+ sites and installation of 10 new sites	$6.0	EPA

[1] $2,000,000 to USGS; $600,000 to EPA; $600,000 to National Park Service; $400,000 to U.S. Forest Service.

2. Clean Air Status and Trends Network (CASTNet)

The CASTnet program, administered by the EPA, measures the component of atmospheric deposition that enters the environment in dry forms such as particles and gases. Monitoring dry deposition is critical to determining the total pollution load across the U.S. In some areas, dry deposition contributes as much as 59% of the total sulfur deposition. At present, CASTnet is a sparse network with only 70 sites nation-wide and none in the central U.S.

Proposal:	Amount:	Agency:
Federal contribution to annual operating costs	$5.0 million	EPA
Installation of 30 new sites	$1.5	EPA
Modernization of existing 79 sites	$3.1	EPA

3. Mercury Deposition Network (MDN)

The mercury deposition network is a patchwork of sites, occurring mostly in the Northeast, that is funded through contributions by state agencies. Some of the highest mercury emitting states, such as Ohio, Kentucky and West Virginia, have no deposition monitoring. Given the tremendous public importance of mercury pollution, it is essential that monitoring be established to develop a mercury deposition baseline and to t rack changes over time.

Proposal:	Amount:	Agency:
Federal contribution to annual operating costs	$1.0 million	EPA
Installation of 60 new sites and upgrade existing sites	$2.0	EPA

4. Temporally Integrated Monitoring of Ecosystems (TIME) and Long-Term Monitoring

The TIME/L TM program monitors lake and stream chemistry and documents changes in response to changing emissions and acid deposition. This program is administered through the

EPA. TIME/LTM is the only national network that directly measures the impact of atmospheric deposition and quantifies the affect of emissions controls. Funding for the TIME/LTM program is both inadequate and unstable. Funding has been cut 50% over the past two years and the program appears to be slated for discontinuation. Funding needs to be reestablished.

Proposal:	Amount:	Agency:
Federal contribution to annual operating costs	$2.5 million	EPA

5. Atmospheric Integrated Research Monitoring Network (AIRMon)

The AIRMon program provides high resolution precipitation and dry deposition chemistry using daily sampling methods operated by the National Oceanic and Atmospheric Administration (NOAA). Funding for this program has been flat for ten years resulting in the unfortunate closure of 3 AIRMon dry deposition sites (Sequoia. CA; Panola, GA; and Burlington, VT). Without an increase in annual operating funds, more site closures are inevitable. Moreover, AIRMon equipment dates to 1984 and has exceeded its life expectancy.

Proposal:	Amount:	Agency:
Federal contribution to annual operating costs	$1.5 million	NOAA
Modernization of existing 20 sites	$1.0	NOAA

In addition to the resources required to stabilize and improve these networks, there are other monitoring needs. For example programs are needed to assess the recovery of soils, trees and aquatic biota in response to decreases in acidic deposition. There are few sites at which mercury in studied in soil, surface waters and fish. None of these have been established as long term monitoring sites to determine baseline conditions and to monitor ecosystem response to changes in atmospheric mercury deposition following emission controls. Finally there is a need for comprehensive ecosystem study sites, such as the National Science Foundation, Long-Term Ecological Research (LTER) program. Comprehensive ecosystem study sites should be viewed as a national resource. At these sites, many environmental measurements are made and ecosystem processes are studied in detail. Through research at comprehensive study sites, cause and effect relationships between air pollution and ecosystem health can be determined, and environmental models used to predict ecosystem response to policy and management strategies can be tested.

Current Air Quality Monitoring Networks and the Pollutants Monitored

The following two tables (Tables 1 and 2) summarize key aspects of the monitoring networks. These tables as well as other more detailed air monitoring information are available from a 1999 report, "The Role of Monitoring Networks in the Management of the Nation's Air Quality" (http:bqs.usgs.gov/acidrain/).

Table 1.

Network	Visibility	Nitrogen species	Sulfur species	Lead	Particulate matter	Carbon monoxide	Ozone or precursors	Stream chemistry	Atmospheric deposition
NADP/NTN*									*wet
AIRMon		*	*						*dry and wet
CASTNet	*	*	*		*		*		*dry and wet
MDN									*wet mercury
TIME/LTM								*	

Prepared by: Hubbard Brook Research Foundation, May 1, 2001, Tel. 603-651-0930, hbrook@hbresearchfoundation.org

Table 2.

Network	**Lead Federal Agency**	**Website**	**Estimated FY99 Federal funding (millions $)**	**Proposal (millions $)**		**Non federal partici-pation?**	**Number of sites**	**Year initiated**
				Annual	Upgrade			
NADP/NTN	USGS	http:llnadp.sws.uiuc.edu/	$3.50	$3.6 Inter-Agency	$6.0 EPA	Yes	220	1978
AIRMoN	NOAA	http://www.arl.noaa.gov/research/programs/airmon.html	$0.60	$1.5 NOAA	$1.0 NOAA	Yes	23	1984
CASJNET	EPA	http://www.epa.gov/acidrain/castnet/	$4.50	$5.0 EPA	$4.6 EPA	No	80	1988
MDN	None	http://nadp.sws.uiuc.edu/	*	$1.0 EPA	$2.0 EPA	Yes	36	1996
TIME/LTM			$1.6	$2.5 EPA			80 lakes 125 streams	1982
TOTAL			$10.2	$13.6	$13.6			

* Total funding approximately $0.6M, nearly all from State and local sources.

Testimony by Ellis Cowling

University Distinguished Professor At-Large
North Carolina State University

Question 1: How has acid rain affected forest, rivers, and other ecosystems in the Southeast and have those ecosystems shown signs of recovery since the Clean Air Act Amendments of 1990?

Question 2: How important are monitoring networks for understanding acid rain and related air pollution, and is federal support for monitoring adequate?

Acid rain, ground-level ozone, regional haze, particulate matter, eutrophication, and climate change continue to be major matters of public worry and industrial concern in the southeastern United States. Although Los Angeles is famous for its historically high ozone concentrations, the Houston-Galveston area of Texas is now rivaling Los Angeles as the most remarkable place to observe rapid accumulation of ozone in the U.S. today. Although air concentrations of sulfate aerosol are decreasing in most states in the eastern US, this is not true in the southern Appalachians and most notably not true within the Great Smoky Mountains National Park. Although the number of urban areas designated non-attainment for the 1-hour ozone standard has decreased in recent years, the total number of counties included within ozone non-attainment areas of the south will increase markedly as soon as the 8-hour standard for ozone is implemented. Many of these newly designated ozone non-attainment counties will be rural rather than urban in character. Although total emissions of sulfur oxides have decreased in most states east of the Mississippi River, emissions of nitrogen oxides have remained about the same or even increased. This difference is especially noteworthy in the southeastern states where emissions of ammonia from animal agriculture also have increased substantially in recent years. The visual range over scenic vistas has decreased in most Class I areas in the southern Appalachians. Although a trend of decreasing ozone concentrations has occurred in the US as a whole, during the 10-year period from 1990-1999 ozone concentrations increased in the southeastern states – the 2nd highest daily maximum one-hour average ozone concentration increased by 13 percent and the 4th highest daily maximum 8-hour average increased by 17 percent. Eight of nine US National Parks showing increased ozone concentrations during 1990-1999 were located in southeastern states (AR, FL, KY, NC, SC, TX).

All these differences and trends provide substantial justification for continuing investments in research and pollution-management activities aimed at discovering and implementing cost-effective means for decreasing these impacts in the southeast.

Biography

Ellis Cowling is a forest biologist at North Carolina State University who became a world leader in air pollution research. Beginning in 1975, he led a group of 200 scientists in creating the National Atmospheric Deposition Program (NADP). This network measures the amounts of nutrients and injurious substances transferred in rain and snow from the atmosphere to forest and agricultural land and surface waters at 200 research sites throughout the US. In some parts of our country, these substances are having positive effects on the productivity of crops and forests. In other parts, some of these same substances are having negative effects on forests, fish, and surface and ground water quality. Cowling will discuss how scientists and engineers in NADP (and in the Southern Oxidants Study which he now leads) are increasing scientific and public understanding about environmental change and the sustainability of ecosystems. As Chair of the session on Regional Impacts, he will also lead our discussion of how society can adjust the procedures by which air and water quality are maintained. Cowling has two earned Ph.D. degrees-one from the University of Wisconsin and the other from the University of Uppsala in Sweden. He has served as major professor for 66 graduate and postdoctoral students at Yale and at NC State Universities. He has been a member of the National Academy of Sciences since 1973.

Themes to be Addressed:

1. Celebrating the achievements of NADP's first twenty years
2. Ensuring the long-term stability of the NADP network
3. The growing importance of nutrient nitrogen in determining the pollution climate of North America
4. Prospects for developing area-wide estimates of total (wet plus dry) deposition through collaboration with other organizations in North America and Europe

What Are the Most Notable Achievements of NADP's First Twenty Years?

1. Building and maintaining for 20 years, a uniquely successful partnership among many disparate federal, state, industrial, and university research communities
2. Developing a high quality environmental database that is trusted by both scientific and policy leaders throughout North America and around the world
3. Achieving a remarkable degree of personal and professional satisfaction through collaboration and cooperation among atmospheric, agricultural, forest, aquatic, and terrestrial scientists on a continental scale

Celebrating the Achievements of NADP's First Twenty Years

1. Using the bottom-up traditions of the Regional Project System of the State Agricultural Experiment Stations as an organization paradigm
2. Selecting, developing, and maintaining a single Central Analytical Laboratory at Illinois State Water Survey (ISWS)
3. Winning the confidence of federal agencies (USDA, USGS, USFS, EPA, USF&WS, NPS, BLM), electric utilities, and other organizations to assemble the present 200-station NADP/NTN network under NAPAP
4. Maintaining cooperation and intercomparisons with the Canadian CANSAP and CAPMoN networks
5. Certification of network-wide QA/QC through site visits by EPA and laboratory intercomparison by the World Meteorological Organization
6. Periodic peer reviews of NADP by NAPAP and the Regional Associations of Directors

Celebrating the Achievements of NADP's First 20 Years

1. Undertaking special studies of atmospheric transport and deposition of pesticides and heavy metals and creation of NADP's Mercury Deposition Network (MDN) and Atmospheric Integration Research Monitoring, Network (AIRMoN)
2. Publishing isopleth maps showing spatial gradients in deposition of major nutrient cations and anions
3. Learning to use the Internet as a means of data dissemination and information transfer to major users of NADP data and information
4. Consolidation of the NADP Coordination Office in Colorado with the Central Analytical Laboratory in Illinois
5. Demonstration that Title IV of the Clean Air Act Amendments of 1990 is working
6. Learning to survive periodic budget crises – a continuing challenge of organization self-education, future planning, and proactive marketing

Ensuring the Long-Term Stability of the NADP Network

The financial stability and longevity of the NADP program depends critically on the <u>number, geographical distribution, and organizational or political</u> clout of satisfied "customers" that include:

- Many data users who care about the data, use the data often, and willing to speak out publicly about the important values they derive from analysis and interpretation of the data.

Hence the following recommendations to:

- NADP Marketing, Executive, and Technical Committees
- NADP Coordinator and his colleagues at the ISWS
- NADP Administrative Advisors within CSREES
- Regional Associations of Directors within the SAES

Recommendation I

Use the NADP Web Site and other outreach venues (see list below) to build an enduring support base of "satisfied and vocal customers:

- "Inside Rain: A Look at the National Atmospheric Deposition Program,"
- "NADP/NTN Wet Deposition in the United States 1995,"
- "NADP/NTN Wet Deposition in the United States 1996,"
- "NADP/NTN Wet Deposition in the United States 1997,"
- "Uses of National Atmospheric Deposition Program National Trends Network Data for Science and Education and Environmental Problem Solving,"
- "Trends in Precipitation Chemistry in the United States, 1983-94: An Analysis of the Effects in 1995 of Phase I of the Clean Air Act Amendments of 1990, Title IV."

Recommendation II

Cultivate and education the NADP Administrative Advisors in CSEERS and the SAES Directors within the four Regional Associations of Directors – Northeast, Southern, North Central, and Western.

This education is essential to maintain the confidence of current leaders within the Regional Research System of Land Grant Universities, the State Agricultural Experiment Stations (SAEA), and the Cooperative State Research, Education, and Extension Service (CSREES).

Recommendation III

Develop and maintain a standing group of current and former NADP leaders who are prepared on short notice to mount an educational campaign for reaffirmation of support whenever one of the major NADP support organizations is threatening to withdraw or decrease its support.

Current members of the NADP Executive Committee and a carefully selected, well connected, and articulate spokespersons including:

- Current and former Chairmen of NADP – Stan Coloff, Mike Kelley, Steve Lindberg, Bill McFee, Jim Lynch, Ellis Cowling, etc.
- Current and former CSREES Administrative Advisors to NADP
- Current and former Directors of NAPAP – Mike Uhart, Dereck Wnstanley, Patricia Irving, James Mahoney, and Chris Bernabo
- Current and former scientific leaders in NADP and closely related organizations including Jim Galloway, Jim Gibson, Eville Gorham, Dick Simonin, Rick Linthurst, Walter Heck, David Shriner, Jim Lodge, Rona Birnbaum, Gene Likens, Jerry Millilo, Bruce Hicks, Dan Albritton, Christina Bierbaum, Milton Russell, Senator Patrick Moynihan, Congressman David Prices, etc.

Recommendation IV

Identify within every state within NADP, a member of the NADP Technical Committee who is willing, to serve as the NADSP Designated State or University Representative for his or her particular state.

The responsibility of these State or University Representatives should be to actively, state by state, cultivate, communicate with, and maintain a roster and current mailing list of "NADP customers" (data users) within his or her own particular state. In this way customers can be informed of NADP activities and enlisted as supporters.

Recommendation IVa

NADP data users for each state should include leaders and/or outreach persons in the following organizations:

- NADP Site Operators

Leaders in various university organizations:

- Botany
- College of Agriculture
- College of Forest Resources
- Crop Science or Agronomy
- Earth and Atmospheric Sciences
- Fisheries and Wildlife Management
- Forestry
- Horticultural Science
- Marine Biology
- Multidisciplinary Studies
- Outdoor Recreation
- Park, Recreation, and Tourism Management
- Political Science and Public Administration
- Public Health and Environment
- Soil Science
- State Agricultural Experiment Station
- Water Resources Research Institute
- Zoology

State Divisions of:

- Agriculture
- Air Quality
- Environmental Conservation
- Environmental Statistics
- Fish and Game
- Parks and Recreation
- Public Health
- Water Quality

Environmental officials of major industries and trade associations:

- Animal waste research groups
- Cattlemen's association
- Electric utilities
- Fish producers
- Forest products companies
- Municipal waste managers association, etc.
- Pork producers
- Poultry producers

State and/or Regional District Offices:

- Fish and Wildlife Service
- Forest Service
- Soil Conservation Districts
- U.S. Geological Survey

Regional and local chapters of environmental groups

- Conservation Foundation
- Ducks Unlimited
- Environmental Defense Fund
- Natural Resources Defense Council
- Nature Conservancy
- Sierra Club

Regional research groups

- Albemarle and Pamlico Sounds Res. Prog.
- Chesapeake Bay Research Program
- Regional Associations of air quality managers such as NESCAUM, MAEAMA, SESARM, etc.
- Tampa Bay Research Program

County and municipal public health and parks and recreation departments
Members of the National Teachers Association

Recommendation IVb

At least once each year (and occasionally in between), the NADP Designated State or University Representative should use the mailing list of data users within each state for at least three general purposes:

1. Call attention to the time, place, and major presentations during at least the public-information portion of the NADP Annual Meeting
2. Ask if there are additional interests that could be satisfied by available NADP data

and the information products or through additional studies or products needed within each state

3. Ask these persons to continue to speak out vocally (and in appropriate institutional publications) about the values they derive from use of the NADP data and information

Recommendation IVc

Every five or so years, the NADP Designated State or University Representative should organize state-wide meetings with organizational leaders from the state to:

1. Communicate NADP appreciation for continuing multi-organizational support and participation in the program
2. Determine if NADP operations are continuing to be adequately representative of current nutrient-, air quality-, water resource-, natural resource-, and fisheries-management and other environmental concerns within the state
3. Identify improvements that may be needed in the location and/or operation of NADP Collection Sites in the state
4. Identify additional uses that might be made of existing NADP data an information products or other data and information needed within the state
5. Discuss contemporary financial and other aspects of the NADP program during past years and the need for continuing use of the data and participation in the program during future years

Recommendation V

The NADP Executive Committee, Regional and CSEERS Administrative Advisors, and all NADP support organizations should explore the possibility of building a contingency ("rainy day") fund to provide temporary support for high quality NADP sites that are in jeopardy of losing their support.

One means of doing this is to add a "contingency surcharge of 2-5% to the analytical service fees of CAL" which would be administered by the NADP Coordinator with the approval of the Executive Committee to provide time-limited (12-24 months) support for sites that lose support by sponsoring organizations.

Recommendation VI

Circulate a Resolution of Mutual Commitment {A "Gentleman's Agreement"} among NADP leaders

As responsible leaders within various universities, federal and state agencies, and industrial organizations in the United States, we are pleased to pledge our continuing participation and financial support of the National Atmospheric Deposition Program (NADP).

We believe that continuing multi-organizational support of NADP, collect5ively totaling less than $4.0 million dollars per year, is by far the most cost effective means by which our country can continue to provide high-quality, scientifically-valid information on contemporary changes in the chemical climate of the United States.

If, at some future time, our organization finds it necessary to decrease our support for this valuable program, we promise to do our best to provide a minimum of 2-years' advance notice of the necessity to decrease our participation. We make this promise out of respect for other organizations which will have to make up for our decreased participation.

Rational For a "Resolution of Mutual Commitment"

The amount of precipitation (rain, snow) occurring in urban, suburban, rural, and remote regions in the United States is currently measured on a daily basis at about 4,000 locations. The longevity of these valuable records of precipitation amount varies across the United States from about 150 to 250 years (since about 1750 in many pars of the eastern US and since the early 1800s in other parts of the nation). The total cost of these measurements and routine reporting of results is currently more than $25 million dollars per year.

For only the past 20 years, the nature and amount of beneficial nutrient substances, toxic substances, acidic and acidifying substances, and growth-altering substances dissolved or suspended in precipitation has been measured by NADP on a weekly basis at about 200 locations in the United States. The cost of these measurements and their public reporting is currently borne by a voluntary association of nearly 100 organizations affiliated with about 20 state and private universities, 7 federal and 6 state regulatory and natural resource management agencies, and 5 commercial firms which have voluntarily joined together to support NADP for the past 20 years.

Testimony by Jill S. Baron

Ecologist
U.S. Geological Survey

Mr. Chairman and distinguished Members of the Committee, thank you for this opportunity to present, on behalf of the U.S. Geological Survey, this statement regarding sources, transport, and fate of atmospheric deposition in the Western United States.

I am an ecologist with the U.S. Geological Survey (USGS). In 1981, I began research into the susceptibility of Rocky Mountain National Park lakes to acidic atmospheric deposition. I, and my colleagues, now have over 20 years in understanding alpine and sub-alpine ecological processes and the influence of atmospheric deposition on these natural environments. The USGS has been quite active in studying the ecological and biogeochemical influences of airborne pollutants throughout the United States, and has produced cutting-edge science from our long-term watershed research and monitoring efforts. I am confident of the results I will present to you today regarding atmospheric deposition and ecological effects for the specific areas we have studied. The response to atmospheric nitrogen deposition in Southern California has been pronounced, and it is well documented by scientists working there. In Colorado, we see significant changes from nitrogen deposition that tell us our ecosystems are just beginning to respond, and our results are also well-documented in the scientific literature. However, the West is very large, and our studies and monitoring are limited in geographic scope. There is also a great amount we do not yet know about combined interactions between many environmental stresses and the plants, animals, and microbes that make up our ecosystems.

Definition of the Issue

The history of atmospheric deposition to the Western United States beyond the 100th Meridian (hereafter referred to as West) differs from that in the Eastern half of the country. Because the West has been settled more recently than the East, it has not experienced emissions of sulfur and nitrogen oxides for as long. Although population in the West is increasing rapidly, human density is still far below much of the rest of the country. This has resulted in a lower density of power plants and industrial facilities that have been a large source of air pollutants elsewhere. Compared to the East, the West is fortunate to have energy sources that generally do not emit high amounts of sulfur dioxide, one of the major precursors to acid rain. Low sulfur coal, hydropower, nuclear, and solar and wind energy have helped the West suffer less the effects of atmospheric deposition to ecosystems and historical structures common to the East. The major sources of sulfur oxide emissions in the Intermountain West were refineries and smelters for metals such as copper, and many of these smelters have closed over the past 20 years. As they have closed, both emissions and atmospheric deposition of sulfur compounds have decreased significantly (Epstein and Oppenheimer 1986). A final important difference between West and East is that the complex topography and localized arid climates of the West combine to create places where most atmospheric deposition occurs as dry particles and aerosols, such as in Southern California and in the deserts of the Southwest.

Emissions and atmospheric deposition of nitrogen oxides and ammonium, however, are increasing rapidly and significantly in the West, creating their own set of environmental concerns (Lynch et al.1995, Campbell, unpub. data). While we consistently find acid rain, especially in the summer, in Rocky Mountain National Park, it has not yet had a measurable effect on lake chemistry in the Park (NADP/NTN 2000, Baron et al. 2000, Campbell et al. 1995).

The major anthropogenic source of nitrogen oxides is combustion of fuels at high temperatures. Combustion sources include automobiles; trucks, trains, and heavy farm and construction machinery, the utility and industrial sectors, and increasingly, energy development from coal-bed methane, natural gas. Additional nitrogen comes from agricultural emissions, both from fertilized croplands and from large manure piles of confined animal feeding operations. These are regionally important, and in some areas, such as the Colorado Front Range, account for more than 20 percent of the nitrogen emissions (Baron and Mosier, in prep).

Where are the Regions of Elevated Wet Nitrogen Deposition?

The National Atmospheric Deposition Program/National Trends Network (NADP/NTN) provides our best coverage of precipitation chemistry in the United States, and its records began in 1978. As measured at NADP/NTN sites, wet deposition of nitrogen in the West is lower on an annual average than other more industrialized parts of the world, such as Europe and the Northeast United States (NADP/NTN 2000). Nevertheless, there are hotspots of elevated wet nitrogen deposition in Southern California and along the Colorado Front Range when compared with the rest of the West (Fenn et al. 1998). These are regions of high population and agricultural activity, and sufficient precipitation to receive the deposition in rain and snow. Wet deposition measurements in Southern California reflect only 10 percent of the total atmospheric nitrogen deposition (Bytnerowicz et al. 1998). Because of its Mediterranean climate, most deposition occurs as dry deposits in Southern California, and is not measured by the NADP/NTN network. Wet deposition to the high mountain areas of the Colorado Front Range are perhaps an order of magnitude lower than those from California, but they are high enough to have caused chemical and ecological change, as described below (Baron et al. 2000, Fenn et al. 1998, NADP/NTN 2000). Other parts of the West, such as near Tucson, Phoenix, and Las Vegas, may also have high nitrogen deposition, but it falls as dry deposition due to the arid climate.

Chemical and Ecological Responses to Elevated Nitrogen Deposition

Nitrogen is an essential fertilizer, and atmospheric nitrogen deposition is a boon to croplands and commercial forests around the world. In natural forests and parks, however, it can affect both plant communities and aquatic environments. Some species are better able to use nitrogen for enhanced growth, allowing them to out-compete their neighbors. This can lead to changes in plant and animal community structure. In many parts of the world with long histories of nitrogen additions, grasses now dominate meadows and fields that once had much greater species diversity (Vitousek et al. 1998). This is well-documented in England and the Netherlands, in experimental plots in the Great Plains, and increasingly, in native coastal sage scrub vegetation of Southern California (Padgett and Allen 1999). Experiments have shown that nitrogen additions have the same effect on alpine tundra (Bowman et al. 1993,1998,2000), enhancing grass production at the expense of the wildflowers.

Trees grow and respond to nitrogen additions at a slower pace, and there are few forest species that can out-compete those adapted to harsh western environments. At high elevations of the Rocky Mountains all plants live a hard life, and the cold, short growing season, and limited water help to create a situation called nitrogen saturation, with excess nitrogen flushing into streams and lakes. As nitrogen moves through soils, chemicals essential for forest fertility are stripped into the water, enhancing lake and stream acidification (Driscoll et al. 2001).

Before acidification occurs there will be lake fertilization effects. Many high elevation lakes do not get enough nitrogen for algal growth, so additions from atmospheric deposition cause increased productivity and changes in community composition (Morris and Lewis 1992, Elser et al. 1990). Excess nitrogen deposition influences plant growth and community composition. Leaching of nitrate through watersheds leads to eutrophication of lakes and streams, and ultimately, to acidification.

In summary, the best evidence for effects of nitrogen deposition occur in Southern California and the Colorado Front Range. In Southern California, ozone damage, eutrophication of streams, and loss of soil fungi essential to endangered species are evident. On the Colorado Front Range, terrestrial plant communities and algal lake communities are altered and increased nitrogen cycling is poised to create acidification similar to that in the East.

Examples of the Effects of Nitrogen Deposition in the West

Southern California:

- Wet plus dry deposition in Southern California is very high, around 35 kilograms per hectare per year (comparable to deposition in northern Europe and the eastern United States, Bytnerowicz et al. 1999). It, combined with high concentrations of ozone in the San Bernardino Mountains, has caused ponderosa pine growth to be severely reduced (Fenn et al. 1998, Grulke et al. 1999, Arbaugh et al. 1999).
- Perennial streams in the San Bernardino Mountains have average concentrations greater than 10 milligrams of nitrate-N per liter. This average concentration is greater than U.S. Environmental Protection Agency drinking water standards (Fenn and Poth 1999).
- The coastal sage scrub is a unique environment with a great number of native species, many of them sensitive or listed as threatened or endangered. Researchers have found a direct relationship between the increase of nitrogen deposition and the loss of soil microorganisms such as fungi and bacteria. These microorganisms are essential to the sage scrub community, in that they enhance nutrient uptake by plants, and help decompose dead material (Edgerton-Warburton and Allen 2000). Without them, native species are weaker and open to invasion by non-native grasses. The non-native annual grasses perpetuate themselves by providing the fuel for frequent fires (Callaway and Davis 1993). Increased fire, perhaps helped along by the nitrogen deposition, poses risk to people and property as well as the native ecosystem.

Colorado Front Range:

- Alpine tundra with added nitrogen shows a shift in plant community composition that favors grasses over flowering plants (Bowman et al. 2000, 1998, and 1993).
- In the harsh high elevation environments of the Rockies, nitrogen deposition has significantly increased rates of forest and soil nitrogen cycling—the first stage of expected effects on the road toward nitrogen saturation (Baron et al. 2000, Rueth and Baron, accepted).
- Alpine lakes in areas with elevated nitrogen deposition have significantly higher nitrate concentrations than lakes in low deposition areas-another indication of expected effects from excess nitrogen deposition (Baron et al. 2000, Baron, 1992).
- Lake algal communities have changed significantly in the past 50 years. These microscopic plants are the base of the lake food chain. Algae in some alpine lakes is more abundant, and now dominated by species representative of pollution. These species have been shown to dominate in lakes allover the world that have been fertilized or otherwise disturbed, and their presence indicates a profound change in their environment. The communities now found in these remote and otherwise undisturbed lakes are different than any other throughout lake history (Baron et al. 2000, Wolfe et al. 200 I).

I would like to spend the remainder of my time emphasizing the need to develop better research and monitoring information throughout the West.

Research and Monitoring Needs

Monitoring:

- There are hundreds of square miles in the West without any wet deposition monitoring, so our knowledge of atmospheric deposition is incomplete. High elevation regions, especially, are under-represented, but this is where the greatest amount of wet deposition occurs.
- Dry deposition monitoring is even more sparse, but dominates deposition through much of the West.
- Long-term monitoring of chemical and biological trends in watersheds is greatly needed. There are only a handful of long-term study areas in the West, operated by the U.S. Geological Survey, the National Science Foundation Long- Term Ecological Research program, USDA Forest Service, and some universities. In a region undergoing rapid change in population and energy development, we are very poorly set up to detect the consequences. Managers of public lands depend greatly on objective scientific reports of the condition of their resources. Long-term monitoring is highly effective at detecting trends, and when coupled with research, give early indications of the causes of change.

Research:

- Our confidence in atmospheric transport models is poor in the West, where the complex topography makes modeling extremely difficult. More effort in this area would yield great benefits in understanding source-receptor relationships.
- The mechanisms of dry deposition are still not well understood; more research into transport and deposition mechanisms is needed.
- Our understanding of the complex ecological interactions that occur with nitrogen deposition is incomplete. Communities that evolved with low nitrogen availability are changing in novel ways that we do not understand. More research is needed to tell us about possible effects on disturbance regimes such as fire, changes in nutrient cycling and leaching, increased opportunities for invasions and insect outbreaks, and how much nitrogen causes acidification in Western watersheds. Better understanding of ecosystem functions will help managers of public lands, and regulatory agencies make decisions on critical thresholds of ecological change.

Mr. Chairman, this concludes my remarks. I am happy to respond to questions Members of the Committee may have.

Testimony by Gerald J. Keeler

Assoc. Prof. of Environmental Health Science, Assoc Prof. of Atmospheric, Oceanic, and Space Science
University of Michigan

The Problem of Mercury

In the U.S. more of our surface waters are impaired for fishing due to mercury contamination than for any other toxic contaminant (U.S. EPA). Fish consumption is most often the route of human exposure to toxic contaminants, and the U.S. Food and Drug Administration (FDA) routinely issues guidelines to states on levels that warrant closures of fishing waters to recreational and commercial fishing. Currently, thirty-nine states have mercury fish advisories, which likely impact both sport fishers and subsistence fishers (USGS). Other counties around the world, including our neighbor to the north-Canada, and several European countries also issue fish consumption warnings due to elevated levels of mercury in fish. The form of mercury found in fish is Methyl mercury (MeHg), which is a neurotoxin. Those most at risk for brain and nervous system damage are children and unborn babies of mothers who eat mercury contaminated fish during pregnancy. Subsistence fishermen, such as many Native American Tribes, are particularly affected by these bans on consumption of mercury-contaminated fish because of the problems associated with alternative foods. There are numerous Alaskan tribes that depend on fish and wildlife for their source of food. Recent research has revealed that the Arctic ecosystem may be particularly at risk due to unusual meteorological conditions and atmospheric chemistry observed there that lends itself to greater deposition of mercury to the surface of the frozen ecosystem.

Today the dominant source of fish contamination is believed to be emissions to the atmosphere from fossil-fuel burning power plants (primarily coal-fired), municipal waste incinerators, and other industrial sources. Some forms of the emitted mercury compounds (e.g. mercuric chloride) directly deposit to water bodies, or to the watershed where they runoff to the adjacent lakes and streams. To date we have a limited data set for the emissions of Hg in its various forms from the major emissions source types.

It should be noted here that in parts of the U.S. significant amounts of mercury reach surface and ground water due to indirect inputs such as urban runoff and leaching from mines and waste disposal sites. Elevated mercury concentrations have also been measured in the influent of municipal wastewater treatment facilities in urban areas that might accept discharges of industrial wastewaters as well as domestic waste.

Natural emissions of mercury to the atmosphere from abandoned mine tailings and from natural geological materials are also significant in parts of the western U.S. Understanding the magnitude of these natural emission sources is made more difficult due to the observation that previously deposited mercury can also undergo chemical transformations that convert it back to the elemental form that readily leaves the earth's surface (land and water) to re-enter the global background of mercury. The rate at which mercury is re-emitted from the surface varies with the type of surface, forest vs. farm field for example. Research to understand the factors that influence the rate of surface emission is ongoing and will meet a critical data need for the development of emissions inventories needed for sounds environmental management of impaired ecosystems.

After it enters the environment, biological and chemical processes in the soil, surface waters, and sediments can transform mercury into methyl mercury (MeHg), the most toxic form of Hg, which can be taken up by plants and animals. The levels of MeHg tend to build up or bioaccumulate in organisms on up through the food chain. While mercury can transform from one chemical form to another, it does not degrade or disappear like some other persistent organic pollutants.

Mercury and Acid Deposition

The problems of acid rain and mercury contamination are closely linked. Research has revealed that the levels of Hg in fish tissue tend to be higher in more acidic water bodies. The same sensitive ecosystems that have been the recipient of acidic compounds from the burning of fossil fuels also receive elevated concentrations of mercury and other toxic trace elements. With the Vice President's announcement of an energy policy that continues to focus heavily on fossil fuels, there is some urgency to understand the linkage between atmospheric emissions from these sources and the risks to humans and the environment. Research aimed at understanding the sources and cycling of mercury in the environment will also help us in understanding the fate and risks from the other 15 "pollutants of concern" identified in the *Third Report to Congress* of the Great Waters Program. Since the atmospheric fate and transport of mercury is so inter-twined with the chemistry of atmospheric acids and ozone, federal agencies should aim to develop integrated high-temporal and spatial monitoring sub-networks in select ecosystems that have comparable aquatic and biological monitoring.

Environmental Monitoring

The importance of high quality monitoring data for development of sound environmental policy can not be understated. At present, the proper balance of long-term atmospheric deposition monitoring sites and integrated process level research sites does not exist. While there are currently several national monitoring networks for acid deposition *(NADPINTN,* CASNet, AIRMON) as well as the National Air Monitoring Stations (NAMS) and State and Local Air Monitoring Stations (SLAMS) for assessment of the National Ambient Air Quality Standards, there is not a national effort to properly assess the trends in ambient mercury or the dry deposition of mercury, which can equal the deposition of mercury deposited by precipitation. Several states have initiated weekly mercury monitoring as part of the NADP sub-network referred to as the Mercury Deposition Network (MDN). However, the spatial coverage at present is not adequate to assess emissions trends or to assess the efficacy of reduction programs. The highest deposition (wet and dry) is found in and downwind of our large urban areas where there are, at present, few monitoring sites. Support from the federal government will be required to improve the spatial coverage as well as to improve the quality of the monitoring equipment presently being utilized. Current efforts by the U.S. EPA and USGS need to be expanded to allow for a continued improvement of monitoring networks and utilization of modem technological breakthroughs in microelectronics that are now becoming more widespread. Improvements in monitoring methods will also be required for attainment of the goals of the bi-national (U.S.–Canada) and tri-national (U.S.–Canada-Mexico) environmental programs and action plans.

Source Attribution

The question of whether one can determine the source or source types contributing the Hg in a particular water body or in the fish in a particular lake is one asked often. Monitoring conducted for source allocation would require the use of short-duration measurement techniques that are now available as well as event-based wet and dry deposition methods. The use of proper techniques and complete chemical and elemental characterization of the deposition samples allows the estimation of the contribution of mercury from the major sources (See Dvonch et al. 1999, ES&T for more details). Detailed measurements that are needed for determining source-receptor relationships can also be used to better understand the basic processes that are controlling the fate of mercury emissions from various sources in different regions of the U.S.

A well-defined network of monitoring sites together with the chemical and elemental characterization would allow for regional and national assessments of the major sources contributing to the mercury deposition. This network would provide the needed data to track changes in mercury deposition and assess the efficacy of regulatory programs. Funding for the development and implementation of this type of network would also need to include both wet and dry deposition measurements. In addition, the enhanced deposition data would allow linkage to whole ecosystem studies tracking changes in the biological and physical environment.

Appendix A – Conference Executive Summary

FOREWORD

The purpose of this May, 2001 conference held in Washington, D.C., was to put the acid rain problem squarely back on the forefront of the public agenda.

The 1990 Clean Air Act Amendments, despite a cost effective emissions trading program, have not proved adequate to combat the adverse impacts of acid deposition on the nation's lakes, streams, fisheries, soils, forest, public health, materials, cultural resources, and ultimately, the public purse.

Acid rain is linked to other serious environmental problems

Scientific research in the past ten years has documented linkages between acid deposition and other serious environmental problems, including mercury accumulation, particulate matter, over-fertilization of ecosystems, ozone formation, and climate change.

The acid rain problem is continental in scope

The conference was co-sponsored by fifty four federal and state agencies, organizations, associations, institutions, and companies. Attendees from 24 states, the District of Columbia and Canada participated in the conference. From Maine to California, from Alabama to Wisconsin, they came from all regions of the U.S. and from Canada.

Highlights

Conference speakers included some of the nation's leading scientists, economists and policy analysts whose presentations are described in this Executive Summary. Highlights of the conference are captured in the Excerpted Quotes of the speakers which precede the Executive Summary and include:

- information about new research findings since 1990, including the impacts of nitrogen compounds;
- economic values of resources at risk;
- the magnitude of health benefits vs. the costs of the trading program;
- the need for a pollution prevention program which accounts for all sources of emissions including transportation and food production.

Credits

The Center for Environmental Information gratefully acknowledges the financial support and the in-kind assistance of the fifty four co-sponsors who made this conference possible. Special thank yous to **James C. White**, professor emeritus, Cornell University, the program coordinator; **John Mimikakis**, staff director of the Environment, Technical, and Standards Sub-Committee of the House Science Committee, for his cooperation; **John Terninko**, the conference and publication coordinator; **Shirley Sherman**, the support staff and registration coordinator; and **Fred O'Hara**, of Fred O'Hara Associates, editor of the Executive Summary.

Elizabeth Thorndike
Center for Environmental Information,
Rochester, New York
July, 2001

Conference Quotes

"I look at what we know today that we didn't know 10 years ago when we last acted on this issue. I look at the progress we have made in both the science and in our ability to manage multi-state air pollution problems We have shown that benefits of additional reduction of SO_2 and NO_X would far exceed additional costs."

Brian McLean
U.S. Environmental Protection Agency

"Scientists, land managers, policy makers, environmental advocacy groups, and the public are going to be increasingly involved in dialogue about the way in which source emissions reductions can be made to mitigate individual and cumulative impacts from all types of air pollutants. These impacts have serious implications for the life support system on which this planet relies, as they impact the web of life in the soil, air and water that sustains ecosystem and humans."

Tamara Blett
National Park Service

"At the current levels for sensitive sites, we are talking many, many decades before we would see significant recoveries in chemical indicators that we would expect to lead to improvement in biological resources."

Charles Driscoll
Syracuse University

"Relative to sulfate, nitrate is growing in importance as a cause of acidic deposition."

Van Bowersox
Illinois State Water Survey

"The nitrogen cycle is being altered more rapidly than the cycle of any other major element globally. 64 percent of the coasts and rivers and bays of the United States are now moderately to severely degraded from nutrient pollution. We are talking about potential losses of resources in the neighborhood of $80 to $100 billion, a very, very large problem.

Ammonia is a major contributor to acidification of soils and surface waters. Although the ammonia is a base, once in soils and waters, it is readily oxidized to nitrate. The process of "nitrification" releases two protons, so that atom-for-atom, ammonia actually has twice the acidifying power as does nitric acid in deposition. Rates of ammonia deposition have increased greatly over the past decade, so that it now probably exceeds nitric acid in importance overall in the US. Ammonia emissions are practically unregulated, and in fact, some control technologies for NO_X end up emitting ammonia instead.

Nitrogen is the biggest pollution problem in the coastal water of the US, with over 60% of the coastal rivers and bays in the country moderately to severely degraded. The sources of the nitrogen are many, including agricultural sources and sewage. However, atmospheric deposition is a major contributor to many coastal systems.

For the northeastern US as a whole, atmospheric deposition is the largest source of nitrogen to the coast. Much of this comes from deposition onto forest lands, which leaves the forests as dissolved organic nitrogen. The importance of this flux pathway has only been discovered within the past few years."

Robert Howarth
Environmental Defense/Cornell University

"The production of food increases the acidity of the atmosphere. The production of food increases the rate of acidification of our ecosystems. The production of food increases ozone concentrations in the atmosphere. The production of food increases particulate matter in the atmosphere. The production of food, primarily through mercury coating on seeds, increases mercury in the environment. So, when we talk about linkages, when we talk about acid rain and what it is related to, we can't just talk about energy production and high temperature combustion. We have to talk about agriculture. To separate the two is a tremendous scientific mistake and a tremendous policy mistake. It has to be a holistic approach to nitrogen, sulfur, ozone, particulate matter and mercury."

James Galloway
University of Virginia

"Reactive nitrogen from food and energy production is accumulating in the environment. Humans, on a daily basis, create more reactive nitrogen than do natural terrestrial processes. We have overtaken nature in making nitrogen mobile in the environment and available for this cascade of effects. So you have ozone, particulate matter with the health effects and visibility effects, then a whole cascade of effects on forests, then on the soil, then groundwater and then streams. Next we have river effects, coastal effects, coastal eutrophication…Linkages among effects require a multi-pollutant strategy for control."

James Galloway
University of Virginia

"When you look at the power plant sources of NO_X, in comparison to other sources, that is going to go below 20 percent. Transportation sources, which are now over 50 percent, will be, as a percentage, even higher. If we really want to focus on nitrogen oxides, it is going to be important to look at more than just power plants."

John McManus
American Electric Power

"Depletion of calcium from the soil is something we weren't aware of in 1990. Calcium is essential for wood formation in trees. If they don't have enough calcium, they don't grow at as fast a rate. Calcium in the soil is the primary element that neutralizes acidity…. As nitric acid is formed and leaches into the surface waters, it is contributing to the calcium depletion problem."

Greg Lawrence
U.S. Geological Survey

"Acid rain doesn't kill the sugar maple trees directly, but it reduces their ability to deal with stresses."

Greg Lawrence
U.S. Geological Survey

"The fine particles, as represented by sulfate, are about 90 percent as high in most indoor environments as outdoors, because fine particles go in through the cracks and the air handling systems. We are not just exposed to outdoor generated fine particles outdoors, but indoors as well…Pollution that you worry about in terms of ecological effects and perhaps material effects, is also clearly related to fairly convincing evidence that there are serious health consequences."

Morton Lippman
New York University
School of Medicine

"We have heard a lot of talk today about recovery. When a sculpture or a building or a stone is altered, it is permanently altered. We can't talk about recovery. It is gone."

Mary Striegel
National Center for
Preservation Technology
and Training

"California's residents drive about 800 million miles per day. This clearly is at the heart of our nitrogen-dominated acidic deposition concerns. Emissions are about 80 percent from motor vehicles and about 20 percent from stationary sources and area sources."

Brent Takemoto
California Air Resources Board

"In Wisconsin we have sandy, loamy soils. We have some base cations left, although we don't want to go any lower than this, if we can help it."

Doug Knauer
Wisconsin Department of
Natural Resources

"Nitrogen saturation in the ecosystems is much greater on the east side of the continental divide than on the west side…One of the scariest things is we just have so few long-term monitoring sites in the western United States."

Don Campbell
U.S. Geological Survey

"This is a picture taken from the summit of Mt. Mitchell. The chemistry of the air that you see, where it is brown, is near the Blue Ridge Parkway. The chemistry of that layer of the air is roughly what you might find over Philadelphia."

Ellis Cowling
North Carolina State University

"There has been significant improvement around Sudbury. Improvements elsewhere have been modest. Some lakes and rivers in the eastern-most part of the country are still acidifying, for example, the Atlantic salmon rivers in Nova Scotia. We have estimated that in order to get rid of this problem, we would need reductions in SO_2 emission from Canada as well as from the eastern United States on the order of about 75 percent from phase II of the program."

Guy Fenech
Environment Canada

"We are all in this together, and I think that is one of the most important things we must understand. No state can do an effective job of managing its own air quality. No province in Canada can do it either."

Ellis Cowling
North Carolina State University

"We are actually talking and have been for several years now, with the administration, the environmental groups and Congress about an approach whereby we could eliminate these many steps in a more streamlined approach which would allow us to eliminate duplication, minimize confusion and reduce uncertainty. We call this the comprehensive approach where, once again, there is an integrated strategy."

John Kinsman
Edison Electric Institute

"That is probably one of the most singular features of this type of trading program. It allows the industry flexibility."

Jeremy Platt
EPRI

"With respect to health, they provide the lion's share of quantifiable benefits and alone, are an order of magnitude greater than costs, and may justify further reductions in the cap on a health basis alone."

Dallas Burtraw
Resources for the Future

"If we summarize phase I, we see cost estimates that came in at about $730 million a year, according to MIT's tally...The future looks highly uncertain, driven by politics and regulation and stringency."

Jeremy Platt
EPRI

"There are health benefits associated with trading, but I am talking $100 million compared to health benefits of over $17 billion per year, by the year 2005, due to the overall reduction in sulfur dioxide emissions with the 50 percent cap."

Dallas Burtraw
Resources for the Future

"I am a firm believer in the idea that sound science drives good policy."

Representative John Sweeney

"Your meeting here today provides even greater impetus and more voices to the extraordinarily urgent message that we can and we must develop a clear set of standards to govern air emissions, deal with acid rain, and have a balanced approach to energy. I urge you to do everything you can, both with your elected members and colleagues back home, to sound the call to action, that we will and we must address acid rain this year, before it ends, before we cause any more damage, and it must be addressed in a bipartisan effective manner."

Senator Hillary Rodham Clinton

"The response to acid rain has thus far been a story of the system succeeding-a story of Congress acting on the basis of the best research, even in the face of uncertainty, and then trying to update policy on the basis of continuing research. That's a pattern we need to perpetuate. We need to keep funding the research and heeding its results."

Representative Sherwood Boehlert

"We can no longer resort to the traditional regional antagonism with regard to acid rain. While New York's position is precarious, it is no longer unique. We need to make it clear that the Virginia trout are worth protecting. That the front range of the Rockies may be in danger. That the futures of Chesapeake Bay and New England maples are not certain. That the historical monuments in Washington, DC are under fire. We have the bills to address these problems. Let us resolve to use this great occasion-this milestone conference-to redouble our efforts to broaden support for these measures."

Senator Charles Schumer

EXECUTIVE SUMMARY

INTRODUCTION

This conference was convened when monitoring data, scientific research, and modeling results indicated that the standards set by the 1990 Clean Air Act amendments, despite a cost-effective emissions-trading program, have not proved adequate to combat the adverse impacts of acid deposition on the nation's lakes, streams, fisheries, soils, forests, public health, materials, cultural resources, and, ultimately, the public purse.

The problem of acid deposition is now recognized as being even more chemically diverse and continental in scope than it was seen to be 10 years ago. In addition, new scientific knowledge has identified linkages between acid rain and other serious environmental problems, including mercury accumulation, particulate matter, overfertilization of ecosystems, ozone formation, and climate change. All of these issues raised new concerns and broadened the scope of the conference.

The aims of the conference were to:

- review scientific progress that has occurred during the 20 years of the National Acid Precipitation Assessment Program (NAPAP) and the 10 years of sulfur dioxide abatement under the Clean Air Act,
- determine the environmental effects of that abatement effort and of the concurrent changes in emissions,
- assess the current trends and needs for future actions, and
- thereby put the acid rain problem squarely back in the forefront of the public agenda.

These objectives were accomplished by reviewing the conduct and results of the scientific programs, legislative initiatives, regulatory efforts, industrial responses, and economic feedbacks that have been brought to bear on the problem. The sources of emissions and the effects of acid precipitation on various ecosystems and different regions of the continent were analyzed. Costs and benefits of the current abatement program and the need for additional reductions in pollutants were examined. The conference ended with a discussion of the needed legislative initiatives and the potential for their enactment.

ADDRESSING THE ACID RAIN PROBLEM – TWENTY YEARS IN RETROSPECT

OVERVIEW

Christopher Bernabo, ***Collaborative Solutions***

In retrospect, the 1980 legislation creating NAPAP, a cooperative federal program to coordinate and conduct research and to report the findings to Congress, was the beginning of a nationwide consciousness not just to conduct scientific research but to harness that research to help make decisions. NAPAP spent $500 million during the next 10 years, and produced its first assessment in 1995. During that period, it brought science to bear on the issue and provided defensible information for reports to the President and Congress, congressional hearings, newspaper articles, and scientific papers.

An initial concern about impacts on agriculture was supplanted with a concern about forestry. The focus on sulfur was enlarged to include nitrogen oxides. Science was translated into legislation and policy, and the program continues today. This conference is, in part, an effort to measure whether the nation got what it wanted.

One notable result of the research program was a market-based program of trading and banking emission allowances. Such a market-based program could not be effective without a fundamental understanding of the source-receptor relationship because the most cost-effective sources to control must be linked to the location of the benefits, results, and receptors. NAPAP provided the scientific understanding to make such judgements and links.

FEDERAL LEGISLATION AND REGULATION

Brian McLean, ***U.S. Environmental Protection Agency***

The 1980s opened with two polarized views about the issue of acid rain. One view held that acid rain was a serious environmental problem. The other held that it was premature to do anything about the issue, and the scientific uncertainties should first be reduced. Research was driven by two basic questions: How bad is the problem? What should we do about it? A national lake and stream survey was conducted, forests were surveyed, and monitoring systems were set up, increasing our understanding of (1) the response of ecosystems to deposition, (2) atmospheric chemistry and transport, and (3) emissions control. That knowledge and understanding was captured in models.

In 1980, the principal instrument to address environmental problems was command-and-control regulation. There was little use of economic instruments. In 1989, a new approach was introduced that relied on a national emissions cap with trading of emission allowances. The goal was to reduce sulfur dioxide emissions by 10 million tons, the bulk of that (8.5 million tons) from electric utilities. As a result of that program, emissions declined sharply by 1995, and they have declined more since then.

Wet sulfate deposition also declined significantly throughout the east. A few years after the program began, it was evaluated, and cost estimates were reduced sharply, from $7-9 billion down to $1-1.5 billion. Competition across the different control options pushed prices down. The market improved as new techniques and banking provisions removed disincentives to over-controlling, and actual cost data were accumulated against which to judge options. When the program is fully implemented, health benefits in the range of $50 billion per year are expected.

Deregulation of the railroads, demand for western coal, and increased competition in western railroads contributed to the decreasing costs of emission reductions. Scrubber costs then had to compete with the reduced prices for lower-sulfur coal. Energy efficiency may have contributed to lower emissions, but the impact is difficult to measure or detect and is indirect. The energy-conservation and renewable set-aside provisions of the Clean Air Act were not attractive because of the large cost differential between renewables and conventional power.

STATE AND LOCAL LEGISLATION AND REGULATION

Richard Valentinetti, ***State and Territorial Air Pollution Program Administration (STAPPA)***

Early discussions of emission reductions focused on cost. Then one week, both *National Geographic* and *Sports Illustrated* ran stories on acid rain and showed fish belly up. That produced a real shift in the debate. Since 1990 and the Clean Air Act Amendments, sulfur deposition rates have declined. Early responses included a lot of fuel switching from coal to oil and from high – to low-sulfur coals. Engineering approaches included the installation of precipitators, bag houses, and scrubbers. Emission reductions have resulted in chemical improvements in some lakes. On the other hand, there are a number of lakes in which no chemical improvement and no biological recovery has occurred.

The past 10 years have shown the importance of nitrogen. There have been no improvements in nitrogen deposition, and chronic acidification is occurring in areas with nitrogen-saturated soils. Sulfur and nitrogen concentrations in the soil are linearly related to their respective deposition rates. Once those elements are deposited in soil, sulfuric and nitric acids tend to accelerate the loss of critical nutrients from forest soils, resulting in lower productivity. Soils can be viewed as a trust fund for forest productivity, and a major question arises: How much of the principal of this trust fund has been used up?

The Clean Air Act has been too model-pollutant-centric and has dealt with air pollution piecemeal. Reductions of both sulfur and nitrogen oxides need to be included. A holistic approach is needed to come up with the control systems that best reduce all pollutants, and a national trading program is needed to deal with all these issues.

COMPLIANCE EXPERIENCE OF UTILITIES:

John McManus, ***American Electric Power***

Once Title IV was passed and the regulations began, the initial focus was on a self-sufficient compliance plan. Every company looked at ways to comply just within their system, so the market itself did not play a significant role as a compliance tool. Each state had a different perspective on how allowances should be handled and accounted for.

Compliance steps taken in Phase I were pollution controls, fuel switching, and allowances. The allowance program has been a tremendous success. It provides flexibility, reduces costs, and is easy to administer. In Phase II, compliance will be achieved through fuels. Nitrogen oxides are part of the program, and utilities have installed low-nitrogen oxide burners.

The coal market and the allowance market are directly interrelated. As allowance prices dropped, it became cost effective to look at higher-sulfur coals. A large bank of sulfur dioxide allowances was built up industry-wide during Phase I.

The Title IV Phase II cap is more stringent than people think it is; converting every coal unit to burning Powder River Basin coal is not practical or economical. There will have to be additional scrubbers installed and additional controls on nitrogen oxides. The real question is, how much reduction do we want and how fast do we want it?

Every coal plant in the country is now regulated for sulfur dioxide and nitrogen oxides. The NO_X SIP (nitrogen oxides state implementation plan) Call Program is a fairly aggressive program. Granted, it only operates in the summer for five months, but it significantly caps emissions in those five months.

Transportation sources, which now produce more than 50% of the emissions, will be, as a percentage, even higher producers as emissions of nitrogen oxides from powerplants are reduced under the SIP Call Program. If we really want to focus on nitrogen oxides, it is going to be important to look at more than just power plants.

INTRODUCTORY REMARKS

As a prelude to the first paper in the second session, "Research and Analysis: What the Experts Say," **Bernard Melewski** reminded the audience that a good deal of scientific research and analysis has been conducted during the past 20 years, and he listed the key reports on acid deposition that have been issued since the Clean Air Act Amendments of 1990: the General Accounting Office's 1994 report on air pollution and allowance trading; the EPA's 1995 acid deposition standard feasibility study report; the U.S. Geological Survey's 1997 report on the effects of acid rain on buildings and monuments in our nation's capital; the National Acid Precipitation Assessment Program's 1998 biennial report to Congress; the Adirondack Council's 1998 report, *Acid Rain, a Continuing National Tragedy*; Trout Unlimited's 1998 report, *Acid Rain, Current and Projected Status in the Coldwater Fish Communities of the Southeastern United States*; the U.S. Geological Survey's 1999 report, *Soil Calcium Deposition Linked to Acid Rain in the Eastern United States*; the Ecological Society of America's 2000 report, *Acid Deposition, the Ecological Response*; the General Accounting Office's 2000 report, *Acid Rain, Emissions, Trends and Effects in the Eastern United States*; the EPA's 2000 analysis of the acid deposition in the Ozone Control Act; and the Hubbard Brook Research Foundation's 2001 report, *Acid Rain Revisited.*

NATIONAL ACID PRECIPITATION ASSESSMENT PROGRAM ANALYSIS

Michael Uhart, ***National Acid Precipitation Assessment Program***

NAPAP in the 1980s was a focused monitoring and research program that resulted in Title IV of the 1990 Clean Air Act Amendments. In Title IX of those amendments, NAPAP was reauthorized to continue, only its role was changed. It still must monitor, but now it must also assess the effectiveness of Title IV and the costs and benefits by (1) coordinating federal acid rain research and monitoring, (2) evaluating the status and trends of certain variables, and (3) conducting periodic assessments (every two years).

A NAPAP report to Congress in 1998 focused on the economic benefits and other effects of Title IV. The assessment's general findings were that sulfur dioxide reductions were as good as expected and reductions in nitrogen oxides were very limited. The variables looked at were emissions, amine concentrations, depositions, and the response of sensitive receptors.

Those measures of effects come from monitoring. But monitoring is jeopardized. The National Atmospheric Deposition Program (NADP) is under constant threat of being reduced, and aquatic monitoring networks are already being reduced. At the same time, the observable ecosystem responses associated with Title IV are limited, largely because of confounding factors, the lack of monitoring information, and the lack of information on anything but the local scale. The Hubbard Brook Research Foundation has described five major monitoring programs related to acid rain that are needed, and estimated their funding needs. [Presented at the House Science Committee hearing on May 3, 2001.]

The compliance costs have been less than expected. Emissions trading has been effective, and many lakes and streams have experienced decreases in sulfate concentrations. However, additional reductions of sulfur and nitrogen deposition would be necessary to bring about the recovery of sensitive lakes.

High-elevation spruce fir forests are the most sensitive and are seeing quite adverse consequences directly related to acidic deposition, but most forest ecosystems in the east, south, and west are not *currently* adversely affected. A gradual leaching of soil nutrients has been observed and documented. A close relationship exists between ambient concentrations and visibility. Sulfate concentrations were 15% lower in the eastern United States, but because data were available for only limited areas, this result may be misleading. The expected reduction of 8% in haze was not seen.

Cultural resources are at greater risk than purely operational assets (buildings, bridges, and cars).

NAPAP helped bring human health effects into the forefront and made connections between fine particles and human health based on epidemiological studies. Those studies strongly indicate a direct relationship. Human-health effects are a significant part of the benefits of reducing acidic deposition. In agriculture, nitrogen and sulfur are nutrients, but the biggest impact is ozone damage. The economic benefits are much greater than what the compliance costs are, but future values are unknown.

The next assessment is currently being drafted. Tropospheric ozone, fine particles and health, sulfates and radiated forcing as a climate issue, energy production, and the deposition of nutrients and heavy metals are linkages that have to be considered. They are not being considered because that is not in NAPAP's mandate.

Sources, Impacts, and Recovery

Paulette Middleton, ***RAND Corporation*****; Charles Driscoll,** ***Syracuse University*****; and James Galloway,** ***University of Virginia***

[Middleton's paper was originally the first paper in the fourth session, "Acid Rain Linkages"; Driscoll's paper was originally the third paper in the first session, "Addressing the Acid Rain Problem, 20 Years in Retrospective"; and Galloway's paper was originally the second paper in the fourth session, "Acid Rain Linkages."]

SOURCES

Middleton

Acid rain is one of many manifestations of how actions of society can have adverse effects on human health and welfare. The relationships among human activities, natural processes, chemical emissions, the formation of secondary harmful chemicals from these emissions, and the effects of the chemical mix on various ecosystems, human health, and materials are highly complex. Identifying the source magnitudes and spatial and temporal patterns for all of the key chemicals [i.e., sulfur and nitrogen oxides, volatile organic compounds (VOCs), ammonia, particulate matter, and other toxics] is an essential step in developing near and long-term strategies for addressing multiple air-quality concerns. To determine the most effective emission-management strategies, it also is essential to characterize chemical connections linking multiple sources to multiple chemicals to multiple impacts.

The scope of sources includes point sources, area sources, and mobile sources. Natural sources are important contributors to VOCs, ammonia, particulates and climate-change-related gases. However, it is important to note that the recent reductions in man-made sulfur dioxide emissions are small relative to the large difference between natural and anthropogenic emissions. Likewise, ammonia emissions are large relative to what occurs naturally.

The United States has definitely reduced sulfur dioxide and sulfate deposition. However, it still has acid rain because of nitrogen oxides and the role that ammonia emissions play in stabilizing nitrogen oxides. Nitrogen oxides and VOCs interact to form ozone, and nitrogen oxides form nitric acid and nitrate aerosols (which cause haze) and ozone. The key point with the VOCs is that they vary in reactivity. Because of this difference in reactivity, some emission-management strategies may have to vary. Some classes of VOCs are more important than others.

These connections among the pollutants make it necessary to look at the whole pollutant mix to meet many of our major environmental goals, such as acid-deposition reduction. On the positive side, the connections among the various pollutants and their sources make it possible to address multiple problems through the same kind of emission-management strategies (e.g., dealing with nitrogen oxides will address ozone, acid rain, and particulate-haze).

Driscoll

An assessment initiated by the Hubbard Brook Research Foundation showed that sulfur dioxide emissions peaked around 1973, that there have been marked reductions (about 35%) since that time, and that additional reductions are expected.

The pattern for nitrogen oxide is a little bit different. Increases ramped up after World War II and have remained relatively flat since about 1980, but are well above background values. On a global basis, about 80% of the reactive nitrogen created by human action is caused by food production, and only 20% is caused by energy production.

Generally, one can see a pattern of west-to-east migration for nitrogen deposition and a similar pattern for sulfate emissions. This transport is affected by mountains; the higher elevation areas, such as the Poconos, Catskills, Adirondacks, Green Mountains, and White Mountains, show higher deposition. There is also a temporal component: statistically significant decreases have occurred in sulfate in precipitation during the past decade in many areas of the eastern United States. That decrease has coincided with a modest improvement in the acidity of precipitation, but it is a long way from background levels. In contrast, neither nitrate concentrations nor ammonia concentrations have exhibited a statistically significant temporal change.

Precipitation data show a clear, linear relationship between sulfur dioxide emissions for the source area and sulfate concentrations in precipitation. A unit decrease in sulfate concentrations results from every unit change in sulfur dioxide emissions.

Galloway

Agrisystems have four areas of nitrogen leakage. One is the seepage of nitrate into groundwater, producing a cascade of effects. The second is ammonia entering the atmosphere (with its subsequent effects on particulate-matter concentrations and visibility) and the terrestrial ecosystem (where it can be quickly oxidized to nitric acid and then cascade through these systems). The third is emissions from fertilized fields in the form of nitrogen oxides, impacting atmospheric ozone. And the fourth is nitric oxide emitted into the atmosphere with similar effects.

IMPACTS

Galloway

Once sulfur is deposited to an ecosystem, the major effects are increases in soil or water acidity and changes in productivity. Acidification of soil and freshwater decreases the biodiversity of those systems. After being stored in the soil or freshwater and then traveling through the headwater stream, sulfate is essentially locked up in the oceans with perhaps a very small recycling via dimethyl sulfoxide.

Reactive nitrogen from food and energy production is also accumulating in the environment. Humans create more reactive nitrogen than do all natural terrestrial processes. A cascade of nitrogen effects occurs on forests, soil, groundwater, streams, and coastal waters. About 90% of the nitrogen (in its myriad chemical forms) that enters coastal ecosystems is converted to elemental nitrogen, and very little reaches the open ocean.

Other impacts of air pollution include increased aerosol loading in the atmosphere, which has regional impacts on visibility. It does not just affect the urban areas. Depending on location, it is driven by organics, nitrates, or sulfates. Increased mercury emissions to the atmosphere from incineration and fossil-fuel combustion impact human health indirectly through the consumption of food and perhaps directly through inhalation. Other trace metals also impact lungs, forests, freshwater bodies, and agricultural crops. Agricultural production increases the acidity of the atmosphere, the rate of acidification of our ecosystems, ozone concentrations in the atmosphere, particulate matter that is at least 2.5 microns in diameter in the atmosphere, and mercury in the environment.

Driscoll

Impacts arise because some areas have soil characteristics that make them sensitive to acid rain. Sensitive or not, soils are experiencing elevated accumulations of nitrogen and sulfur. When deposition is reduced, sulfur bleeds out of the soil. The sulfur that accumulated during the past 150 years represents a legacy that will endure. With the uptake of sulfur and nitrogen, soils exhibit a significant loss of nutrient cations, principally calcium and magnesium. The soil at Hubbard Brook, has lost more than half of its exchangeable, plant-available calcium. As the soil becomes more acidic, significant quantities of aluminum are leached out of the soil. That aluminum is toxic to plants and aquatic organisms.

In terms of the forest effects of acid precipitation, the Adirondacks and the Green Mountains have seen about a 50% loss of red spruce since 1980; and the White Mountains, about 25%. Acid rain causes the loss of membrane-bound calcium, which makes red spruce more susceptible to freezing damage. In the sugar maple, a loss of calcium and magnesium from the soil leads to less being taken up by the tree, making the tree more susceptible to insect infestation and drought stress.

Lakes and streams in high-elevation areas that have very shallow soils (and, subsequently, limited neutralization) are sensitive to acidification. Even lakes whose catchments have soils that can buffer the acid precipitation may become acidic during short periods, such as storm events or snowmelt.

Some lakes, though, are chronically acidic. For the Adirondacks, about 40% of the chronically acidic lakes have been impacted by acid rain and, for the New England area, about 15%. With acidification of the lake water, the condition of the fish deteriorates, their numbers diminish, and species richness decreases. In the Adirondacks, about a quarter of the lakes are fishless. At least 20% of those are fishless because they are too acid to support any fish.

RECOVERY

Driscoll

The process of recovery will have certain chemical indicators: base saturation in soils, increased calcium-to-aluminum ratios in soils, rising pH, or falling aluminum concentrations of lakes. Once the chemical-recovery process improves, biological recovery can proceed. The fastest growing and the least sensitive organisms would be expected to recover the fastest. Those organisms with very long life cycles and the most sensitive species would take longer. Accumulation of sulfur in soil and the loss of bases (calcium and magnesium) in soil will create additional delays in the chemical recovery process.

On an encouraging note, soil sulfate has decreased in response to decreases in atmospheric deposition. But the acidity of soils has shown relatively little recovery, and significant damage has been done, particularly the loss of bases (e.g., calcium and magnesium) within soils, which will delay the recovery process.

Emission reductions made under Title IV of the 1990 Clean Air Act amendments are not going to be adequate to achieve the desired degree of recovery. Model projections suggest that the greater the level of reduction, the faster the rate of recovery.

A recovery of the calcium and the magnesium baselines in soil will take a long time. Model calculations suggest that some sites in the Adirondacks are very sensitive and are still losing base resources. Except for maybe the most sensitive sites, we have, by and large, turned the corner and are starting to move toward recovery. But at the current rates, it will be many decades before chemical indicators show the significant recoveries that would lead to improvement in biological resources.

Galloway

Linkages among effects require a multi-pollutant strategy for control. We cannot just focus on ozone or sulfur dioxide or mercury alone. A multi-pollutant approach must be taken to these materials, our science, and our policy. And the next Clean Air Act must include all sources of nitrogen and take into account the cascade of effects for sulfur and nitrogen.

Acid Rain Impacts: State of the Science

RESULTS FROM MONITORING DEPOSITION

Van Bowersox, *Illinois State Water Survey*

NAPAP I data from 1979 through 1987 suggest that all the chemical species analyzed except the hydrogen ion (the measure of acidity) exhibited a decreasing but not significant trend in deposition and concentration for the eastern half of the United States. Sulfate and nitrate decreased at nearly every station in the eastern United States, but acidity did not. The decrease in these acid species was offset by a decrease in base cations (e.g., calcium and magnesium).

The National Trends Network within the National Atmospheric Deposition Program (NADP) measures acidic species, nutrients, base cations, mercury, and precipitation. Comparing the 1993 data (just before Phase I began) with the 1996 data shows a clear decrease. Mean annual sulfate concentration in the rainwater shows a clear decrease between 1985 through 1994 and from 1994 to 1999. From Pennsylvania through New England, concentrations of sulfate depositions were level to decreasing with similar patterns in the Midwest and Southeast. The northeast saw a step-function drop in sulfate deposition during that period. In a number of regions, the deposition increased, but bear in mind that the total sulfur dioxide emissions in the eastern United States are 92 to 93% of the nation's total.

The amount of sulfur deposited in Ohio by wet deposition is about 10% of the amount that is emitted in Ohio. The sulfur emission reductions in Ohio were about 28%, and the reduction in deposition was 13%.

The emissions of sulfur in New York are only about one-third as large as they are in Ohio. In the state of New York, sulfur emissions decreased 8% from before to after Phase I. About a third of those emissions are wet deposited, so New York is also a sulfur exporter. Vermont increased sulfur emissions by 61% from before to after Phase I (it had only about 1% of the emissions of Ohio). The sulfur deposition, however, decreased substantially.

Since NAPAP (1990), sulfate deposition has decreased at most stations in the east, but not at all. The average change in annual sulfate deposition decreased about 12% averaged over the entire east. The largest decreases (10 to 25%) occurred in and downwind of the major source areas, Illinois, Indiana, Pennsylvania, New York, and Maryland. Over the whole east, about a fourth of the sulfur that is emitted is deposited by precipitation.

In the west, the sulfate deposition appears to have increased at most sites.

Nitrate emissions and deposition did not change much. There were year-to-year variations, but no consistent trend. There is a band from Illinois to New York that received the highest concentrations. Temporal changes exhibit no clear trend.

Relative to sulfate, nitrate is growing in importance as a cause of acidic deposition. Ammonia emissions are a growing problem, also. When the ammonium ion is deposited on the Earth's surface, it acts as an acid because it contributes its extra proton. Ammonia sources are largely from fertilizer, automobiles with catalytic converters, power plants, senescing crops, and animal waste. Ammonia emissions are increasing from the Texas panhandle to North Dakota, and from Kansas east to Ohio. The highest concentrations occur in South Dakota and Minnesota, a clear source area.

Hydrogen ion deposition decreased at 88% of National Trends Network sites. The largest decreases generally occurred in the same area as the largest sulfate decreases, and there was evidence that cation increases are causing hydrogen ion decreases. In the west, hydrogen ion deposition has not changed.

TERRESTRIAL SYSTEMS

Greg Lawrence, *U.S. Geological Survey*

In the past 20 years, we have developed a better understanding of soil processes and how changes in soils are linked to changes in surface waters. Acid rain produces three primary changes in soil:

- depletion of calcium,
- mobilization of inorganic aluminum, and
- accumulation of sulfur and nitrogen.

Calcium is essential for wood formation in trees, and trees have a very high demand for calcium. Calcium in the soil also neutralizes acidity.

Because soil science in this country is a fairly young discipline, there was really no pre-acid rain data available for comparison. The Russians, on the other hand, have soil museums where archived soil samples are housed and are available for analysis. By sampling soils in Russia today at locations that were sampled between 1893 and 1895, calcium concentrations between the two dates could be compared, showing that there is much less calcium now.

In the top layer of soil, the amount of calcium is very strongly related to the weathering potential of the local rocks. In the mineral horizon directly below the top layer, the correlation is much weaker. In the upper organic layer, very intense root activity recycles the calcium, preventing the calcium from leaching out. In the deeper soil, this process is not nearly as strong. As calcium is lowered, aluminum begins to be mobilized. The aluminum starts to control the chemistry and buffers any further decreases in calcium.

Aluminum can be mobilized by organic acids, forming organic aluminum, which is generally not toxic to aquatic life and does not impair tree growth. As the soil pH decreases, however, inorganic aluminum enters soil solution. At an aluminum-to-calcium ratio of greater than one, the growth of the tree is potentially impaired.

Sulfur is accumulated within the soil but does not necessarily stay there indefinitely. It seems to be gradually working its way out, particularly when an extended dry period is followed by rain. That releases some of the sulfur, producing a very severe acidification episode. The outputs of sulfur are currently exceeding the inputs. If all inputs of sulfur and acid rain were shut off tomorrow, some sulfur would still be lost into surface waters as sulfuric acid, continuing to put a drain on base cations.

Nitrogen also is accumulating in forest soils to the point where some forests are not very effective at retaining nitrogen. Again, a dry period followed by a wet period releases a lot of nitrate in the form of nitric acid, producing an acid episode. As the nitric acid leaches into the surface waters, it contributes to the calcium-depletion problem.

Factors other than acid rain can also change soils very significantly over decadal time scales. The changes in soils have reduced the stress tolerance of red spruce trees and contributed to the die-off of current-year needles during extremely cold periods as a result of the amount of calcium incorporated into the wood as the tree grows.

At the early stages of acid rain, calcium actually was more available. The hydrogen ion was freeing up the calcium, making it easier for the tree roots to take it up. The higher calcium availability was reflected in the calcium in the tree tissue. With continued acidification of the soils, the calcium has leached out at a higher rate than the rate at which it was being added through the breakdown of the rocks and minerals. This situation has produced lower levels of available calcium, which is also reflected in the chemistry of the tree tissue.

Magnesium responds to acid rain leaching very similarly to calcium. If trees do not get enough calcium and magnesium, they are much more susceptible to stresses, such as defoliation from insects. If there is sufficient calcium and magnesium available to the foliage, the trees can recover from the defoliation. If there is not, they die. Acid rain does not kill the trees directly; it reduces their ability to deal with stresses.

FRESHWATER AQUATIC SYSTEMS

Arthur Bulger, *University of Virginia*

Pure rainwater in the eastern United States is acutely lethal to all native fish species; not long-term lethal, but acutely lethal. Not even a goldfish could survive a day in pure rainwater. The modeling study, Fish in Sensitive Habitats, reconstructed the past and future statuses of fish streams. The negative effects of acidification on fish are clear and unambiguous. Hydrogen ion, measured by low pH, is directly toxic to fish. Moreover, low-pH waters leach aluminum from soils. Aluminum is quite lethal to fish and other aquatic organisms under acidic conditions, eliminating the most sensitive species first and lowering biodiversity.

A dose-response relationship exists between acidity and fish (and large aquatic insects) because of the differences in sensitivity among species. Virginia and the southern Appalachians have had a delayed response to acidification because the soils are thicker and retain sulfur. When that sulfur-retention capacity is exceeded, the amount of sulfur that comes in equals the amount that goes out.

In the southeast, the primary sensitive aquatic receptors are streams. In southern Appalachia there are about 30,000 miles of potential trout streams.

The site of the toxic action of acid and aluminum is fish gills, which perform respiration (exchanging gases) and regulate blood electrolytes. Fish can tolerate lower-pH water if aluminum is absent, but acid and aluminum together disrupt electrolyte balance and cause a significant reduction in the concentration of electrolytes in fish blood. Electrolytes are lost from the blood directly to the stream. Osmosis then drives water into the red blood cells, causing them to swell. The volume of the cells increases dramatically, from 40 to 70% of the blood volume. When the red blood cells swell to twice their normal volume, they burst. The serum becomes pink from all the red blood cells that have burst, and the blood is noticeably stickier. The heart cannot pump this very viscous blood or even to supply its own tissues with sufficient oxygen to sustain cardiac activity, and the fish has a lethal heart attack.

With sublethal stress, you can tell months or years in advance when a population may be likely to disappear. Once you get below a pH of 5.8, there is basically a dose-response relationship: The more acid the water, the smaller and skinnier the fish are. The Shenandoah National Park effects study showed that acid rain causes chronic and acute stress to fish, with the young usually more sensitive than the adults. Fish in acidifying streams were found to be in poorer physical condition, and more biodiversity losses are likely at current sulfur-deposition levels. About 30% of the probable historic trout streams in Virginia have become too acidic to support any fish.

For the Southern Appalachians, reductions in sulfur greater than 70% will be needed just to stop further losses, and very substantial reductions will be needed for recovery. For the northeast, the 1990 Clean Air Act amendments will not achieve chemical recovery; deeper cuts in emissions will be needed. Model results indicate that substantial recovery of acidic streams like Hubbard Brook will occur in 25 years if power-plant sulfur dioxide reductions are lowered by 80%.

How long soil recovery will take is dependent on the rate at which sulfur deposition is lowered. Once soil recovery has occurred and stream conditions have recovered to a pH of about 6, the big stream bugs may recover in as little as three years. Zooplankton communities will recover their functionality in about 10 years. Fish, will probably take 10 to 20 years. The recovery of animals in a stream might be given a jump start by liming. The most sensitive species, the first to go, will be the last to come back.

COASTAL ECOSYSTEMS

Robert Howarth, ***Environmental Defense/Cornell University***

Forests do a good job of retaining nitrogen deposition until they reach their saturation point. Once you add so much nitrogen that the forest is no longer limited by nitrogen, that nitrogen starts to leach out as nitrate. As more nitrogen is deposited into a forest, some proportion leaves the forest as organic nitrogen. As that organic nitrogen hits the coastal waters, it is converted back to a plant nutrient, contributing to coastal nutrient pollution.

Additional nitrogen pollution in the coastal zone occurs through atmospheric transfer. Seawater systems are limited in their production of ionic nitrogen, so adding more nitrogen produces a very, very strong nutrient effect.

A 2000 report from the National Academy of Sciences, *Clean Coastal Waters: Understanding and Reducing the Effects of Coastal Nutrient Pollution*, concluded that atmospheric deposition is very much underappreciated and under-regulated. Added nitrogen produces more growth of a variety of things. Algal blooms can lead to low oxygen levels that are unable to support fish and other higher forms of life. Dead zones are clearly growing in size off coastal Louisiana, in the Chesapeake Bay, in Long Island Sound, and in several other systems. Adding nitrogen to coastal waters also increases the incidence and duration of *harmful* algal blooms (producing brown tides, red tides, and pfiesteria blooms).

When coastal rivers and bays are fertilized with nitrogen, biodiversity is lost in a very obvious way, largely by the degradation or destruction of seagrass habitat. Two-thirds of the coastal rivers and bays of the United States are moderately to severely degraded from this problem, roughly one-third of them are moderately degraded, and one-third are severely degraded.

Nitrogen deposition comes from fossil-fuel combustion, from fertilizer, from nitrogen fixation by crops, or by movement of nitrogen in and out of a region in foods. Globally, fertilizer is the largest contributor; but in the northeastern United States, deposition from fossil-fuel sources is the major source of nitrogen under human control.

For the United States, 92% of the ammonia flux comes from animal waste resulting from industrial farm-animal production. It is a major contributor to atmospheric deposition and to the coastal waters. The ammonia flux onto the landscape with subsequent runoff into coastal waters at least equals the problem from the deposition of nitrogen oxides and probably exceeds it. Collectively, these airborne pollutants make up about 40% of the total flux of nitrogen to the coast.

For the northeastern United States, humans have probably increased this nitrogen flux by at least tenfold and, in parts of Europe, by even more. These changes have greatly accelerated during the past four decades. As a result, the nitrogen cycle is being altered more rapidly than the cycle of any other major element globally; 64% of the coastal waters, rivers, and bays of the United States are now moderately to severely degraded from nutrient pollution. The potential losses of resources amount to $80-100 billion.

Most models and reports have underestimated the importance of the deposition of nitrogen oxides to coastal nutrient pollution by a factor of 2 or 3. Guidelines for fertilization tend to encourage slight over-fertilization, about a 10 to 15%. More than half the nitrogen in the fertilizer is not only taken up by the crop, but goes into the harvestable product and is removed from the field. Over-fertilization is small compared to the animal wastes and to the deposition of nitrogen oxides. We *are* seeing massive over-fertilization on golf courses and lawns. These uses are relatively small, but they damage local coastal ponds.

HUMAN HEALTH

Morton Lippmann, *New York University School of Medicine*

Until the mid-1980s, most airborne health impacts were defined by the experience in London and the United Kingdom. In December 1952, 4000 people died in London in an acid-fog episode when one could not see more than a few feet in front of one's face. A royal commission report led to the Clean Air Act that banned soft-coal burning, replacing it initially with coke and now with natural gas. What was killing people was the strong acidity associated with sulfur dioxide.

In the mid-1980s, the United States started conducting some epidemiology that served as the basis for standards on air pollution. The sulfur dioxide gas emitted forms aerosols after oxidizing to sulfites, combining with water, and producing very fine droplets of sulfuric acid that rapidly coagulate and produce fine particles that have a long atmospheric lifetime. Those fine particles are about 90% as high in most indoor environments as outdoors.

The clearly identifiable human health effects of ambient-air acidity follow their inhalation into the respiratory tract. In eastern North America, most of the acidity is associated with sulfuric acid aerosol, which is formed in the atmosphere as ultrafine particles (less than 0.1 microns in diameter) through the oxidation of sulfur dioxide emitted during fossil-fuel combustion. Those aerosols are gradually agglomerated into fine particles (between 0.2 and 1.0 microns in diameter) and also are neutralized to ammonium sulfate by reaction with ammonia gas, a product of anaerobic decay at ground level. Short-term peak exposures to acidic aerosols can cause asthma exacerbation and can decrease the lung's ability to clear itself of inhaled particles. Long-term chronic exposures are closely associated with premature mortality, increased rates of emergency-room and hospital admissions, respiratory symptoms, and lost time from work and school.

The landmark health-research study is that sponsored by the Health Effects Institute at Johns Hopkins University and Harvard University. It looked at daily excess mortality in 90 U.S. cities and provided an overwhelming indication that there is a statistically significant association between the particulate-matter (PM) content of air and mortality. That study indicated that including sulfur dioxide or carbon monoxide or nitrogen dioxide into a multiple-pollutant regression does not shift the particulate matter coefficient very much. That is why fine particles need to be regulated.

Studies of hospital admissions indicate that the highest excess risk per unit of particles was for asthma. Pneumonia was next, and chronic obstructive pulmonary disease is significantly excessive. Studies of population cohorts indicated that the mortality rates in different communities varied in relation to fine particles and sulfate.

In the economic benefit analyses of the Clean Air Act, 90% of the benefits that could be both quantified and monetized were associated with mortality. The people affected lost perhaps 15 years of life span. Averaged over the entire U.S. population, this is a loss of about 1.5 years in life span because irritation to the cardiopulmonary system caused victims to age faster. So, the 15-microgram PM 2.5 limit has been confirmed by reanalysis, and there are serious health consequences to this kind of pollution. The data do not indicate any threshold; one cannot set a standard to fully protect the public health.

The study also showed that, if the effects of particulate matter and ozone are separated, ozone has a bigger effect on *hospital admissions* than particulate matter does. That is unlike *mortality*, where particulate matter drives it and ozone has a much smaller effect.

MATERIALS AND CULTURAL RESOURCES

Mary Striegel, ***Center for Preservation Training and Technology, National Park Service***

Outdoor man-made air pollution is affecting our cultural resources and causing decay. Sulfur dioxide is one of the most important pollutants affecting stone, reacting with stone through dry and wet deposition. Once the pollutant has deposited onto the stone, it can interact to create an alteration crust, usually calcium sulfate, which is more water soluble than the stone itself and is washed away by subsequent rains.

Stone is heterogeneous, contains many different minerals, is porous, and has a wide variety of properties. Damage may occur in the form of chemical changes or physical changes. A dose-response relationship cannot be devised because of the complexity of the system. Moreover, stone has a very long response time (up to 100 years).

Studies were conducted on the effects of texture on pollution uptake for several types of limestone. Samples were exposed eight times to 24-hour exposures of simulated acid rain. The first and second exposures were found to produce a much higher deposition of air pollution onto the stone than subsequent runs did. The rougher stone has more deposition that the smoother stone, but the surface roughness itself is not the controlling or dominant factor. It may affect the deposition, but only within stone types. Porosity and the number of valleys on a stone surface will enhance deposition, while protrusions will limit deposition. That is important because each cleaning treatment changes the stone surface and sets up a different type of deposition.

The National Park Service has been researching treatments and mitigation and how to prevent the damage to the stone. Damage is going to happen; what is unknown is the time frame and the level of damage. No exposure threshold is believed to exist; any sulfur dioxide is ultimately going to lead to some problems. Furthermore, once sulfur dioxide is deposited onto the stone, one has a lot of trouble getting it off. Therefore, even if all the atmospheric sulfur dioxide was eliminated, problems would still exist.

REGIONAL OUTLOOKS

NORTHEAST

Richard Poirot, ***Conference of New England Governors and Eastern Canadian Premiers***

In 1985, the New England governors and Eastern premiers developed a plan to deal with acid precipitation. Recently, they took stock of the progress made and the improvements needed.

They found that the kind of ecological recovery desired was not being seen and that a number of additional issues had emerged. They concluded that the big picture needs to be studied and the multiple effects associated with these pollutants need to be understood. The acid rain issue is very closely tied to the visibility issue, the fine-particle health issue, the effects on ozone, the eutrophication of coastal estuaries, and so on. The multiple benefits from reducing emissions in all these different areas must be emphasized.

Some of the future needs that would be especially helpful to the northeast and in which it could use some assistance from the federal government include the development of emission inventories that are consistent across the United States and Canada. In addition, better agreement and coordination are needed on the development of reasonable emission-reduction scenarios for the future. Comprehensive models are needed to evaluate the multiple effects and multiple benefits in not only deposition but aerosol concentration, fine particles, etc. These models also need to look at long-term concentrations over at least continental-scale domains. Alternative controls are needed to maximize reductions of mercury and carbon dioxide. And finally, monitoring data is absolutely critical to moving forward.

Aerosol sulfate concentrations have exhibited a fairly long-term improvement. But episodes still occur. On July 17, 1999, the highest sulfate level ever recorded was logged at Acadia National Park. This was a massive sulfate episode. Fine-particle pollution was certainly worse just downwind of the Ohio River Valley, but it is extending well up into New England and into eastern Canada.

These concentrations were well below the U.S. fine-particle short-term standard. On the other hand, they were well above the Canadian short-term standard.

MIDWEST

Douglas Knauer, *Wisconsin Department of Natural Resources*

In the upper Midwest a major concern is related to seepage lakes with no surface inputs. The inputs to such lakes are what falls out of the sky and what comes in from the groundwater system. The concern arises because the Midwest has sandy, loamy soils with a base saturation index of about 30%. (The Adirondacks' is around 10%.) So, some base cations are left, but we do not want to go lower than this. The pH of the soil solutions at a depth of 8 cm is about 5.6 to 6.1. Down about 60 cm, it is 6.5 to 6.7. These values are not as acidic as in the Adirondacks.

In modeling seepage lakes, no surface runoff was found to come into these lakes. The sandy soils absorb any water. The groundwater-contributing basin was the major source of water for the lakes. If the groundwater was taken out of the model, the lake would go acidic from the direct rainfall. Groundwater is extremely important as far as buffering capacity goes.

A small percentage of lakes sit above the groundwater table and receive everything out of the sky. These are the ones that are acidic.

In 1986, statewide legislation was passed and, by 1993, a 50% reduction in sulfur dioxide emissions was achieved with a resulting decrease in sulfate deposition to 10 kg per hectare per year, producing much lower loading levels. There have been some slight increases also in the pH. If there is a drought, it decouples the groundwater from the lakes, and the alkalinity declines considerably. Therefore, there is great concern over just how climate changes will impact the lakes.

The pH is related to the amount of mercury seen in the fish. Mercury is significantly elevated in an acidified lake. The sulfate-reducing bacteria are the methylators that produce methyl mercury that is taken up by the fish. If the pH is lowered, the Hg(II) that comes in has more opportunity to bump into the bacteria and to be methylated.

SOUTHEAST

Ellis Cowling, *North Carolina State University*

The southern United States has a continuing concern about visibility effects, continuing concerns about acidification effects, and a growing concern about eutrophication effects. White pines continue to suffer from ozone exposure.

The area is a substantial contributor to its own pollution: The ammonia emissions of North Carolina now exceed that state's emissions of nitrogen oxides, and Tennessee is a big emitter state. No state can do an effective job of managing its own air quality; cooperation is essential.

INTERMOUNTAIN/WEST

Don Campbell, *U.S. Geological Survey*

Quite a bit of the west (beyond the 100th Meridian) and the intermountain region is not worried about acid rain. In the high deserts and high plains, there is not enough precipitation for significant wet deposition. For sulfate deposition and hydrogen ion deposition, the highest depositions are in northern Colorado. But overall, the deposition is not very large in the west. The sulfate in the Rockies is decreasing.

Lakes in this region are very sensitive and very responsive to sulfur deposition. On the other hand, sulfate concentrations have been going down.

Nitrogen is a different story. Again, the patterns in deposition are driven by the precipitation amounts that vary widely over short distances in the west. Northern Colorado has some of the highest nitrogen deposition along with a small area on the front range of Colorado. Relatively high deposition also occurs in Utah and around Yellowstone, where there are a lot of agricultural sources.

Many of these areas do not have any long-term monitoring of the terrestrial or the aquatic ecosystems. The long-term data from the NADP network show a fairly strong increasing trend in sulfur from the mid-1980s through 1999, but the amounts (0.5 to 1.5 kilogram per hectare) are very low compared to almost all the other areas on the continent. In the Rockies, a pulse of pollutants during snowmelt delivers half the year's pollutants in a very short time.

Nitrogen is coming into the lakes, even in Wyoming, because there is so little soil and vegetation, and that may have some implications for the aquatic ecosystems there. Nitrogen may be being stored in the soil and vegetation, but there are no data on any of the terrestrial effects. Sampling indicates that nitrogen saturation in the ecosystems is much greater on the east side of the continental divide than on the west side, but many sites show no significant trend. However, a number of sites in northern Colorado are seeing increasing trends in nitrate along with other sites throughout the northwest. Quite a few sites throughout the Rocky Mountains and far west show increasing trends in ammonia concentrations. At most of the sites, ammonia constitutes roughly a third of inorganic nitrogen deposition. This is a big concern.

Electric power generation is a big source of nitrogen oxides in the west and mountain states. Some plants are putting controls on, but a lot more plants are being built, so that may cancel out any control benefits.

Transportation is a concern; the increase in vehicle miles is going to overwhelm any hope of reduced emissions from cleaner cars. Agriculture produces a lot of ammonia emissions, but the contribution that it is making has not been well characterized. Crop production and livestock production are big on the eastern plains; a lot of hog farms have moved to eastern Colorado, increasing the ammonia emissions.

Two unknowns are mercury and pesticides, which may undergo long-range transport and then get trapped in these cold systems because they cannot get re-volatilized. Mercury and pesticides (including DDT, which has not been used in many years) are observed in sediments in some wilderness lakes.

CALIFORNIA

Brent Takemoto, *California Air Resources Board*

California's acidic deposition is dominated by nitrogenous inputs stemming from its heavy use of motor vehicles. This nitrogen is largely deposited as gases during the dry half of the year. Because of geographic and elevational ranges, the total range in deposition varies considerably across the state.

About 11 times as much of nitrogen oxides is emitted than sulfur dioxide. The emissions of nitrogen oxides are about 80% from motor vehicles and about 20% from stationary sources and area sources. About 55% of the sulfur dioxide comes from mobile sources, and about 45% from area and stationary sources. Going from urban areas to remote forest sites, total nitrogen drops from about 23 to 3 kg of nitrogen per hectare per year, while sulfur drops from about 2.5 to 1.5 kg. Dry nitrogen deposition accounts for 80 to 90% of total nitrogen deposition in urban sites, but only about 40% in forest sites.

Human-health research has focused on the effects of acidic fog and nitric acid vapor alone and in combination with other pollutants in both human and animal systems. Acute-exposure studies using nitric acid vapor indicated slight changes in airway reactivity and inflammation with above-ambient exposures. Chronic-effect studies found increased risks of chronic respiratory illness in adulthood, stemming from a reduction in the maximum lung function that can occur as a young adult.

Forest research assesses the combined effects of ozone and atmospheric nitrogen on a mixed conifer forest and of acidic fog on agricultural crops. Ambient ozone has an adverse effect on pines in Southern California. A forest-model study found that the potential for chronic soil acidification and aluminum toxicity was fairly low, but that rates of soil nitrogen emissions were comparable to those that can be measured in agricultural fields.

Aquatics research has focused on wet-deposition effects to high elevation systems in the Sierra Nevada. Nitrate levels in streams were observed to exceed the highest atmospheric nitrogen loads. Although many lakes were surveyed, no indication of chronic acidification was found, although episodic acidification does occur during snowmelt in the spring. Episodic acidification was linked to reductions in the populations of some invertebrates as well as some spring-spawning fish species.

More measures will be adopted to reduce emissions of nitrogen oxides, focusing on their role in ozone and PM10 formation. It is anticipated that there will be concurrent reductions in many other effects resulting from acidic pollutants. One issue under study is ammonia from feedlots in the San Joaquin Valley; that concern was driven by particulate matter associated with ammonium nitrate.

CANADA

Guy Fenech, ***Environment Canada***

Like everywhere else, Canada's emissions of sulfur dioxide have declined fairly rapidly in the seven easternmost provinces. Emissions of nitrogen oxides have stayed just about constant. Deposition on certain lakes has been reduced by 80 to 90%, and there has been chemical as well as biological recovery.

The improvements elsewhere have been rather modest. Some lakes and rivers in the easternmost part of the country (Nova Scotia) are still acidifying, and salmon runs have decreased in all those rivers. Several of those rivers do not support salmon any more, and predictions are that only two of the rivers may have a chance to support salmon in the long run.

Of extreme concern are the chemical alterations of forest soils. Starting with the critical load or the tolerance level, it has been estimated that, to eliminate this problem, reductions in sulfur dioxide emissions from Canada as well as from the eastern United States on the order of about 75% would be needed from Phase II of the program. The national strategy also commits the government to assess the role of nitrogen in acidification.

A calcium budget for lakes indicates that they are losing approximately 5 kg per hectare per year. Dendrogeochemistry has indicated that, from 1920 to the mid-1950s, iron and aluminum concentrations in the wood did not change; but since 1950, iron and aluminum have increased.

The major conclusions are that there has been significant progress in addressing the acid rain problem in both countries, but much more is needed in terms of significant emission reductions. Monitoring must continue to verify that the programs are having the desired effects. The return rate of the salmon has dropped significantly in the past few years, but the reason is unknown; it may be competition by genetically modified salmons, or it may be because the fish that reach the ocean are not in good enough shape to survive the return trip.

Perspectives on Additional Reductions

NEEDS

Rona Birnbaum, ***U.S. Environmental Protection Agency***

[Originally the second paper in the second session, "Research and Analysis: What the Experts Say"]

In 1995, an EPA report to Congress stated that, based on improved models, the recovery expected when the 1990 amendments were passed may not come to pass. The Ecological Society of

America (ESA) report in 1999 documented that (1) some regions are seeing recovery but others are not and (2) reductions under the Title IV program may not be sufficient. ESA also released a report in 1997 that raised concerns whether further nitrogen reductions might be necessary to address the coastal-waters issue. A Government Accounting Office report came out just last year that focused primarily on the most sensitive ecosystems, indicating that the expectation for recovery may not be as apparent as it was once thought to be and emphasizing nitrate levels in particular.

In 1990, the expectation was that a 50% reduction in sulfur dioxide emissions from the utility sector would do something rather substantial and that the effect would be rapid for surface waters. Now, the results of long-term monitoring and research like that at Hubbard Brook are solidifying the case that further reductions may be necessary.

We have seen significant emission reductions. There was about a 40% reduction just in the state of Ohio. And no regional emission shifts have occurred because of trading. The overwhelming majority of allowances were retired in the same state in which they were allocated, and no significant in-flows of emissions have resulted.

An additional 50% reduction in sulfur dioxide and about a 60% reduction in nitrogen oxides beyond Title IV would produce reductions with broad impacts and would affect sensitive ecosystems not only in the high-elevation complex terrain but also in the lower-elevation terrain. Similar impacts would be produced by improvements in annual particulate matter emissions. Some of the highest benefits are realized in the highest-emitting states. States in the Ohio valley are, in fact, achieving the greatest health benefits from the current program.

INDUSTRY'S RESPONSE

John Kinsman, ***Edison Electric Institute***

By 2004, emissions of nitrogen oxides from the electric utility sector will be less than 20% of total man-made emissions of nitrogen oxides. In 2010, emissions of sulfur dioxide will be at their lowest mark in a century, except for two years during the Great Depression. In the past 30 years, electricity from coal has doubled. In terms of emission rates per unit of energy produced, nitrogen oxides are down 50%, sulfur dioxide is down 70%, and PM10 is down 94%. Although coal use, total electricity, electricity from coal, and electricity from gas will all virtually triple from 1970 to 2020, the absolute tons of emissions from electricity will still be down by about 20% for nitrogen oxides, about 50% for sulfur dioxide, and down 80% or so for PM10.

Mercury reductions in 2007 will bring the associated benefit of additional sulfur dioxide reductions. PM2.5 emissions will decrease to zero over 60 years, and nitrogen oxides and ozone will also see further reductions.

To maximize benefits while minimizing costs, the utility industry has been discussing a comprehensive approach or integrated strategy that would allow it to eliminate duplication, minimize confusion, and reduce uncertainty.

ECONOMICS

COMPLIANCE COSTS AND MARKETS

Jeremy Platt, ***Electric Power Research Institute (EPRI)***

In 1995, a cap was imposed, and sulfur dioxide emissions by electric utilities took a step jump down because there were incentives to add scrubbers. Between 1995 and 1997, the industry was emitting well below the cap. The use of Powder River Basin coal helped drive down sulfur dioxide emissions. EPA auction results indicated that, in 2000, sulfur dioxide emissions came down a million tons from the previous year, but not all of that came from the electric utility industry. Now the nation is facing coal shortages, and the price of coal has run up dramatically because of low inventories and an inability to get enough supply. That means companies have to burn anything that they can get, regardless of the sulfur content. Emissions will probably be up a little bit, because the industry will use allowances that are in the bank. Another economic influence is the cost of scrubber operation. In 1966, the variable costs for running a scrubber were very low. Those costs jumped up to around \$200 per ton of coal burned in 1998, and in 2010 prices may be \$350 to \$500 a ton, depending on the discount rate.

Having legislation that provides the flexibility

to deal with these changes in market fundamentals is one of the most stellar features of this type of regulation. The banking and trading system was widely embraced by the electric utility industry. It permits internal efficiencies that are very important within units under ownership of any one organization. It provides operating flexibility. It has also had a big impact on the design of scrubbers because allowances eliminated the need for a spare. As a result, a more cost-effective, larger scrubber could be built with greater throughputs of gas and true engineering savings. Trading and banking allowances also remove uncertainties about when to build scrubbers because any costs incurred by overspending on scrubbers can be offset through the excess allowances that those scrubbers produced. Finally, allowances provide intricate intertemporal flexibility.

The cost of compliance fell when standards were defined, allowing a number of additional types of coal, largely from Indiana and the Powder River Basin, to be used. Changes in infrastructure also permitted more coals to come to bear and more blends to take a role, and that gave companies options to switch. About half the reductions were produced by scrubbing, and about half by fuel switching.

Tremendous reductions took place in the Midwest; even though there was trading going on, they ended up reducing their emissions considerably in the upwind states. To achieve these reductions, companies adopted quite a range of strategies. Most companies opted to control their emissions on their own account and not to be net buyers of allowances. One exception was Illinois; it bought allowances to be able to continue to use local coal. Now they are going to switch to Powder River Basin coal and are going to go way down in actual emissions.

Phase I produced control costs estimated from \$0.5-1.3 billion a year for the industry; \$730 million a year, according to an independent study done at the Massachusetts Institute of Technology. Some of the factors that have helped drive down compliance costs have been the Powder River Basin coal phenomenon, additional fuel switching, lower flue-gas desulfurization operating and maintenance (O&M) costs, the technical capability to remove greater amounts (to more than 95%) of the sulfur dioxide, and the heightened competition among multiple options.

Cutting the sulfur dioxide emissions cap by 50% would require building and operating scrubbers on 60 to 100 GW of generators; that cost would be significant. Key factors in assessing the future include the new source review, which produces some regulatory and interpretative uncertainty. Now people are saying the price range in Phase II is on the order of \$100 to \$400 per ton. One company projects the cost stream to be about \$450 per kW, with a huge fraction of that going to sulfur dioxide removal because those scrubbers will be a big capital cost. However, O&M and disposal costs will be offset because of the higher-sulfur coal that they can use.

All in all, the future looks highly uncertain, driven by politics and regulations and stringency. For an allowance market to work, some source of over-compliance at a relatively low cost is needed to create a supply of allowances for trading or for consumption. It would be very difficult to generate excess allowances under strict regulations for nitrogen oxides or mercury.

ECONOMIC BENEFITS OF CONTROLS

Dallas Burtraw, *Resources for the Future*

Use values stem from changes in well-being through direct exposure or experience, like swimming, boating, fishing, angling, and bird watching. Nonuse values stem from the willingness to pay for an environmental improvement, even if you never plan on visiting the resource. The economics and integrated-assessment literature indicate that:

1. The science and economics are still not adequate to really value most of the ecological effects. The use values alone do not justify the costs of Title IV emission reductions. However, nonuse values have not been valued, and there is ample evidence that they could be very large.
2. Health benefits alone are an order of magnitude greater than costs and may justify further reductions in the cap.
3. Trading has not undercut benefits.

An integrated-assessment model was used to

link changes in emissions to changes in transport, to changes in concentrations and deposition, to changes in a number of environmental end points, and then to use values. The study found that health mortality and nonuse value for ecosystem health are the highest-valued benefits. Costs are estimated at about $5 per person across the nation for electricity consumers. A number of benefit categories are in the same ballpark, including visibility improvements and morbidity effects. Mortality effects are an order of magnitude greater. Recreational lake fishing is not highly valued. The model results indicated that it would be difficult to justify the costs of the program from a benefit-cost perspective.

The epidemiology literature now suggests that the dose-response coefficients are double what they were when premature mortality from changes in nitrates and sulfates was calculated in 1995 and 1996. On the other hand, there were a lot of philosophical and conceptual problems with the approach used to value premature mortality, and some adjustments were made, offsetting the changes produced by the doubling of the dose-response coefficients. So the valuations in the 1990s that suggested an order of magnitude difference between health benefits and the cost of the program still stand.

The model also showed that, in the Midwest, the greatest changes (but still very small changes) are in health status as a consequence of slightly greater emissions there, arising because of the trading process. However, the evidence is not compelling that the pattern of trading has harmed health or the environment. Indeed, the whole eastern seaboard has realized a very slight improvement in health status as a result of the pattern of trading. However, the effects of trading are very small, producing health benefits on the order of $100 million a year. Compare that to more than $17 billion per year in health benefits by the year 2005 because of the required 50% cap plus a 50% reduction of emissions. The overall cap is what was really important in achieving the benefits that have resulted from the program.

Trading has been important to realizing cost savings and flexibility for the industry, and it has had close to a trivial effect in terms of the benefit estimations. Stated another way, transparency and simplicity are the primary reasons for why sulfur dioxide trading programs' costs have been low and it has been such a success.

Current controls on nitrogen oxides use a summer cap in 19 states; however, an annual cap would more than double the improvement in health benefits from reducing particulate matter while increasing the compliance costs by only about 20%. Because nitrogen oxides are worse contributors to deposition and environmental damage in the winter, an annual nitrogen oxides reduction program would dominate a seasonal nitrogen oxides reduction program. Therefore, the EPA and the eastern states should consider revising the initiative to aim at annual reductions.

The Legislative Outlook

As part of the conference, the Committee on Science of the U.S. House of Representatives, chaired by **Representative Sherwood Boehlert**, conducted a hearing to (1) examine the latest scientific findings regarding acid rain's impact on sensitive ecosystems and those ecosystem's ability to recover, (2) explore the continuing threat posed by acid rain, and (3) assess the adequacy of our nation's air pollution monitoring systems. Testimony was provided at that hearing by four researchers. In addition, several legislators addressed the conference attendees and the committee members.

Congressman Boehlert (R-NY) noted that Title IV of the Clean Air Act of 1990 has made a real, if insufficient, difference. The continuing research has, as expected, demonstrated that the problems caused by acid rain are truly severe, varied and widespread. The impacts of acid rain on lakes, aquatic species, trees, and soil can now be documented in ways that we could only guess at in the early 1980s. The response to acid rain has thus far been a story of the system succeeding, a story of Congress acting on the basis of the best research, even in the face of uncertainty, and then trying to update policy on the basis of continuing research. That pattern needs to be perpetuated through continued funding and the heeding of research results.

Jill Barton of the U.S. Geological Survey testified that hundreds of square miles in the West do not have any wet-deposition monitoring, so our knowledge of atmospheric deposition is incomplete. High-elevation regions especially are underrepresented, and those regions are where the greatest amount of wet deposition occurs. Dry-deposition monitoring is even more sparse, even though that mechanism dominates deposition throughout much of the West. Research concerns that need to be addressed include:

- Long-term monitoring of chemical and biological trends in watersheds is greatly needed.
- Confidence in atmospheric transport models is poor in the West, where complex topography makes modeling extremely difficult.
- Mechanisms of dry deposition are still not well understood.
- Understanding of the complex ecological interactions that occur with nitrogen deposition is incomplete.

Charles Driscoll of Syracuse University also stressed the need for additional monitoring and called for the upgrading of existing equipment and techniques so researchers could produce accurate data on which government could formulate reasonable policies. The National Atmospheric Deposition Program system needs substantial modernization and a modest number of new sites. The NADP network currently relies on World War II vintage equipment. The Clean Air Status and Trends Network (CASTNet) at present is a sparse network with only 70 sites nationwide and none in the central United States. The installation of 30 new sites and the modernization of existing 79 sites is proposed. The Mercury Deposition Network (MDN) is a patchwork of sites funded through state agencies. It is essential that monitoring be established to develop a mercury-deposition baseline and to track changes over time The installation of 60 new sites and the upgrading of existing sites is proposed. The Temporally Integrated Monitoring of Ecosystems (TIME) and Long-Term Monitoring Program monitor lake and stream chemistry and document changes in response to changing emissions and acid deposition. Funding is inadequate and unstable and has been cut 50% during the past two years; the program appears to be slated for discontinuation. Funding needs to be reestablished.

Gerald Keeler from the University of Michigan testified that, in the United States, more of our surface waters are impaired for fishing because of mercury contamination that for any other pollutant. The problems of acid rain and mercury contamination are closely linked. Research has revealed that the levels of mercury in fish tissue tend to be higher in more-acidic water bodies. The dominant source of mercury contamination in fish is believed to be emissions to the atmosphere from fossil-fuel-burning power plants, municipal waste incinerators, and other industrial sources. Currently, no national effort properly assesses the trends in ambient mercury or the dry deposition of mercury. Support is required to improve the spatial coverage as well as to improve the quality of the monitoring equipment presently being used.

Ellis Cowling of North Carolina State University summarized, in a terse but extensive list, the environmental effects of air pollutants. (For example, the Houston-Galveston area is rivaling Los Angeles as the most remarkable place to observe the rapid accumulation of ozone in the United States today.) He cited these effects as substantial justification for continuing investments in research and pollution-management activities aimed at discovering and implementing cost-effective means for decreasing these impacts.

In addition, he called for building a support base of satisfied NADP customers, developing a group of NADP leaders to mount educational campaigns to encourage NADP support, maintaining a mailing list of NADP customers so they could be informed of NADP activities and enlisted as supporters, organizing state-wide NADP meetings, building a contingency fund for high-quality NADP sites that are in jeopardy of losing their support, and formulating a resolution of mutual commitment for NADP leaders.

Senator Charles Schumer (D-NY) expressed satisfaction that this problem is coming to be viewed not in regional terms but as a national challenge that requires a coordinated and cooperative response. The Adirondacks should not be thought of as just New York's tragedy, but as a cautionary tale to those who would do nothing about acid rain. Maryland, for example, receives some of the highest concentrations of sulfate and nitrate deposition in the United States.

Along with Senator Clinton, he has sponsored two bills that would combat acid rain throughout the nation. The Acid Rain Control Act would lower the current caps on sulfur dioxide, create a cap and trade program for nitrogen-oxide emissions, and regulate mercury emissions from utilities. The Clean Power Act would also reduce carbon dioxide emissions.

Schumer noted that President Bush has recognized the importance of reducing sulfur and nitrogen levels, opening the door for possible legislative action. He expressed hope about the prospects for such action in the 107th Congress.

Congressman John Sweeney (R-NY) is the lead sponsor of the Acid Rain Control Act in the House of Representatives. He called attention to the fact that loons are adversely affected by higher levels of mercury. Spruce forests are dying because the trees, made weak by acid rain, are threatened by insect infestation and are susceptible to winter injury. The current acid rain program has shown measurable success and is putting the environment within reach of a recovery. He advocates further emissions reductions by using the framework of the current acid rain program. He also noted that President Bush and EPA Administrator Whitman have made public commitments to further reductions in sulfur, nitrogen, and mercury.

Senator Hillary Rodham Clinton (D-NY) observed that, on the Senate Environment and Public Works Committee, Chairman Smith of New Hampshire and the ranking member, Harry Reid of Nevada, have said that they are willing to move a multi-pollutant bill in Congress. The delegations from the east are working very hard to find common ground with colleagues in the Midwest. The nation's energy problem cannot be addressed at the expense of the environment, but a balanced energy program can emphasize both supply and demand and can focus on conservation and the development of cleaner sources of energy. Some who have not been interested in, or support-

Appendix B – Conference Chairs, Speakers, and Guest Speaker Biographies

Session Speakers and Chairs

Mary C. Barber

Mary C. Barber is the Director of the Sustainable Biosphere Initiative (SBI) and Science Programs for the Ecological Society of America (ESA). The SBI focuses on issues of ecosystem sustainability, global change and biodiversity. Since coming to ESA in 1994, Barber has undertaken a range of activities that use ecological information to inform the public debate on issues as diverse as land use planning, ecosystem services, habitat classification, and ecological forecasting.

Prior to her current position she was Senior Environmental Scientist with Science and Policy Associates, Inc., and has held positions with the National Oceanic and Atmospheric Administration, the Oceanic Society, and the National Science Foundation. Other activities include: Women in Science and Engineering, Association of Women in Science, and Women's Aquatic Network.

She received her B.A. from Vassar College and her Ph.D. in Ecology and Evolutionary Biology from Johns Hopkins University.

Chris Bernabo

Chris was the founding Director of the National Acid Precipitation Assessment Program (NAPAP) from 1980 to 1985. He is now President of Collaborative Solutions, which designs and implements policy-relevant programs using multi-stakeholder approaches.

His other experience in linking science to decision-makers' needs includes serving as: Director – RAND Environmental Science and Policy Center, President – Science and Policy Associates, Inc., Senior Policy Analyst – NOAA, NRC Research Fellow on Climate Change, and Congressional Science Fellow – House Science Committee.

Chris has a Ph.D. in Earth Sciences from Brown University.

Rona Birnbaum

Rona Birnbaum is Chief of the Assessment and Communications Branch in the Office of Atmospheric Program's Clean Air Markets Division. She has been with the Program since 1991, soon after the Acid Rain Program was created. Throughout her time in the Air Office, Rona has focused on the science-policy interface relating to atmospheric deposition, program assessment and communication of results. Much of Rona's work involves linking research and environmental monitoring to air pollution policies such as market-based mechanisms. She has also been providing support to the US-Canada Air Quality Agreement since 1991. Rona has been with the EPA for over 14 years. She holds a Masters Degree in Environmental and Natural Resource Policy from George Washington University.

Tamara Blett

Tamara joined the National Park Service as an Ecologist in the Air Resources Division in October 2000. She focuses on the protection of natural resources from the effects of air pollution by developing, synthesizing and interpreting research and monitoring information on ecological effects and applying these analyses to air quality policy issues.

Previously Tamara was an Air Resource Management Specialist with the Rocky Mountain Region of the USDA Forest Service, an Ecologist with the USDA Forest Service Riverside Forest and Range Experiment Station in California, and a researcher studying physiological plant ecology projects for UCLA and U.C. Irving.

Tamara has a B.S. degree in biology from U.C. Irvine and a M.S. degree in Fire Ecology from Colorado State University in Fort Collins, Colorado.

Jill S. Baron

Jill Baron is an ecosystem ecologist with the U.S. Geological Survey, and a Senior Research Ecologist with the Natural Resource Ecology laboratory at Colorado State University. She is principal investigator of a long-term ecological research and monitoring program that works toward understanding Rocky Mountain biogeochemical cycles, and the influences of land use, climate change, and atmospheric deposition on ecosystem processes. Her recent interests lie in applying knowledge of mountain environments, freshwater ecosystems, and human-dominated basins such as the South Platte Ricer of Colorado. Baron received her Ph.D. from Colorado State University.

Van Bowersox

Van Bowersox is Coordinator of the U.S. National Atmospheric Deposition Program (NADP), a multi-agency program that provides regional and national-scale data and information on the amounts, temporal trends, and geographic distribution of chemical deposition by precipitation. The NADP Program Office is located at the Illinois State Water Survey, which is affiliated with the University of Illinois. Van joined the Illinois State Water Survey in 1979 and has been a Senior Scientist since 1990. His research interests are in the scavenging of gases and particles from the atmosphere by precipitation and in understanding the relationships between pollutant sources and deposition. Van has a M.S. in Meteorology and a B.S. in Chemistry from Penn State Univ.

Arthur J. Bulger Jr.

Arthur J. Bulger earned his BA at Harvard University and his Ph.D. in Zoology (fish Ecology and Physiology) at the University of Connecticut. He works in the Department of Environmental Sciences of the University of Virginia as a Fish Ecologist/Physiologist, studying problems relating to environmental biology of fishes. He is one of the nation's leading experts on acid rain effects on fish.

Bulger has collaborated on many projects related to acid deposition with a number of organizations including: the Hubbard Brook Research Foundation, the National Park Service, the US Geological Survey, the US Environental Protection Agency, the Southern Appalachian Mountain Initiative, and Trout Unlimited. He was also the recipient of a Senior Scientist Research Fellowship from the Royal Norwegian Council for Scientific and Technical Research

Dallas Burtraw

Dallas Burtraw is a Senior Fellow at Resources for the Future, a non-profit, non-partisan research institute that specializes in environmental, energy and natural resource economics. His research interests include the restructuring of the electric utility market, the social costs of environmental pollution, and benefit-cost analyses of environmental regulation, and the design of incentive-based environmental policies.

Recently, Burtraw has investigated the effects on electric utilities of the sulfur dioxide emissions-permit trading program legislated under the 1990 Amendments to the Clean Air Act. He has also helped to evaluate benefits of emission reductions resulting from the 1990 Amendments. Current projects: Integrated Approaches to Pollutant Control in the Electricity Sector; and the Valuation of Natural Resource Improvements in the Adirondacks. Burtraw has a Ph.D. in economics, Master in public policy, University of Michigan.

Don Campbell

Don Campbell is a research hydrologist with the US Geological Survey in Denver, Colorado. He received an undergraduate degree in Biology at Penn State, and worked at the Virginia Institute of Marine Science for 3 years. He moved west for graduate studies in watershed science at Colorado State University, and has been studying the hydrology and biogeochemistry of mountain lakes and streams for 18 years with the USGS.

Ellis Cowling

Ellis Cowling is a forest biologist at North Carolina State University who became a world leader in air pollution research. Beginning in 1975, he led a group of 200 scientists in creating the National Atmospheric Deposition Program (NADP). This network measures the amounts of nutrients and injur-ious substances transferred in rain and snow from the atmosphere to forest and agricultural land and surface waters at 200 research sites throughout the US.

Cowling has two earned Ph.D degrees – one from the University of Wisconsin and the other from the University of Uppsala in Sweden. He has served as major professor for 66 graduate and postdoctoral students at Yale and at NC State Universities. He has been a member of the National Academy of Sciences since 1973.

Charles T. Driscoll

Charles T. Driscoll is Professor of Environmental Systems Engineering in the Department of Civil and Environmental Engineering at Syracuse University. Driscoll has worked on the effects of acid rain and mercury in New York and New England since the mid 1970's. His research interests include the chemistry of soils and drainage waters , environmental modeling, and the long-term biogeochemical patterns in forest and aquatic ecosystems. Driscoll served as Chair of the Gordon Conference on Forested Catchments, was designated by the National Science Foundation as a Presidential Young Investigator and has been honored with the Syracuse University Chancellor's Citation of Academic Achievement. He holds a Ph.D. in Environmental Engineering from Cornell University.

Guy Fenech

Guy Fenech is Senior Science Advisor in the Science Assessment and Policy Integration Branch of Environment Canada. The Branch's main products are science assessments of major atmospheric issues such as "Acid Rain", "Smog" and Stratospheric Ozone". Guy Fenech directed the 1997 Canadian Acid Rain Assessment and he is currently setting up the science program to support the Canada-Wide Acid Rain Strategy. He has degrees in mathematics and meteorology.

James N. Galloway

James N. Galloway earned his BA from Whittier College with a double major in Chemistry and Biology. He was awarded his PhD from the University of California, San Diego in Chemistry and was a Postdoctoral Fellow at Cornell University.

He is currently Professor and Chair of the Department of Environmental Sciences at the Univ. of Virginia. His research is in the area of biogeochemistry and includes work on the natural and anthropogenic controls on chemical cycles at the watershed, regional and global scales. He is the author of over a hundred scientific papers. He is internationally recognized for his work on acid deposition effects on soils, waters and forests, watershed biogeochemistry and the influence of Asia on the global environment.

Robert Howarth

Howarth has a B.S. in biology from Amherst College and a Ph.D. in oceanography issued jointly by MIT and the Woods Hole Oceanographic Institution. He has been on the faculty at Cornell University since 1985, and has been the David R. Atkinson Professor of Ecology and Environmental Biology since 1993. He currently is a senior marine scientist and manager of the Oceans Program at Environmental Defense, and is an adjunct research scientist at the Marine Biological Laboratory in Woods Hole, MA. Howarth has been co-chair of the International SCOPE Nitrogen Project since 1994 and has been Editor-in Chief of the journal Biogeochemistry since 1983. From 1998-2000, he chaired the Committee on Causes and Management of Coastal Eutrophication of the U.S. National Academy of Sciences.

Gerald J. Keeler

Gerald. Keeler has been on the faculty at the University of Michigan (UM) in Ann Arbor since 1990. He holds a joint appointment in the Department of Environmental health Sciences in the School of Public Health an in the Department of Atmospheric, Oceanic, and Space Sciences in the College of Engineering. He serves as the Director of the UM Air Quality Laboratory, an Interdisciplinary research group which is internationally recognized as a leader in atmospheric mercury research. Keeler served as a Technical Reviewer of the EPA Mercury Report to Congress. He and his team are currently working on mercury projects in Florida, New England, Michigan, Great Lakes Region, and the Arctic. Keeler has a Ph.D. in Atmospheric Sciences from the University of Michigan.

John D. Kinsman

John Kinsman is the Director, Air Quality Programs at the Edison Electric Institute in Washington, D.C. Mr. Kinsman's environmental career has spanned 20 years, including the last 13 years at EEI, where he works on the issues of acid rain, ozone, particulate matter, regional haze, mercury and global climate change. He works with different constituencies to obtain reasonable environmental laws and regulation, and then assists the industry in compliance with laws and regulations. His degrees in environmental science are from the University of Virginia and George Mason University.

Douglas Knauer

Douglas Knauer is the Section Chief of Environmental Contaminants Research for the Wisconsin Department of Natural Resources. He has been associated with acid rain research projects in Wisconsin since 1983. He was involved with a whole-lake acidification experiment at Little Rock Lake, an acid lake neutralization experiment at Max Lake, the development of an acid rain model through the Regional Integrated Lake/Watershed Acidification Study, and a long-term survey of acid sensitive lakes in northern Minnesota, Wisconsin and the upper peninsula of Michigan.

Since 1988, Knauer has participated in a variety of mercury studies ranging from the biogeochemical fate of mercury in the aquatic ecosystem of inland lakes, Lake Superior and the Florida Everglades to a mercury risk assessment of the common loon.

Greg Lawrence

I began studying acid rain in 1980 when I began work on my M.S degree at the State University of New York College of Environmental Science and Forestry (completed in 1982). In 1987 I completed a Ph.D. in Civil Engineering at Syracuse University after studying the effects of acid rain and clearcutting on stream chemistry at the Hubbard Brook Experimental Forest in New Hampshire. After 3 years as a research professor at the University of Maine, I joined the Water Resources Division of the U.S. Geological Survey in Troy, New York, where I have continued my research on the effects of acid rain on soil and water chemistry. To date, I have published over 50 technical articles and reports related to acid rain.

Morton Lippmann

Morton Lippmann holds a Bachelor of Chemical Engineering from the Cooper Union (1954), a master's degree in industrial hygiene from Harvard (1955) and a Ph.D. in environmental health sciences from NYU (1967). He is a diplomate of the American Board of Industrial Hygiene, with certification in the engineering aspects and comprehensive practice of industrial hygiene. He the recipient of numerous awards.

Lippmann is a Professor of Environmental Medicine at the New York University School of Medicine. He directs a research program on Human Exposure and Health Effects, and the EPA supported Particulate Matter Health Effects Research Center. His publications include 260 research and review papers in the scientific literature and two reference texts on environmental health science.

Brian McLean

Brian McLean is the Director of the Clean Air Markets Division, part of EPA's Office of Air and Radiation, which is best known for the development and use of the "cap and trade" emissions trading approach to pollution control. Presently, the Division is responsible for the management of trading programs to control emissions of sulfur dioxide (SO2) and nitrogen oxide (NOX) including the NOX Budget Program for the Northeast Ozone Transport Commission. Additionally, Mr. McLean is involved in efforts to develop emis-sions trading programs to address other U.S. and international pollution issues.

While with the EPA Mr. McLean has helped develop the Administration's acid rain legislative proposal which was enacted in 1990 as Title IV of the Clean Air Act, and was a principle negotiator of the 1991 U.S. – Canada Air Quality Accord.

Mr. McLean holds a Bachelor's degree in Electrical Engineering from Lafayette College, a Master's degree in City and Regional Planning from Rutgers University, and a Doctorate in City Planning from the University of Pennsylvania.

John M. McManus

John M. McManus is the Manager of Environmental Strategy & Compliance Planning for the American Electric Power Company located in Columbus, Ohio. In this position, Mr. McManus has responsibility for evaluating environmental issues that have the potential to affect the electric utility industry and formulating responses to such issues. In addition, he is the primary interface with the U.S. Environmental Protection Agency on matters related to compliance with the requirements of the acid rain control program for coal-fired units on the AEP system.

Mr. McManus has been with AEP since 1977. He holds a Bachelor of Science degree in Environmental Engineering from Rensselaer Polytechnic Institute (1976) and is a registered Professional Engineer in the State of Ohio.

Bernard C. Melewski

Bernard C. Melewski has been Legislative Director of the Adirondack Council since 1990, directing the Council's legislative and public policy initiatives toward Adirondack Park protection. Mr. Melewski has been involved in most of the major environmental initiatives in New York in the last two decades. He is the former counsel to the State Commission On Solid And Hazardous Waste Management, and past executive director of the Environmental Planning Lobby (now Environmental Advocates), a statewide advocacy organization.

Mr. Melewski participated in the drafting and passage of the first acid rain legislation in the country in 1984 and represented the Adirondack Council in successful negotiations with the US EPA in the revision of its acid rain regulations. He currently directs legislative efforts on acid rain in both the New York State Legislature and in Congress. He is a frequent guest speaker on the topic.

Paulette Middleton

Paulette Middleton, Co-Director RAND Environment, has 25 years of experience as an atmospheric chemist, policy analyst and program director, applying cutting-edge research and strategic thinking to environmental policy making. Before coming to RAND in 1998, she worked at the National Center for Atmospheric Research, the Atmospheric Sciences Research Center at the State University of New York at Albany and Science & Policy Associates, Inc. Currently she is also involved with the Global Emissions Inventory Activity Center, as chair of the Air Quality Modeling Subcommittee of the EPA Science Advisory Board, the Environmental Models Committee currently reviewing the National Air Toxics Assessment, and as an Associate Editor of the Journal of Air and Waste Management Association. Middleton has authored over 50 peer-reviewed articles.

She has a Ph.D. in Chemistry, University of Texas

Jeremy Platt

Mr. Platt manages research on power and fuel markets for EPRI (founded as the Electric Power Research Institute in 1973) where he has been responsible for research including fuel supply analysis and procurement; energy market behavior; electric industry restructuring; techniques for analyzing forward markets in energy and ancillary services; and the fundamentals driving the SO2 emission allowance market.

Mr. Platt is also involved with the Energy Minerals Division of the American Association of Petroleum Geologists and the Potential Gas Committee. He has lectured occasionally at Stanford University, and has made presentations on EPRI research to a variety of industry and academic groups.

Mr. Platt received an A.B. degree with concentration in economic geology from Harvard University (1971) and an M.S. in Geological Sciences from Stanford University (1974).

Richard Poirot

Rich graduated from Dartmouth College in 1972. For the past 24 years, he's been employed as an Air Quality Planner with the Vermont Agency of Natural Resources. He currently serves on the Technical Committee of the National Atmospheric Deposition Program, the Monitoring and Assessment Committee of the Northeast States for Coordinated Air Use Management (NESCAUM), the Steering Committee of the Interagency Monitoring of Protected Visual Environments

(IMPROVE), the Subcommittee on Scientific Cooperation for the US/Canada Air Quality Agreement, and on the Acid Rain Data Exchange Workgroup for the Conference of New England Governors and Eastern Canadian Premiers.

Fred Stoss

Fred Stoss is the Biological Sciences Librarian in the Science and Engineering Library at the University at Buffalo. In addition to his Master in Library Science Degree, Fred holds undergraduate and graduate degrees in biology and zoology. He has more than 10 years of research experience in the area of environmental toxicology and nearly 20 years of experience in the area of information and library science. He is the past-chair of both the Environmental Information Division of the Special Libraries Association and the Task Force on the Environment of the American Library Association. He is on the adjunct faculty of the Department of Library and Information Studies at the University at Buffalo.

Mary F. Striegel

Mary is the Environmental and Materials Research Program Director for the National Center for Preservation Technology and Training (NCPTT) Materials Research Program. Mary came to NCPTT in 1995 from the Getty Conservation Institute. Her past work has included studies of the effects of formaldehyde on inorganic materials, uses of Thin-layer Chromatography for the analysis of binding media, and applications of digital imaging and technical photography in the analysis of works of art. Mary earned her PhD in inorganic chemistry from Washington University in St. Louis, where she pursued interdisciplinary research on residual stresses in numismatics.

Brent Takemoto

Brent Takemoto is an Air Pollution Research Specialist in the California Air Resources Board's Office of Community Health. His primary responsibilities include work related to the Board's Neighborhood Assessment Program, environmental justice issues, air quality in the eastern Sierra, and ozone effects on plants. Prior to joining the Office of Community Health, Brent was a contract manager in the Board's Research Division where he organized the forest ecosystems research program to evaluate the effects of acidic deposition and ozone on mixed conifer forests in southern California.

Brent received a Ph.D. in biology from Bowling Green State University in Bowling Green, Ohio. Prior to joining the Air Resources Board, he was a Postgraduate Researcher at the Statewide Air Pollution Research Center at the University of California, Riverside.

Elizabeth Thorndike

Elizabeth Thorndike is the founder of the Center for Environmental Information where she served as Executive Director from 1974-1992. She remains a member of the Board of Directors. She has served in public service positions under three New York governors: as a commissioner of the Adirondack Park Agency for fifteen and a half years; a member of the governors's Environmental Advisory Board for 10 years; and since 1997 as a member of the board of the NY State Energy Research and Development Authority (NYSERDA), in the designated environmentalist appointment. She is currently a consultant on collaborative environmental problem-solving, a member of the advisory board of the Cornell University College of Agriculture and Life Sciences, a Visiting Lecturer there in the Department of City and Regional Planning, and a trustee of the Association for the Protection of the Adirondacks. She has degrees from Stanford and Harvard and a Ph.D, with a concentration in Natural Resource Policy and Management, from Cornell.

Michael Uhart

National Acid Precipitation Assessment Program

Richard A. Valentinetti

B.S. in Environmental Health, University of Massachusetts; Master of Public Health in Public Health Administration, University of Michigan.

Mr. Valentinetti is the Director of the Air Pollution Control Program for the State of VT where he manages the state's air resources issues associated with an integrated natural resources/ environmental protection program, with particular emphasis on the effects of municipal waste combustion on ambient air; monitoring and projections of acid deposition; chlorofluorocarbons; greenhouse gases; and emissions from wood stoves.

Previously Mr. Valentinetti has worked with the U.S. EPA, State of Vermont Air and Solid Waste Programs; National Air Pollution Control Administration (NAPCA), State and Territorial Air Pollution Program Administrators (STAPPA); and North Eastern States for Coordinated Air Use Management (NESCAUM).

Guest and Luncheon Speakers

Representative Sherwood L. Boehlert

Sherwood L. Boehlert (R-New Hartford), chairman the House of Representatives Science Committee, is a senior member of Congress proud of his accomplishments at the national and local levels.

Boehlert was first elected to the House in November 1982 and is currently serving in his tenth consecutive term representing Central New York. In the 2000 election, he again won all nine counties and received a convincing 60% of the vote.

Boehlert has served on the Science Committee since 1983, and was elected Chairman in January 2001. The Committee has jurisdiction over all federal nonmilitary scientific and technology research and development programs, on which the federal government spends more than $30 billion a year. The Committee has jurisdiction over NASA, the National Science Foundation, and research and development initiatives within the Environmental Protection Agency, the Department of Energy, and the Department of Commerce. In addition, the Committee has jurisdiction over civil aviation research and development and marine research.

Boehlert is the third-ranking member of the House Transportation and Infrastructure Committee, serving as Chairman of its Subcommittee on Water Resources and Environment from 1995 to 2000. He remains an active member of that Subcommittee. Boehlert also sits on the Subcommittee on Highways and Transit, and the Subcommittee on Railroads.

Boehlert was reappointed by House Speaker J. Dennis Hastert as a member of the Select Committee on Intelligence, where he is on the front line of important intelligence decisions faced by Congress. Boehlert is a delegate to the NATO Parliamentary Assembly, also at the appointment of the Speaker, where he serves as chairman of the Assembly's Scientific and Technology Committee.

Born on September 28, 1936 in Utica, New York, Boehlert is a graduate of Whitesboro Central High School and Utica College (Bachelor of Science, 1961). Before serving as Oneida County Executive (1979-83), he was manager of public relations at Wyandotte Chemical (1961-64) and served two years in the U.S. Army (1956-58).

Boehlert served as chief of staff for two area Congressmen, Alexander Pirnie (1964-72) and Donald Mitchell (1973-79), where he became intimately familiar with the people, places and issues of the 23rd District. In honor of his former boss, Boehlert was able to secure passage of legislation in 2000 to rename the Veterans' Outpatient Clinic in Rome as the "Donald J. Mitchell Department of Veterans Affairs Outpatient Clinic."

An avid New York Yankees fan and movie buff, Boehlert and his wife, Marianne (Willey) Boehlert, make their home in New Hartford, New York. They have four grown children and five grandchildren. When Congress is in session, he returns home each weekend to stay in touch with people he feels fortunate to represent in Washington.

The 23rd includes all or parts of Broome, Chenango, Delaware, Herkimer, Madison, Montgomery, Oneida, Otsego, and Schoharie counties.

Senator Hillary Rodham Clinton

Hillary Rodham Clinton was elected United States Senator from New York on November 7, 2000. She is the first First Lady elected to the United States Senate and the first woman elected statewide in New York.

Senator Clinton has been an advocate for children and families for more than thirty years. She brings to the Senate the same commitment and energy that took her to each of New York State's 62 counties during her 16-month campaign.

To build a better future for working families,

Hillary Rodham Clinton supported policies to expand the economy, raise the minimum wage and the earned income tax credit, increase tax deductions for children, and make credit more available, including microcredit loans for women entrepreneurs. Senator Clinton is dedicated to bringing jobs to Upstate New York. Through tax credits for small businesses, investments in telecommunications infrastructure, creating technology extension programs, skills training, restructuring utilities and lowering airfares to increase regional accessibility, she proposes to make it possible for the economy in upstate and all regions of New York to flourish and to stem the outmigration of young New Yorkers and their families.

Senator Clinton serves on the Budget, Environment and Public Works, and HELP (Health, Education, Labor and Pensions) Committees. These important committees are responsible for many issues that affect the residents of New York State. Addressing the Senate on the subject of the 2002 Federal Budget, Senator Clinton commented that, "We must pass a budget that keeps paying down the debt, provides sensible tax cuts and invests in priorities that matter to the people we represent."

Appointed by President Bill Clinton in 1993 to chair the Task Force on National Health Care, she and her task force members worked for months meeting with families and health care professionals. Their efforts culminated in the Health Security Act of 1994. Disappointed that the Task Force was unable to make more progress, she has said that the experience brought her to the "school of smaller steps," adding that "we must continue to make progress. It's still important that we increase access to quality health care for working families." As First Lady, she led the fight to pass the Children's Health Insurance Program that provides health insurance for millions of working families. She worked to increase funding for breast cancer research and treatment for breast cancer, prostate and colon cancer, osteoporosis and juvenile diabetes. She worked to pass strong anti-crime measures, including the Brady Bill and the Assault Weapons ban. Senator Clinton is a strong supporter of the HMO Patient's Bill of Rights and of action to strengthen Medicare and include prescription drug benefits. She advocates expanding federal funding for childhood vaccinations and diseases such as asthma and epilepsy. Health care was the topic of Senator Clinton's maiden speech in the Senate chamber.

Senator Clinton is recognized around the world as an advocate for democracy, religious tolerance and human rights, and as a champion for women and girls, emphasizing access to education, economic opportunity, family planning and women's right to choose. With her husband, former President Clinton, she has worked for peace in Northern Ireland, the Balkans and the Middle East. Her Vital Voices program has brought women together in Asia, Africa, Latin America, and Europe to encourage their increased participation in economic and political decisions.

Born in Chicago, Illinois on October 26, 1947, Senator Clinton is the daughter of Dorothy Rodham and the late Hugh Rodham. She grew up in Park Ridge, Illinois and attended public school. She attended Wellesley College. A 1973 graduate of Yale Law School, Hillary Rodham Clinton was named one of the National Law Journal's 100 Most Influential Lawyers in America in both 1988 and 1991. She was appointed chair of the Legal Services Corporation by President Jimmy Carter in 1977, and served as chair of the American Bar Association Committee on Women in the profession in 1987. From 1986-1989, she chaired the board of the Children's Defense Fund.

Senator Charles E. Schumer

Charles E. Schumer is in his first term representing New York State. He currently serves as a member on the Committee on Banking, Housing and Urban Affairs, the Judiciary Committee, the Energy and Natural Resources Committee and the Rules Committee. Before his election to the Senate, Schumer represented the Ninth Congressional District in Brooklyn and Queens for nine terms.

For the past two decades, Chuck Schumer has been a leader on national issues and a champion for New York State. For his efforts, the Rochester Democrat & Chronicle called Schumer "an accomplished, far-sighted legislator," while The New York Times wrote that Schumer "is a more serious lawmaker with more rooted values, sounder policy positions and a deeper commitment to the common good."

Armed with an impressive resume of accomplishments and an 18-year House record, Chuck Schumer was elected to the Senate in 1998 – one of only two Democrats to defeat a Senate incumbent that year.

Since his election to the Senate, Schumer has made improving New York's economy his top priority. He has been particularly successful in bringing affordable air service to Upstate New York, helping deliver new airline JetBlue to Buffalo, Rochester and Syracuse and working with low-cost carrier Southwest Airlines to expand service to Albany.

Schumer is also working to ensure that all Americans have quality health care and access to affordable prescription drugs. He is currently fighting to restore hospital cuts inflicted by the 1997 Balanced Budget Act, provide seniors with a prescription drug benefit under Medicare, and knock down the barriers that delay low-cost generic medications from coming to the marketplace.

A member of the Banking committee in the House and the Senate, Schumer worked for a decade to pass the 1999 Financial Services Modernization legislation, which modernizes regulations governing the US banking, securities and insurance industries. He played a key role in drafting language to ensure that financial companies serve traditionally underserved areas and has exposed unequal lending practices of banks and predatory lending practices of subprime lenders in minority communities.

Throughout his 20 years in Congress, Schumer has been a pioneer in the fight against crime. His work in this area led Attorney General Janet Reno, the nation's top law enforcer, to state, "I have never met a public official more dedicated to fighting crime than Mr. Schumer."

Schumer sponsored and helped pass the Omnibus Crime Bill of 1994, which put 100,000 new cops on the street, enforced "three strikes and you're out" sentencing, and created after school programs for troubled teens. As of August 2000, the Crime Bill's COPS program had put 11,461 new officers on New York's streets.

Schumer has also worked to keep guns out of the hands of criminals and children. He authored the 1993 Brady Bill, which instituted mandatory background checks for handgun purchases, and the 1994 Assault Weapons Ban. In the Senate, Schumer won the first federal funding for Albany, Buffalo, Syracuse and Rochester to implement Project Exile, a program that enforces strict sentencing guidelines for illegal gun possession.

To protect a woman's right to choose, Schumer wrote the Freedom of Access to Clinic Entrances Act, which makes blockading family planning clinics a federal crime. He also authored the Violence Against Women Act, the first federal legislation protecting women from domestic abuse, and has been a leader in the fight against hate crimes and terrorism.

A product of the Brooklyn public schools, Schumer, 50, is a graduate of Harvard College and Harvard Law School. He was elected to the New York State Assembly at age 23, one of the youngest members since Theodore Roosevelt, and to Congress at 29. Schumer upset Al D'Amato to become New York's junior Senator in 1998.

Senator Schumer lives in Brooklyn with his wife, Iris Weinshall, and their daughters, Jessica and Alison.

Representative John E. Sweeney

John E. Sweeney was sworn-in as a member of the 106th Congress on January 6, 1999. In his first term as a representative, Congressman Sweeney quickly distinguished himself among his colleagues. He was appointed to the Republican Steering Committee, also known as the Committee on Committees, by his fellow freshman congressmen to appoint Members of Congress to posts on the House of Representatives' 20 committees.

Congressman Sweeney now serves on the House Appropriations Committee. He has been assigned to the Subcommittees on Transportation and Related Agencies, Treasury Postal and General Government, and the District of Columbia.

Before becoming an elected representative, John Sweeney served for more than two years as New York State Commissioner of Labor where he gave the Labor Department a more pro-active role throughout the state and the nation. His efforts focused on reducing burdensome government regulations on businesses and stimulating economic growth while protecting the interests of the state's workforce.

Congressman Sweeney has built a reputation as a pro-active legislator on environmental issues in the House of Representatives. His most recent legislative initiative-the Acid Rain Control Act-represents one of the most comprehensive bills currently seeking to address the acid rain crisis plaguing the northeastern United States. The Acid Rain Control Act is viewed by many as a significant step towards reducing sulfur and nitrogen oxide emissions by power plants and returning forests and waterways to non-acidic levels. In 2000, Mr. Sweeney successfully advanced legislation reauthorizing the Clean Lakes Program at the Environmental Protection Agency and has secured funding for acid rain monitoring.

Congressman Sweeney is a native of Troy, New York where his father was president of a local shirt-cutters union. After working his way through college, he received a Bachelor of Arts degree from Russell Sage College. The following year, he was appointed head of the Rensselaer County DWI prevention program that, under his direction, became the most successful program of its type in New York State. He continued working while studying law, and earned his law degree from Western New England School of Law. After completing his law degree he was selected as Executive Director and Chief Counsel to the New York State Republican Party.

John Sweeney currently resides in Clifton Park, New York. He is actively involved in many civic and community organizations.